AF785280

Effects of Hydrology on Soil Erosion and Soil Conservation

Guest Editors

Xiaojun Liu
Ling Zhang

MDPI

Basel • Beijing • Wuhan • Barcelona • Belgrade • Novi Sad • Cluj • Manchester

Guest Editors
Xiaojun Liu
School of Agriculture
Ningxia University
Yinchuan
China

Ling Zhang
College of Forestry
Jiangxi Agricultural University
Nanchang
China

Editorial Office
MDPI AG
Grosspeteranlage 5
4052 Basel, Switzerland

This is a reprint of the Special Issue, published open access by the journal *Water* (ISSN 2073-4441), freely accessible at: https://www.mdpi.com/journal/water/special_issues/973U0VGNB9.

For citation purposes, cite each article independently as indicated on the article page online and as indicated below:

Lastname, A.A.; Lastname, B.B. Article Title. *Journal Name* **Year**, *Volume Number*, Page Range.

ISBN 978-3-7258-5039-6 (Hbk)
ISBN 978-3-7258-5040-2 (PDF)
https://doi.org/10.3390/books978-3-7258-5040-2

Contents

Preface

Soil erosion remains a critical global challenge, profoundly impacting soil quality, land productivity, ecosystem biodiversity, and ultimately, human well-being and sustainable development. Despite significant worldwide efforts, the intricate role of hydrology in driving soil erosion processes and enabling effective conservation strategies requires a deeper understanding and broader recognition. The following Reprint, compiling peer-reviewed articles from a Special Issue published by *Water*, addresses this vital nexus.

The primary aim of this collection is to present significant new findings and foster a better understanding of the complex processes and mechanisms governing soil erosion and conservation induced by hydrological factors. Specifically, it explores the fundamental interrelationships between soil erosion and key hydrological elements such as rainfall characteristics (intensity, duration, and interception) and water flow dynamics (infiltration, preferential flow, runoff generation, path, and hydrological connectivity). The scope encompasses the full spectrum of soil and ground component disturbance, encompassing destruction, detachment, transport, and deposition, including detailed investigations into specific erosive agents such as raindrop splash, sheet flow, rill formation, gully development, and underground leakage. Furthermore, the Reprint critically examines diverse conservation approaches—biological, agrotechnical, and engineering measures—designed to mitigate erosion by strategically managing hydrological pathways and energy.

The motivation for compiling this work stems from the urgent need to translate scientific advances into practical solutions. By synthesizing cutting-edge research on how water movement governs erosion and how interventions can alter these hydrological pathways, this Reprint aims to provide robust scientific foundations. It seeks to empower researchers, land managers, policymakers, and conservation practitioners with the knowledge necessary to design and implement more effective, hydrologically-informed strategies for protecting our invaluable soil resources from degradation.

This Reprint is dedicated to the international community committed to soil and water conservation, offering consolidated insights to guide future research and action in safeguarding the foundation of terrestrial ecosystems and agricultural systems worldwide.

Xiaojun Liu and Ling Zhang
Guest Editors

water

MDPI

Article

Photogrammetric Technique-Based Quantitative Measuring of Gravity Erosion on Steep Slopes in Laboratory: Accuracy and Application

Jinwen Xia [1,2,3], Lichao Zhang [1,*], Xuhua Huang [4], Xianghui Lu [1], Peilin Ge [4], Yujie Wei [5,*] and Chongfa Cai [5]

1 College of Hydraulic & Ecological Engineering, Nanchang Institute of Technology, Nanchang 330099, China; xiajinwen2020@163.com (J.X.); xianghuilu@nit.edu.cn (X.L.)
2 Jiangxi Key Laboratory of Degraded Ecosystem Restoration and Watershed Ecohydrology, Nanchang Institute of Technology, Nanchang 330099, China
3 Jiangxi Provincial Technology Innovation Center for Ecological Water Engineering in Poyang Lake Basin, Nanchang 330099, China
4 Jiangxi Academy of Water Science and Engineering, Nanchang 330099, China; 13755674959@163.com (X.H.); gpl13755792030@163.com (P.G.)
5 Key Laboratory of Arable Land Conservation (Middle and lower Reaches of Yangtze River) of the Ministry of Agriculture, Soil and Water Conservation Research Centre, Huazhong Agricultural University, Wuhan 430070, China; cfcai@mail.hzau.edu.cn
* Correspondence: lichaozhang2019@163.com (L.Z.); wyj@mail.hzau.edu.cn (Y.W.)

Abstract: Quantitative measuring of gravity erosion contributes to a better understanding of soil-mass failure occurrence and prediction. However, the measurement of gravity erosion requires the continuous monitoring of the objective terrain, due to its occurrence, usually within seconds, and combination with hydraulic erosion. The photogrammetric technique can quickly obtain terrain data and provide a new method for measuring gravity erosion. Based on a continuous high-overlapping image-acquisition equipment, a Structure-from-Motion-Multi-View-Stereo (SfM-MVS)-integrated workflow, and volume calculation, a new working methodology was established for measuring gravity erosion on steep granitic slopes in the laboratory. The results showed a good match between the digital point clouds derived from SfM-MVS-integrated workflow and terrestrial laser scanning (TLS), achieving millimeter-scale accuracy. The mean distance between the point clouds derived from TLS and SfM-MVS was 1.13 mm, with a standard deviation of 0.93 mm. The relative errors among the volumes calculated by SfM-MVS and TLS or the conventional oven-drying method were all within 10%, with a maximum error of 9.3% and a minimum error of 0.2%. A total of 213 gravitational erosion events were measured in the laboratory by using the SfM-MVS method, further confirming its feasibility.

Keywords: SfM-MVS; gravity erosion; volume calculation; steep slopes

1. Introduction

Gravity erosion is triggered by gravitational force and defined as the erosion and transportation processes of soil mass under its own weight. Gravity erosion, a dominant erosion type on steep hilly slopes, is widely located in gully heads, gully side walls or river valleys [1–3]. Gravity erosion usually occurs within seconds in combination with hydraulic erosion, thus posing a great challenge to its site-specific real-time measurement.

Compared with hydraulic erosion, the body of research of gravity erosion is weak because of the phenomenon's mechanistic complexity and limited measurement methods [4,5]. The quantitative measurement of gravity erosion is very important for a better understanding of soil-mass failure occurrence and prediction. At present, soil-erosion models (such as USLE, CSLE, and WEEP) only consider soil loss under hydraulic force, while the sediment yield produced by gravity erosion is usually indirectly analyzed through the relationship between runoff and sediment or determined through the difference between the amount of total erosion and hydraulic erosion [6–8]. Despite the establishment of soil-erosion models involving the influence of gravity erosion by some researchers, these models are empirical equations of soil loss [9]. The study of gravity-erosion models requires a large number of directly observational data sets; most previous studies of gravity erosion involve qualitative description rather than quantitative analysis, which also limits the establishment of gravity-erosion models. Moreover, slope failure is not only a natural hazard but also contributes to landscape evolution and erosion. It is crucial to understand the relationship between mass movements on gully slopes and sediment transport for land and water management.

The traditional direct monitoring methods of gravity erosion, such as the strain probe method, landslide activity maps or drilling-rod method, also have the shortcomings of limited measurement accuracy and an inability to achieve real-time monitoring [10–12]. In recent years, with the rapid development of 3S (GPS, RS, GIS) technology, significant progress has been achieved in soil-erosion monitoring technology and analysis methods. Terrestrial laser scanning, sonar bathymetry, radar altimetry, aerial photography, and an approach combing aerial photographs with satellite imagery have been used to monitor soil erosion and geomorphic evolution [13–16]. However, most of these methods are expensive and mainly applicable to relatively long-term soil-erosion monitoring and geomorphic evolution, and unsuitable for the real-time dynamic monitoring of gravity erosion.

In the decade or so since its emergence, the photogrammetric technique has become a powerful tool for topographic surveying [17–20]. With the development and improvement in Scale-Invariant Feature Transform (SIFT), Bundler, Patch-Based Multi-View Stereo (PMVS), and other algorithms in the computer-vision community, stereoscopic photogrammetry based on the principle of remote sensing image interpretation, such as Structure-from-Motion (SfM) and Multi-View-Stereo (MVS), has made new progress in obtaining high-resolution terrain data [21–25]. Since the first application of SfM-MVS photogrammetry in geoscientific studies, its implementation has increased significantly, leading to an established method to generate 3D point clouds depicting the earth's surface with high resolution [26,27]. The photogrammetric technique can simultaneously and automatically calculate camera pose and scene geometry based on matching features in a series of overlapping, offset images [23]. Additionally, this technique has also been integrated into image 3D-modeling software, such as VisualSFM [28], 123D Catch [29], Agisoft PhotoScan [30], CMP SfM [31], etc., which allows one to build a 3D dense point cloud of target objects from images obtained by ordinary consumer un-calibrated and non-metric cameras or even mobile phones [32–34]. The precision of the model established by this technique could be comparable to that of a terrestrial laser scanner (TLS) [15,35–37], leading to its wide use in geomorphology. Currently, the photogrammetric technique has been used to survey river-bed topography [38,39], coastal retreat [40], and fluvial [41], glacial [42], and gully erosion [20,43,44], as well as map landslides [45]. However, to our knowledge, no report is available about its application in quantitatively measuring gravity-erosion volume, especially soil collapsing on steeper slopes. In most cases, studying the evolution of objects requires a high temporal frequency of data collection with a high spatial resolution. This technique can quickly obtain terrain data and provide new knowledge for the high temporal frequency of data collection, such as gravity-erosion research [36].

Against the above background, this paper aimed (i) to test the quality of point clouds generated by the photogrammetric method based on the SfM-MVS technique; (ii) to evaluate the accuracy of the photogrammetric method in the quantitative measurement of the gravity-erosion volume by comparing TLS with the traditional oven-drying method; and (iii) to apply the SfM-MVS photogrammetric method for estimating the ratio of gravity erosion to the total erosion on the steep granitic slopes.

2. Materials and Methods

2.1. Steep Slope Construction

In order to verify the applicability of the SfM-MVS method proposed in this paper in monitoring the gravity-erosion process, we constructed a steep granite-soil slope as shown in Figure 1. The soil slope was 1m high, 1m wide, and 1.5 m long, with a vertical collapsing wall and a gentle slope of 5°. The soil slope was modeled using granite residual soil collected from a collapsing gully in Tongcheng county, Hubei. This area was dominated by gravity erosion because of thick and loose granite weathered crust [46]. Consistent with the field, the bulk density and the initial water content were controlled at 1.38 g cm^3 and 10%, respectively. The particle size distribution of the collected granite residual soil was determined using the pipette method and is shown in Figure 2. The basic properties are presented in Table 1.

2.2. Camera System and Calibration

The primary requirement for SfM-MVS is highly redundant and overlapping well-exposed images [20]. The imaging sensors of SfM-MVS can be low-grade compact digital cameras, mobile phones, and even video stills [34], with cost and programmable control as the main considerations for camera selection. The maximum number of cameras was determined as eight for one PC based on the program control and the determine interface restrictions, so eight non-metric cameras (ONTOP X2S) with a USB 2.0 interface were selected to acquire the images of the steep slope (Figure 1). The camera has a complementary metal-oxide semiconductor (CMOS) sensor of 8.0 effective megapixels with a size of 12 mm^2 (3 mm × 4 mm) and a resolution of 8 MPa (2448 × 3264). The 35 mm equivalent focal length is 32 mm. In order to control multiple fixed-positions cameras to take continuous images at the same time, we used C# language to implement a camera control program based on Afroge.net in the environment of net framework (4.6.1). Before each experiment, the camera positions were manually adjusted to ensure the cameras' network could obtain highly redundant and overlapping images for the establishment of 3D point cloud density, with the orientation of cameras orthogonal to the steep slope.

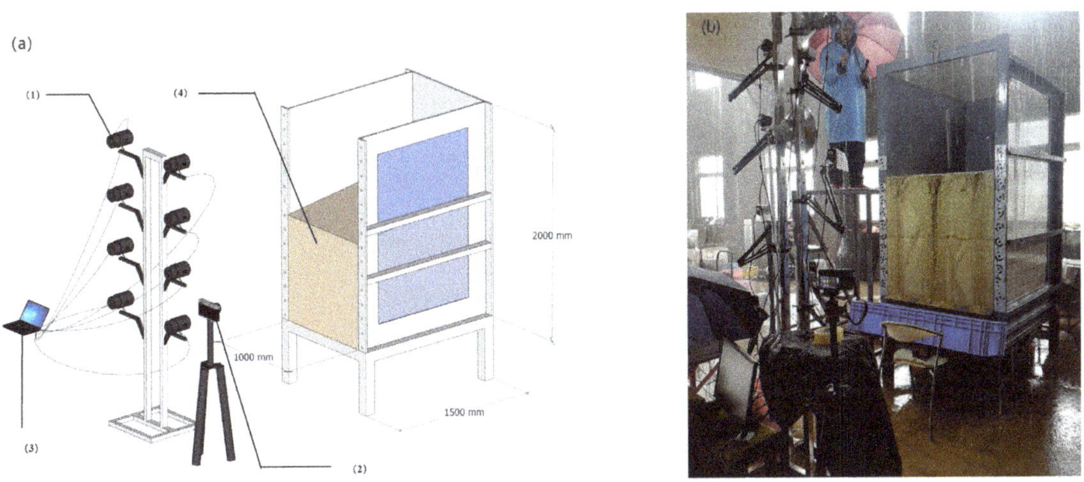

Figure 1. Sketch of experimental equipment. (**a**) Blue print of the experimental system; (**b**) Picture of an experimental site. (1) Image acquisition system; (2) camera, which was used to judge the type and position of gravity erosion after experiments; (3) computer, which was used to control the cameras and store images; and (4) constructed slope.

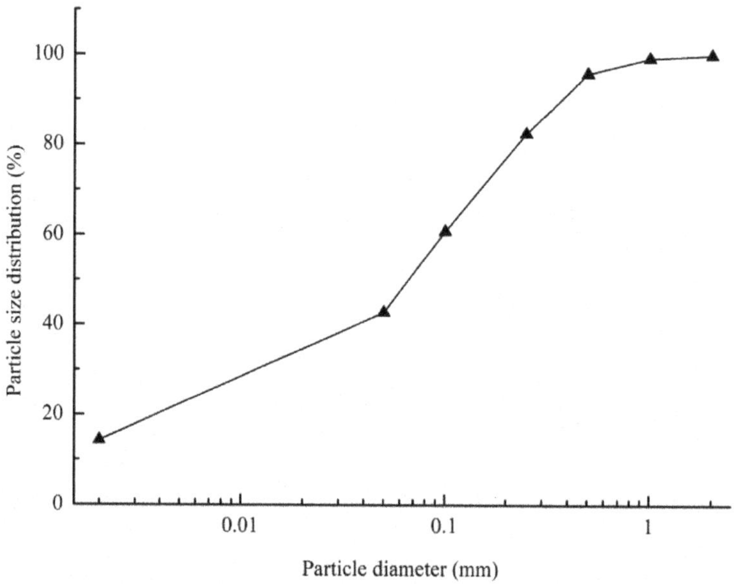

Figure 2. Particle size distribution of sampling soil.

Table 1. Basic properties of sampling soil.

BD ($g\ cm^{-3}$)	PD ($g\ cm^{-3}$)	TP (%)	CP (%)	LL (%)	PL (%)	SOM ($g\ kg^{-1}$)	Fe_d ($g\ kg^{-1}$)	Fe_o ($g\ kg^{-1}$)	Al_d ($g\ kg^{-1}$)	Al_o ($g\ kg^{-1}$)
1.38	2.61	46.31	38.19	36.42	20.35	4.62	5.97	0.08	2.13	1.64

The lens distortion of a non-metric camera can affect the internal camera parameters [47–49], so we needed to calibrate the camera to eliminate the effects of lens distortion. Calibration was performed by using the camera to be calibrated to take pictures of the measured objects with known specifications, followed by comparing the estimated value with the real value to calculate the internal camera parameters. Here, the internal camera parameters (focal length, principal point and distortion estimates) of each camera were acquired through the built-in calibration tool (Agisoft Lens v0.4.1) of photogrammetric softcopy Agisoft PhotoScan. The calculated internal camera parameters could improve the quality of cloud points as initial approximations.

2.3. Referencing Setup

Different from the traditional photogrammetry technique, the SfM-MVS method can solve the geometry of the scene and camera parameters automatically, and the control points (CPs) are not necessary for the construction of 3D point clouds. CPs are only used to scale and georeference the cloud points [50]. The bundle adjustment can be carried out theoretically by only three CPs. However, many researchers concluded that CPs are necessary to compensate for the non-linear model misalignment, and the use of CPs can considerably improve the accuracy of models [48,49]. Here, the relative coordinates of 30 CPs were measured by total station (Trimble M3) to obtain optimal accuracy in the laboratory. The total station had an estimated 3D observation accuracy of 3 mm. The CPs were selected to be evenly distributed around the steep slope model, perpendicular to the camera shooting direction. In order to evaluate the accuracy of 3D point clouds, the 30 CPs were divided into 20 reference points and 10 check points. The check points were not used for spatial registration, so the model accuracy can be evaluated by the distance between check points and their corresponding points in the dense point clouds. The 20 reference points were used to scale and georeference the point clouds.

2.4. Photograph Processing and Model Reconstruction

A lot of software can achieve 3D-model reconstruction of an objective scene from photographs. One of the key techniques is SfM, which was considered as the most promising method to generate 3D models from multiple 2D images through photogrammetric methods and computer visualization techniques. Simultaneously, the development of MVS algorithms improves point clouds by maximizing the description information of the objective scene through the information in the images [36]. In this study, Agisoft PhotoScan (http://www.agisoft.ru/products/photoscan, accessed on 24 August 2020) was used to produce the 3D point clouds from the photographs based on an SfM-MVS integrated workflow. The photogrammetric softcopy includes all the steps in 3D model reconstruction, and the whole procedure is semi-automatic, only requiring camera calibration parameters as initial inputs and the manual identification of CPs. Additionally, a manual interference is required for optimizing the 3D-reconstruction process, such as deleting erroneous peaks and troughs induced by mismatching.

The photographs are first scanned for the identification of matching points using a point-matching algorithm. For automatic matching-point extraction in individual images used for image alignment, PhotoScan implements the SIFT algorithm, as the identified features are invariant to camera rotation and image scaling, and to some extent invariant to illumination changes and 3D camera viewpoint. Then, the network of matching points in multiple images is used to estimate both the camera locations and the camera-lens parameters by bundle adjustment (BA). Before optimizing camera parameters, non-target and stray points were eliminated for sparse point clouds. After this initial scene construction, the CPs were manually marked on multiple images, which could be used to improve the image alignment and the estimation of camera calibration parameters through an iterative process in the workflow of PhotoScan. We could check the photograph-alignment process based on the predicted locations of CPs in the model. Finally, through several iterations of optimization, the reprojection error and CP error were stable. After the alignment process, the final dense point clouds were produced from primary sparse point clouds through MVS dense-image-matching algorithms. Dense-image matching is an important procedure for improving point clouds derived from SfM matching techniques, and could increase point density by two or three orders of magnitude. Here, the 3D models were reconstructed with a moderate-depth filter at high quality. The detailed description of SfM-MVS integrated workflow can be found in the literature [20,25].

2.5. Calculation of Volumes of Gravitational Erosion Events

The detailed objective terrain data could be obtained in real time through the SfM-MVS integrated workflow mentioned above. In order to calculate the volumes of gravitational-erosion events, we needed to identify the timings, types, and areas of each gravitational erosion event from the video of the experimental process (Figure 1). Then, we could select the area of interest by comparing the changes in geometric shape of objective surface before and after each gravitational erosion event. Finally, the volume change in the interested area was defined as the volume of each gravitational-erosion event [51,52]. In this paper, we used Cloud Compare (http://cloudcompare.org, accessed on 5 September 2020), an open-source 3D point-cloud and mesh processing software, to calculate the volume change for the point clouds of the interested area before and after each gravitational-erosion event [19].

2.6. Experimental Setup and Assessment

The accuracy of the models derived from SfM-MVS could be assessed using the residuals of the CPs. The root-mean-square errors (RMSEs) of the CPs could be calculated by comparing the measured coordinates through the total station with the interpolated coordinates of the models generated through the photogrammetric process [20]. Additionally, we could also evaluate the accuracy of the models by directly comparing the distance between the clouds of points derived from TLS and SfM-MVS [50]. In order to calculate the distance between two clouds of points, Cloud Compare was selected for cloud-to-cloud comparisons as it can avoid the complex and risky meshing or gridding processes and is more robust to local noise between two methods [53]. For TLS, we used a Z+F IMAGER 5100C, which has a reported precision of 0.1 mm and an accuracy of 0.3–0.5 mm at a 10 m scan distance depending on the surface reflectance of its objective. Additionally, in order to evaluate the reproducibility of the SfM-MVS method, we performed two exercises to obtain two sets of images of the same steep slope terrain by adjusting the cameras' network [54–56]. For the two exercises, the camera positions were adjusted to gain two highly redundant and overlapping image series. Then, the two sets of images were used to generate the 3D density point cloud through the SfM-MVS integrated workflow mentioned above. In this paper, the reproducibility of the multi-image 3D reconstruction method for the steep slope in the laboratory was evaluated by the cloud-to-cloud comparison between two point clouds. The schematic illustration of the data-acquisition and -processing chain developed for 3D reconstruction with Sfm-MVS is shown in Figure 3.

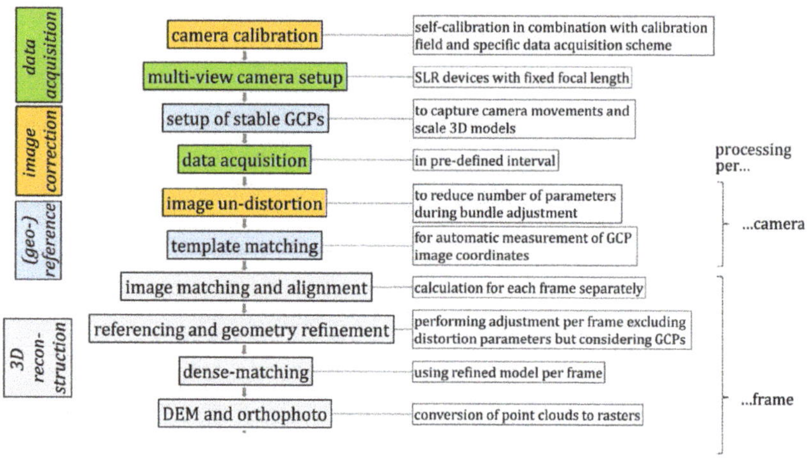

Figure 3. The schematic illustration of the data acquisition and processing chain developed for 3D reconstruction with Sfm-MVS.

The accuracy of the photogrammetric method in calculating the volumes of gravitational erosion events can be evaluated by the accuracy of point clouds derived from SfM-MVS. Generally, the higher the accuracy of point clouds derived from SfM-MVS, the higher the accuracy of the photogrammetric method [19]. However, it is difficult to judge whether the accuracy of point clouds derived from SfM-MVS could meet the requirements for calculating the volumes of gravitational-erosion events. In order to evaluate the accuracy of the photogrammetric method in calculating the volumes of gravitational-erosion events in this paper, manual excavation experiments were performed on the steep slope to simulate gravity-erosion events for the following reasons. One of the most important reasons is that we could not find any other method to measure the volumes of gravitational-erosion events under rainfall conditions, so we conducted several excavation experiments to simulate gravity-erosion events. Another reason is that we could simulate a variety of slope surface types after the occurrence of gravitational-erosion events, such as circular arcs, linear lines, etc., while controlling the scales of gravitational-erosion events simultaneously. In this way, the terrain data could be obtained using TLS and SfM-MVS before and after manual-excavation experiments. Furthermore, the soil volumes of excavation experiments were calculated by the weight of soil after drying and by its bulk density (the conventional oven-drying method). Here, we conducted 20 manual-excavation experiments, and the accuracy of the method proposed in this paper could be verified by comparing the volumes of both the TLS and the conventional oven-drying method.

Finally, this method was applied to monitor the gravitational-erosion processes for the steep granite-soil slope under simulated rainfall conditions in the laboratory (Figure 1). In order to monitor many more gravitational-erosion events, two intense and continuous rainfall conditions were applied, with rainfall intensity designed as 1 and 2 mm min^{-1}, and the rainfall duration set as 5 h. The gravitational-erosion processes were monitored through the SfM-MVS method mentioned above.

3. Results and Discussion

3.1. The Point-Cloud Accuracy Evaluation

After photogrammetric processing, the 3D model of the objective terrain was reconstructed, and the accuracy was assessed in absolute orientation for every CP. Here, the measured coordinates of CPs were compared with the interpolated coordinates, and the RMSEs were calculated. The residuals of reference points were 1.19 mm, 4.08 mm, and 1.91 mm for the X, Y, and Z directions, respectively. The residuals of check points were 1.35 mm, 4.09 mm, and 1.56 mm for the X, Y, and Z directions, respectively (Figure 4). This indicate that millimeter-scale accuracy can be achieved using SfM-MVS-integrated workflow for the steep soil slopes in the laboratory. Simultaneously, the accuracy of the SfM-MVS-integrated workflow method was estimated using TLS and photographs to survey the same steep-slope landscape. The mean distance between the point clouds derived from TLS and SfM-MVS was 1.13 mm, with a standard deviation of 0.93 mm. The cloud-to-cloud distance in the range of −5.0 to 5.0 mm accounted for 98.3%, with 88.0% between −1.0 and 1.0 mm (Figure 5). For the reproducibility experiments, the mean distance between the two point clouds generated through two sets of images of the same steep slope terrain was 0.65 mm, with a standard deviation of 0.51 mm. The cloud-to-cloud distance in the range of −5.0 to 5.0 mm accounted for 99.4%, with 94.2% between −1.0 and 1.0 mm (Figure 5). The repeatability test results showed that the SfM-MVS-integrated workflow method could obtain relatively certain objective terrain data. Furthermore, there was a good match between the 3D model derived from SfM-MVS and that from TLS.

Generally, the SfM MVS method can achieve decimeter scale and even sub-millimeter scale accuracy depending on image characteristics and landform complexity for topographical investigation [15,35–37]. The highest error of a 3D photo-reconstruction method usually occurs in a hidden or low-visibility area, such as steep slopes, vegetation, gullies, and so on [19,42]. In the present study, the orientation of the cameras was orthogonal to the steep slope, which was different from the investigation of aerial images in the field, so the steep slope did not seem to be an important impediment for 3D photo-reconstruction in the laboratory. Additionally, the area of interest was relatively small, and the average point density could reach 2,000,000 points·m^{-2}. The point density was about two orders of magnitude higher than that of the topographical model derived from the SfM-MVS method for the larger areas of terrain in the field [19,42]. Simultaneously, the area of interest was free from vegetation, and the collapsed terrain was not complex. These were all important reasons for the model accuracy in this study to reach the millimeter scale, much higher than the accuracy of models for larger areas of terrain investigations in the field. Furthermore, the cloud-to-cloud distance of models derived from SfM-MVS and TLS was much higher in many field investigations than in our study [19,42,50]. This might also be attributed to the reasons mentioned above. The results of this analysis indicate that the multi-image 3D-reconstruction method is suitable for morphological surface detection in the laboratory due to simple terrain conditions and controllable environmental elements.

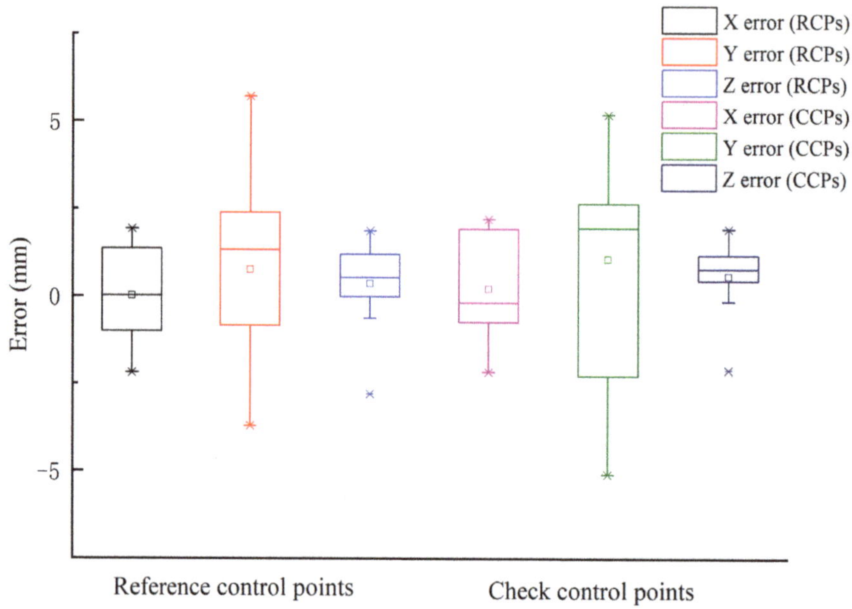

Figure 4. The boxplot of residuals of reference control points and check control points for the X, Y, and Z directions.

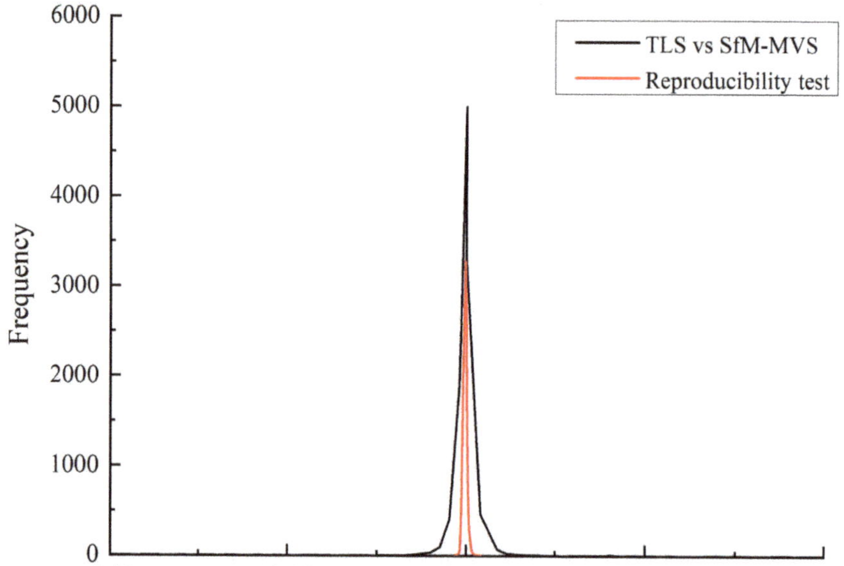

Figure 5. Histogram of cloud-to-cloud distance. Black line refers to the cloud-to-cloud distance between point cloud models derived from SfM-MVS and TLS, and red line refers to the reproducibility test.

3.2. Volume Calculation Accuracy Evaluation

The excavation volumes ranged from ~200 cm^3 to 20,000 cm^3 due to the limitation of model size (Figure 6). The results showed that the relative errors among the volumes calculated by SfM-MVS and TLS or the conventional oven-drying method were all within 10%, with a maximum error of 9.3% and a minimum error of 0.2% for the 20 manual-excavation experiments (Figure 6). This demonstrates that the SfM-MVS method is suitable for the volume measurement of both small-scale gravitational-erosion events (e.g., mudslides) and relatively large-scale collapsing in the laboratory. The linear regression analysis showed that the volumes of gravitational-erosion events calculated by SfM-MVS, TLS and the conventional oven-drying method were closely related ($R^2 > 0.99$, $p < 0.05$) (Figure 6). The measurement accuracy met the general requirements of gravitational-erosion process monitoring.

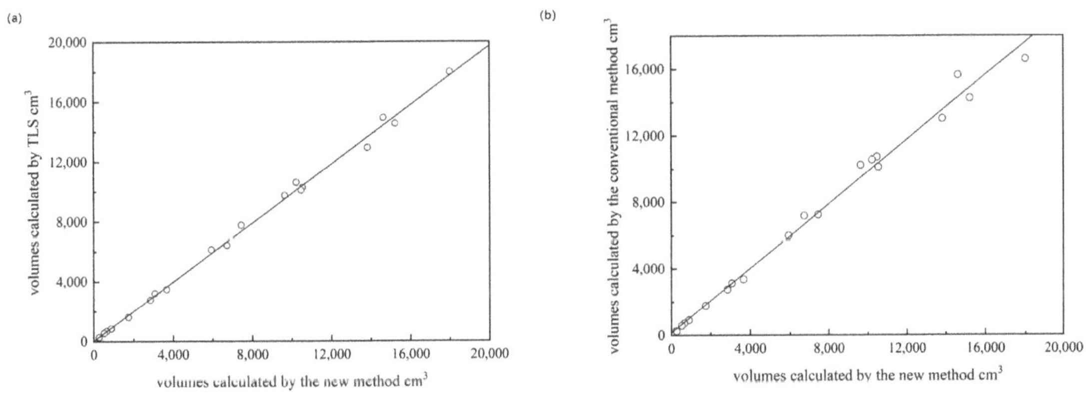

Figure 6. Relationships between gravity-erosion volumes calculated by (**a**) the SfM-MVS method and TLS, (**b**) the SfM-MVS method and conventional oven-drying method. Lines represent the best fit using linear regression.

The 3D photo-reconstruction method is widely applied in soil-erosion monitoring [20,43,44]. Several researchers have concluded that this technique could obtain quite high accuracy relative to the conventional methods in calculating soil-erosion losses. Song et al. (2016) found that the soil-erosion measurement accuracy of close-range photogrammetry could reach 75% using an artificial water washing test, and 83.11% in a natural rainfall experiment. Ming et al. (2019) reported that the average soil-erosion measurement accuracy of close-range photogrammetry could reach 90.67% for free-thaw slopes. However, other researchers argued that close-range photogrammetry was more suitable for surface-elevation change detections in soil-erosion processes than calculating soil-erosion losses directly. They mentioned that the calculation of soil losses from the changes in soil surface may lead to inaccurate results due to alterations in the soil bulk density [5,57]. Even so, the use of close-range photogrammetry techniques in measuring volumes could obtain satisfactory results [58]. In this paper, the relative errors among the volumes calculated by the SfM-MVS method and TLS were all within 10%, indicating that the accuracy of close-range photogrammetry in calculating volumes is quite high, which has been verified by many researchers. Additionally, the relative errors among the volumes calculated by the SfM-MVS method and the conventional oven-drying method were also all within 10%. Therefore, the changes in soil bulk density can be assumed to mainly occur on the soil surface due to the erosion–transportation–deposition processes during rainfall. The soil bulk-density changes might significantly affect the calculation of the soil loss during hydraulic erosion on gentle slopes, but may have little influence on the calculation of the soil loss for gravitational erosion on steep slopes. Furthermore, the volumes of gravity-erosion events were calculated for the interested area before and after each gravitational-erosion event. The calculation error for the volumes in uninterested areas was eliminated, which was also important for improving the calculation accuracy of volumes of gravity-erosion events.

3.3. Application in the Detection of Different Gravitational-Erosion Events

Based on the soil mass failure type, gravitational-erosion events could be divided into avalanche, slide, and earthflow [51,59]. Earthflows have obvious an flow performance and a high water content compared with slides and avalanches. During erosion, the failure block of an avalanche completely separates from the slope surface, whereas that of a landslide slips down as a whole along a weak belt. A total of 213 gravitational-erosion events were monitored in the two simulated rainfall tests in the laboratory, including 8 avalanche events, 14 slide events, and 63 earthflow events under 1 mm min^{-1} rainfall intensity, and 38 avalanche events, 11 slide events, and 79 earthflow events under 2 mm min^{-1} rainfall intensity (Figure 7). The gravitational-erosion volumes ranged from 155 cm^3 to 23,557 cm^3, with the volume of the avalanche and slide much greater than that of the earthflow. The total volume of the avalanche, slide, and earthflow was 37,485 cm^3, 27,119 cm^3, and 27,604 cm^3 for 1 mm min^{-1} rainfall intensity, and 130,648 cm^3, 28,147 cm^3, and 21,876 cm^3 for 2 mm min^{-1} rainfall intensity, respectively. Thus, we could calculate the contribution of gravitational erosion to the total erosion. The proportion of gravitational erosion to total erosion was 69.57% for 1 mm min^{-1} rainfall intensity, while it was 77.04% for 2 mm min^{-1} rainfall intensity. Additionally, the whole gravitational-erosion process could be evaluated through this method. Earthflow existed throughout the whole erosion process with a small value and a large frequency. The avalanche was concentrated in the early stage of rainfall, while the slide mainly occurred in the late stage. The frequency of the avalanche and slide was considerably increased with rainfall intensity (Figure 7). This method could refine and quantify the research on gravitational-erosion processes.

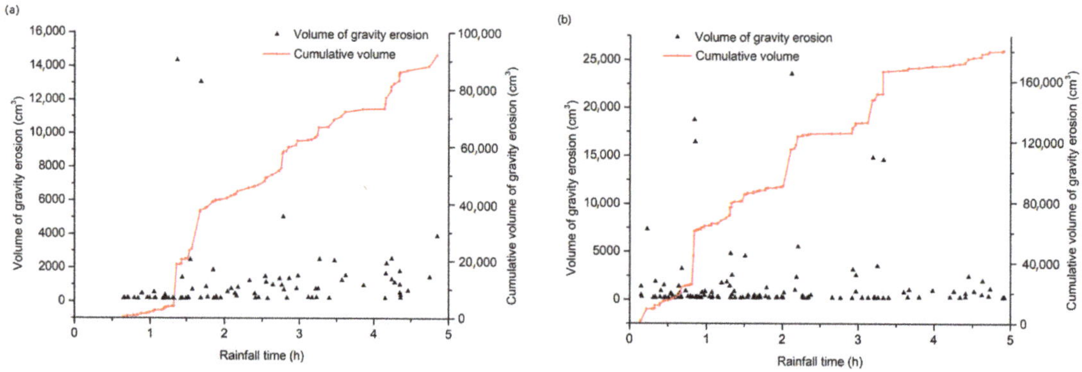

Figure 7. Distribution of gravity-erosion volume with rainfall time. (**a**) The 1 mm min^{-1} rainfall intensity; and (**b**) The 2 mm min^{-1} rainfall intensity.

3.4. Advantages and Limitations

Currently, many devices can be used to monitor changes in surface topography, such as high-precision GPS, high-resolution remote sensing, and airborne/ground laser radar [13–15]. However, none of these devices can be used in the study of gravitational-erosion processes for real-time dynamic monitoring. Mass movements on gully slopes are the main contributors to gully-bank expansion and are a major source of sediment delivered to rivers. However, they are often overlooked because they are either too small or too time-consuming to measure in the field. The SfM-MVS method proposed in this paper can obtain terrain data in real time to quantify each gravitational erosion event. Specially, this method relies on a photogrammetric technique, which has made new progress in obtaining high-resolution terrain data, achieving the millimeter-scale accuracy of the 3D slope model in this paper (Figure 4). Additionally, the volumes of gravity-erosion events were calculated for the interested area before and after each gravitational-erosion event. All of these contributed to improving the accuracy of the SfM-MVS method. Furthermore, this method can be equipped with ordinary consumer un-calibrated and non-metric cameras or even mobile phones [60]. For highly redundant and overlapping images and orthogonal slopes, high-price DSLR cameras are not necessary [33]. Simultaneously, the images obtained by cameras are orderly stored in a computer automatically, and the entire equipment does not require artificial operation in the tests. Therefore, this method has the advantages of a good real-time performance, high accuracy, a low cost, and being labor-saving.

However, this method also suffers some limitations. Firstly, it is only suitable for the detection of gravitational-erosion processes for a relatively small area in the laboratory, and unsuitable for a large area and complex topography in the field, especially for areas with dense vegetation. Secondly, just like any other photogrammetric or laser surface-topography detection techniques, the SfM-MVS method also cannot completely reconstruct the objective terrain in blind or low-visibility areas [19,42]. Gravitational-erosion events occur frequently on those collapsing slopes where very few deep gullies are observed, and such a phenomenon did not occur in our tests, but this limitation did exist in our method.

4. Conclusions

This study introduced the SfM-MVS photogrammetric method to quantify the volume of each gravity-erosion event by controlling multiple cameras to take continuous images at the same time. This method can generate 3D surface models of steep slopes from multiple 2D images based on an SfM-MVS integrated workflow in real time. A millimeter-scale accuracy of 3D models can be achieved using SfM-MVS in the laboratory. The point clouds derived from SfM-MVS matched well with those derived from TLS. Additionally, the SfM-MVS photogrammetric method is suitable for the volume measurement of both small-scale gravitational erosion (e.g., mudslides) and collapsing events in the laboratory, and all the relative errors were within 10%. Finally, this method was used to monitor the gravity-erosion processes under continuous heavy rainfall, and the results demonstrate that this method can refine and quantify these gravitational-erosion processes. Overall, this method has the advantages of real-time data analysis, a low cost, high precision, and being labor-saving, but its application to a large area and a complex topography in the field needs further research.

Author Contributions: Conceptualization, C.C.; methodology, J.X. and Y.W.; writing—original draft preparation, J.X.; writing—review and editing, Y.W., L.Z., X.H., X.L. and P.G.; funding acquisition, L.Z., Y.W. and C.C. All authors have read and agreed to the published version of the manuscript.

Funding: This research was funded by the National Natural Science Foundation of China (No. 41967012; No. 41807065; No. 41630858), the Water Conservancy Science and Technology Program of Jiangxi Province of China (No. 202223TGKT03; No. 202224ZDKT17; No. 202324TGKT08; No. 202324TGKT11), the Science and Technology program of Jiangxi Province of China (No. GJJ2201513) and the China Postdoctoral Science Foundation (No. 2018 M640714).

Data Availability Statement: The data that support the findings of this study are available from the corresponding authors upon reasonable request.

Acknowledgments: The authors thank Hongliang Yu, Jie Wang, Je He and Hao Lu for their help.

Conflicts of Interest: The authors declare no conflict of interest.

References

1. Dong, Y.; Wu, Y.; Yin, J.; Wang, Y.; Gou, S. Investigation of Soil Shear-Strength Parameters and Prediction of the Collapse of Gully Walls in the Black Soil Region of Northeastern China. *Phys. Geogr.* **2011**, *32*, 161–178. [CrossRef]
2. Chen, A.; Zhang, D.; Peng, H.; Fan, J.; Xiong, D.; Liu, G. Experimental study on the development of collapse of overhanging layers of gully in Yuanmou Valley, China. *Catena* **2013**, *109*, 177–185. [CrossRef]
3. Rengers, F.K.; Tucker, G.E. The evolution of gully headcut morphology: A case study using terrestrial laser scanning and hydrological monitoring. *Earth Surf. Process. Landf.* **2015**, *40*, 1304–1317. [CrossRef]

4. Xu, X.-Z.; Zhang, H.-W.; Wang, W.-L.; Zhao, C.; Yan, Q. Quantitative monitoring of gravity erosion using a novel 3D surface measuring technique: Validation and case study. *Nat. Hazards* **2014**, *75*, 1927–1939. [CrossRef]
5. Guo, M.; Shi, H.; Zhao, J.; Liu, P.; Welbourne, D.; Lin, Q. Digital close range photogrammetry for the study of rill development at flume scale. *Catena* **2016**, *143*, 265–274. [CrossRef]
6. Yang, J.S.; Yao, W.Y.; Zheng, M.G.; Li, L. Analysis on gravitational sediment yield in the check-dam controlled basins of Chabagou Watershed. *J. Hydraul. Eng.* **2017**, *48*, 241–245. (In Chinese)
7. Dong, Y.; Xiong, D.; Su, Z.; Duan, X.; Lu, X.; Zhang, S.; Yuan, Y. The influences of mass failure on the erosion and hydraulic processes of gully headcuts based on an in situ scouring experiment in Dry-hot valley of China. *Catena* **2019**, *176*, 14–25. [CrossRef]
8. Liu, X.; Zhang, Y.; Zhang, L.; Fang, X.; Deng, W.; Liu, Y. Aggregate-associated soil organic carbon fractions in subtropical soil undergoing vegetative restoration. *Land Degrad. Dev.* **2023**. *early view*. [CrossRef]
9. Cai, Q.G.; Lu, Z.X.; Wang, G.P. Process-based soil erosion and sediment yield model in a small basin in the hilly Loess region. *Acta Geogr. Sinca* **1996**, *51*, 108–117. (In Chinese)
10. Parise, M.; Wasowski, J. Landslide Activity Maps for Landslide Hazard Evaluation: Three Case Studies from Southern Italy. *Nat. Hazards* **1999**, *20*, 159–183. [CrossRef]
11. Iverson, R.M.; Reid, M.E.; Iverson, N.R.; LaHusen, R.G.; Logan, M.; Mann, J.E.; Brien, D.L. Acute Sensitivity of Landslide Rates to Initial Soil Porosity. *Science* **2000**, *290*, 513–516. [CrossRef] [PubMed]
12. Kirschbaum, D.B.; Adler, R.; Hong, Y.; Hill, S.; Lerner-Lam, A. A global landslide catalog for hazard applications: Method, results, and limitations. *Nat. Hazards* **2010**, *52*, 561–575. [CrossRef]
13. Hu, G.; Wu, Y.; Liu, B.; Zhang, Y.; You, Z.; Yu, Z. The characteristics of gully erosion over rolling hilly black soil areas of Northeast China. *J. Geogr. Sci.* **2009**, *19*, 309–320. [CrossRef]
14. Perroy, R.L.; Bookhagen, B.; Asner, G.P.; Chadwick, O.A. Comparison of gully erosion estimates using airborne and ground-based LiDAR on Santa Cruz Island, California. *Geomorphology* **2010**, *118*, 288–300. [CrossRef]
15. Castillo, C.; Pérez, R.; James, M.R.; Quinton, J.N.; Taguas, E.V.; Gómez, J.A. Comparing the Accuracy of Several Field Methods for Measuring Gully Erosion. *Soil Sci. Soc. Am. J.* **2012**, *76*, 1319–1332. [CrossRef]
16. Ehiorobo, J.O.; Audu, H.A.P. Monitoring of gully erosion in an urban area using geoinformation technology. *J. Emerg. Trends Eng. Appl. Sci.* **2012**, *3*, 270–275.
17. Chandler, J. Effective application of automated digital photogrammetry for geomorphological research. *Earth Surf. Process. Landf.* **1999**, *24*, 51–63. [CrossRef]
18. Carbonneau, P.E.; Lane, S.N.; Bergeron, N.E. Cost-effective non-metric close-range digital photogrammetry and its application to a study of coarse gravel river beds. *Int. J. Remote Sens.* **2003**, *24*, 2837–2854. [CrossRef]
19. Gómez-Gutiérrez, A.; Schnabel, S.; Berenguer-Sempere, F.; Lavado-Contador, F.; Rubio-Delgado, J. Using 3D photo-reconstruction methods to estimate gully headcut erosion. *Catena* **2014**, *120*, 91–101. [CrossRef]
20. Frankl, A.; Stal, C.; Abraha, A.; Nyssen, J.; Rieke-Zapp, D.; De Wulf, A.; Poesen, J. Detailed recording of gully morphology in 3D through image-based modelling. *Catena* **2015**, *127*, 92–101. [CrossRef]
21. Lowe, D.G. Distinctive Image Features from Scale-Invariant Keypoints. *Int. J. Comput. Vis.* **2004**, *60*, 91–110. [CrossRef]
22. Seitz, S.M.; Curless, B.; Diebel, J.; Scharstein, D.; Szeliski, R. A Comparison and Evaluation of Multi-View Stereo Reconstruction Algorithms. In Proceedings of the IEEE Computer Society Conference on Computer Vision and Pattern Recognition, New York, NY, USA, 17–22 June 2006.
23. Snavely, N.; Seitz, S.M.; Szeliski, R. Modeling the World from Internet Photo Collections. *Int. J. Comput. Vis.* **2008**, *80*, 189–210. [CrossRef]
24. Verhoeven, G.; Doneus, M.; Briese, C.; Vermeulen, F. Mapping by matching: A computer vision-based approach to fast and accurate georeferencing of archaeological aerial photographs. *J. Archaeol. Sci.* **2012**, *39*, 2060–2070. [CrossRef]
25. Javernick, L.; Brasington, J.; Caruso, B. Modeling the topography of shallow braided rivers using Structure-from-Motion photogrammetry. *Geomorphology* **2014**, *213*, 166–182. [CrossRef]
26. Matilde, B.; Angel, M.; Jose, L.; Sara, I. Estimation of small-scale soil erosion in laboratory experiments with Structure from Motion Photogrammetry. *Geomorphology* **2017**, *295*, 285–296.
27. Eltner, A.; Kaiser, A.; Abellan, A.; Schindewolf, M. Time lapse structure-from-motion photogrammetry for continuous geomorphic monitoring. *Earth Surf. Process. Landf.* **2017**, *42*, 2240–2253. [CrossRef]
28. Wu, C. Towards linear-time incremental structure from motion. In Proceedings of the 3DV-Conference, International Conference on IEEE Computer Society, Seattle, WA, USA, 29 June–1 July 2013; pp. 127–134. [CrossRef]
29. Chandler, J.H.; Fryer, J.G. AutoDesk 123D Catch: How accurate is it? *Geomat. World* **2013**, *2*, 32–33.
30. Brunier, G.; Fleury, J.; Anthony, E.J.; Gardel, A.; Dussouillez, P. Close-range airborne Structure-from-Motion Photogrammetry for high-resolution beach morphometric surveys: Examples from an embayed rotating beach. *Geomorphology* **2016**, *261*, 76–88. [CrossRef]
31. Jancosek, M.; Pajdla, T. Multi-view reconstruction preserving weakly-supported surfaces. In Proceedings of the CVPR 2011, Colorado Springs, CO, USA, 20–25 June 2011; pp. 3121–3128. [CrossRef]

32. Micheletti, N.; Chandler, J.H.; Lane, S.N. Investigating the geomorphological potential of freely available and accessible structure-from-motion photogrammetry using a smartphone. *Earth Surf. Process. Landf.* **2014**, *40*, 473–486. [CrossRef]
33. Thoeni, K.; Giacomini, A.; Murtagh, R.; Kniest, E. A comparison of multi-view 3D reconstruction of a rock wall using several cameras and a laser scanner. *ISPRS—Int. Arch. Photogramm. Remote Sens. Spat. Inf. Sci.* **2014**, *40*, 573–580. [CrossRef]
34. Mosbrucker, A.R.; Major, J.J.; Spicer, K.R.; Pitlick, J. Camera system considerations for geomorphic applications of SfM photogrammetry. *Earth Surf. Process. Landf.* **2017**, *42*, 969–986. [CrossRef]
35. James, M.R.; Robson, S. Straightforward reconstruction of 3D surfaces and topography with a camera: Accuracy and geoscience application. *J. Geophys. Res. Earth Surf.* **2012**, *117*, 03017. [CrossRef]
36. Fonstad, M.A.; Dietrich, J.T.; Courville, B.C.; Jensen, J.L.; Carbonneau, P.E. Topographic structure from motion: A new development in photogrammetric measurement. *Earth Surf. Process. Landf.* **2013**, *38*, 421–430. [CrossRef]
37. Peter, K.D.; D'Oleire-Oltmanns, S.; Ries, J.B.; Marzolff, I.; Hssaine, A.A. Soil erosion in gully catchments affected by land-levelling measures in the Souss Basin, Morocco, analysed by rainfall simulation and UAV remote sensing data. *Catena* **2014**, *113*, 24–40. [CrossRef]
38. Chandler, J.; Ashmore, P.; Paola, C.; Gooch, M.; Varkaris, F. Monitoring River-Channel Change Using Terrestrial Oblique Digital Imagery and Automated Digital Photogrammetry. *Ann. Assoc. Am. Geogr.* **2002**, *92*, 631–644. [CrossRef]
39. Bird, S.; Hogan, D.; Schwab, J. Photogrammetric monitoring of small streams under a riparian forest canopy. *Earth Surf. Process. Landf.* **2010**, *35*, 952–970. [CrossRef]
40. Gonçalves, J.A.; Henriques, R. UAV photogrammetry for topographic monitoring of coastal areas. *ISPRS J. Photogramm. Remote Sens.* **2015**, *104*, 101–111. [CrossRef]
41. Woodget, A.S.; Carbonneau, P.E.; Visser, F.; Maddock, I.P. Quantifying submerged fluvial topography using hyperspatial resolution UAS imagery and structure from motion photogrammetry. *Earth Surf. Process. Landf.* **2015**, *40*, 47–64. [CrossRef]
42. Westoby, M.; Brasington, J.; Glasser, N.F.; Hambrey, M.J.; Reynolds, J.M. 'Structure-from-Motion' photogrammetry: A low-cost, effective tool for geoscience applications. *Geomorphology* **2012**, *179*, 300–314. [CrossRef]
43. Kaiser, A.; Neugirg, F.; Rock, G.; Müller, C.; Haas, F.; Ries, J.; Schmidt, J. Small-Scale Surface Reconstruction and Volume Calculation of Soil Erosion in Complex Moroccan Gully Morphology Using Structure from Motion. *Remote Sens.* **2014**, *6*, 7050–7080. [CrossRef]
44. Glendell, M.; McShane, G.; Farrow, L.; James, M.R.; Quinton, J.; Anderson, K.; Evans, M.; Benaud, P.; Rawlins, B.; Morgan, D.; et al. Testing the utility of structure-from-motion photogrammetry reconstructions using small unmanned aerial vehicles and ground photography to estimate the extent of upland soil erosion. *Earth Surf. Process. Landf.* **2017**, *42*, 1860–1871. [CrossRef]
45. Niethammer, U.; Rothmund, S.; Schwaderer, U.; Zeman, J.; Joswig, M. Open source image processing tools for low cost uav based landslide investigations. *ISPRS—Int. Arch. Photogramm. Remote Sens. Spat. Inf. Sci.* **2012**, *38*, 161–166. [CrossRef]
46. Xia, J.; Cai, C.; Wei, Y.; Wu, X. Granite residual soil properties in collapsing gullies of south China: Spatial variations and effects on collapsing gully erosion. *Catena* **2018**, *174*, 469–477. [CrossRef]
47. Eltner, A.; Baumgart, P.; Maas, H.-G.; Faust, D. Multi-temporal UAV data for automatic measurement of rill and interrill erosion on loess soil. *Earth Surf. Process. Landf.* **2014**, *40*, 741–755. [CrossRef]
48. James, M.; Robson, S.; D'Oleire-Oltmanns, S.; Niethammer, U. Optimising UAV topographic surveys processed with structure-from-motion: Ground control quality, quantity and bundle adjustment. *Geomorphology* **2017**, *280*, 51–66. [CrossRef]
49. James, M.R.; Robson, S.; Smith, M.W. 3-D uncertainty-based topographic change detection with structure-from-motion photogrammetry: Precision maps for ground control and directly georeferenced surveys. *Earth Surf. Process. Landf.* **2017**, *42*, 1769–1788. [CrossRef]
50. Stöcker, C.; Eltner, A.; Karrasch, P. Measuring gullies by synergetic application of UAV and close range photogrammetry—A case study from Andalusia, Spain. *Catena* **2015**, *132*, 1–11. [CrossRef]
51. Xu, X.-Z.; Liu, Z.-Y.; Xiao, P.-Q.; Guo, W.-Z.; Zhang, H.-W.; Zhao, C.; Yan, Q. Gravity erosion on the steep loess slope: Behavior, trigger and sensitivity. *Catena* **2015**, *135*, 231–239. [CrossRef]
52. Guo, W.-Z.; Xu, X.-Z.; Wang, W.-L.; Yang, J.-S.; Liu, Y.-K.; Xu, F.-L. A measurement system applicable for landslide experiments in the field. *Rev. Sci. Instrum.* **2016**, *87*, 044501. [CrossRef]
53. Lague, D.; Brodu, N.; Leroux, J. Accurate 3D comparison of complex topography with terrestrial laser scanner: Application to the Rangitikei canyon (N-Z). *ISPRS J. Photogramm. Remote Sens.* **2013**, *82*, 10–26. [CrossRef]
54. Clapuyt, F.; Vanacker, V.; Van Oost, K. Reproducibility of UAV-based earth topography reconstructions based on Structure-from-Motion algorithms. *Geomorphology* **2016**, *260*, 4–15. [CrossRef]
55. Smith, M.W.; Vericat, D. From experimental plots to experimental landscapes: Topography, erosion and deposition in sub-humid badlands from Structure-from-Motion photogrammetry. *Earth Surf. Process. Landf.* **2015**, *40*, 1656–1671. [CrossRef]
56. Goetz, J.; Brenning, A.; Marcer, M.; Bodin, X. Modeling the precision of structure-from-motion multi-view stereo digital elevation models from repeated close-range aerial surveys. *Remote Sens. Environ.* **2018**, *210*, 208–216. [CrossRef]
57. Gessesse, G.D.; Fuchs, H.; Mansberger, R.; Klik, A.; Rieke-Zapp, D.H. Assessment of Erosion, Deposition and Rill Development on Irregular Soil Surfaces Using Close Range Digital Photogrammetry. *Photogramm. Rec.* **2010**, *25*, 299–318. [CrossRef]
58. El-Din Fawzy, H. The accuracy of determing the volumes using close range photogrammetry. *J. Mech. Civ. Eng.* **2015**, *12*, 10–15.

59. Guo, W.-Z.; Luo, L.; Wang, W.-L.; Liu, Z.-Y.; Chen, Z.-X.; Kang, H.-L.; Yang, B. Sensitivity of rainstorm-triggered shallow mass movements on gully slopes to topographical factors on the Chinese Loess Plateau. *Geomorphology* **2019**, *337*, 69–78. [CrossRef]
60. Chandler, J.H.; Fryer, J.G.; Jack, A. Metric capabilities of low-cost digital cameras for close range surface measurement. *Photogramm. Rec.* **2005**, *20*, 12–26. [CrossRef]

water

MDPI

Article

Composite Factors during Snowmelt Erosion of Farmland in Black Soil Region of Northeast China: Temperature, Snowmelt Runoff, Thaw Depths and Contour Ridge Culture

Haoming Fan, Yunqing Hou, Xiuquan Xu *, Caihong Mi and Hao Shi

College of Water Conservancy, Shenyang Agricultural University, Shenyang 110866, China; fanhaoming@163.com (H.F.); houyunqing@syau.edu.cn (Y.H.); micaihong@syau.edu.cn (C.M.); shihao@syau.edu.cn (H.S.)

* Correspondence: xuxiuquan1986@126.com

Abstract: Snowmelt erosion could cause serious damage to soil quality and agricultural production conditions of slope farmland in the black soil region of northeast China. Contour ridge tillage is a traditional and effective measure to mitigate soil loss on slope farmland. However, the characteristics and influence factors of snowmelt erosion of slope farmland with contour ridge culture and the effect of this measure on the snowmelt process have not been comprehensively investigated, especially at the field scale. To bridge the gap, in situ observation was conducted on the snowmelt erosion process of a typical farmland in Baiquan County, Heilongjiang Province, China. The results revealed that during the snowmelt erosion period, the average daily snowmelt runoff volume and sediment concentration exhibited a trend of first increase and then a subsequent decrease. In the early stage, although the sediment concentration was large, limited discharge and soil thaw depths led to minimal soil loss. In the following stage, due to increased runoff and thaw depths, 94% of the total soil loss amount was obtained with an obvious erosion path formed. For each event, when soil thaw depths were shallow, sediment concentration had a high and early peak, whereas a reverse trend was observed when thaw depths increased. The hysteresis relationship of discharge–sediment indicated that the location where snowmelt erosion primarily occurred would change, under the influence of variations in runoff, freeze and thaw action, thaw depths, and micro-topography. The results could provide a guide in the control of soil erosion in seasonal snowmelt-erosion-prone areas.

Keywords: snowmelt runoff; soil-thawing depth; rill; sediment concentration; freeze and thaw action

1. Introduction

Snowmelt erosion occurs in areas with high latitudes and high altitudes [1–4]. In these ecologically vulnerable regions, snowmelt erosion could make severe damage to land health, water quality, crop yield, and ecological balance [5–8]. Compared with rainfall-induced soil erosion, snowmelt-induced erosion has its own distinct characteristics, such as snowmelt runoff, which is sensitive to variations in radiation and air temperature, surface soil being affected by freeze and thaw action, the frozen soil layer reducing infiltration, and low vegetation coverage during the snowmelt period [9–11].

In Nordic countries, snowmelt erosion is particularly severe combined with rainfall for partially thawed soil as infiltration is restricted, leading water pollutants to recipient water bodies [12]. In the Tibetan Plateau region, spring melt water erosion is affected by multiple factors on an alpine meadow slope, and fast melting at high flow rates could promote soil erosion [13,14]. Spatially distributed soil erosion should focus on water infiltration, the influence of ice and snow on runoff, and variations in soil surface structure caused by freeze and thaw action [15]. Snowmelt erosion could result in micro-topography changes in space and over time due to frost heave, snowpack, erosion, and deposition by overland flow [16].

Northeast China is the region with the largest seasonal snow amount in the country. Snow accumulation is a prerequisite for snowmelt erosion. The spring thawing period is a time when soil freeze–thaw processes occur strongly, which is generally prone to soil erosion. The theoretical research on soil erosion processes under the influences of snowmelt runoff and freeze and thaw action in China still requires extensive field observations and data. The related research mainly focuses on glaciers and frozen soils, with an emphasis on hydrological processes.

Snowmelt erosion in northeast China accounts for a non-negligible proportion of total soil erosion throughout the year [17]. To some extent, the harm of snowmelt erosion is severely underestimated. Recent research on soil erosion in northeast China has mainly focused on observing rainfall-induced soil erosion and simulation experiments under artificial conditions. In experimental designs, an external water supply method is usually used to simulate snowmelt runoff. However, under natural conditions, there is still snow cover during snowmelt periods, which would gradually disappear. Incompletely melted snow may preserve snowmelt runoff to some extent. At the same time, local micro-topography reconstructed by soil and water conservation measures could also block water from snowmelt and soil-thawing processes and increase infiltration. Consequently, the release of snowmelt water could be influenced by various factors. Thaw depths of surface soil change across different points on slope because many factors could change soil freeze and thaw process, such as air temperature, snow ablation, snowmelt discharge, and local terrain [18,19].

The black soil region in northeast China is an important commodity grain base in the country. During the spring thaw period, the combined effects of freeze–thawing action, snowmelt discharge, and low vegetation cover could lead to serious snowmelt erosion, which might be more severe than rainfall-induced soil erosion in some locations. Snowmelt erosion has a significant impact on agricultural production and regional economic safety [20]. Snowmelt erosion on a slope has a sediment transport and deposition process, which could be revealed by the sediment–discharge hysteresis relationship by quantifying time-scale dynamics [21]. Moreover, potential sediment sources could also be confirmed. Tillage methods could reduce soil erosion [22–24]. Contour ridge tillage is a traditional measure for soil erosion control in this area with small gradients but long slope lengths [25]. Historically, there has been an underestimation of the detrimental impact of snowmelt erosion. While gully erosion resulting from snowmelt erosion lies beyond the scope of this observation, it is crucial to acknowledge that the morphological alterations caused by snowmelt erosion may exacerbate subsequent soil erosion processes.

However, there has been limited research focused on the snowmelt erosion process of slope farmland with contour ridge culture at the field scale. To address this gap, we conducted an observation on typical slope farmland with this measure. The primary objectives of this study were (a) to monitor the snowmelt erosion process, including snowmelt discharge and soil loss amounts and identify critical influencing factors and (b) to analyze the effects of influencing factors such as air temperature changes, thaw depths of surface soil, and micro-topography reshaped by contour ridge culture on the snowmelt erosion process. The findings of this study will provide a scientific and theoretical foundation for addressing snowmelt erosion and designing effective soil conservation measures in areas with seasonal snow cover prone to snowmelt erosion.

2. Materials and Methods

2.1. Study Site

The study site is situated in the Jiusheng small watershed (47°26′54″ N, 126°18′8″ E), Baiquan County, Heilongjiang Province, China. The region is characterized as a combination of over-flood plain and hill areas, with a large proportion of low mountains and hills. The climate belongs to the mid-temperate continental monsoon climate zone, with a feature of cold and dry winters and hot and rainy summers. The annual average rainfall is 490 mm, with 70% of the total amount concentrated between July and September. The annual average temperature is 1.5 °C. The spring snowmelt period is generally from middle to late March, when air temperatures rise rapidly during the daytime. Temperatures rise above 0 °C in the daytime and drop below 0 °C at night, leading to freeze and thaw action, which makes the soil susceptible to snowmelt erosion. The main land use patterns in the region consist of cultivated land and forest land. The major cultivation form on slope farmland is contour ridge tillage or horizontal ridge farming. The main forms of soil erosion are surface erosion and gully erosion occurring on slope farmland.

The investigation site represents a typical slope farmland catchment that has been cultivated over 50 years in a small watershed. It spans approximately 330 m from east to west and 105 m from north to south, with the middle position lower than other parts. The land use type is contour ridge culture. The cultivation measures are corn ridges, planted in May and castrated in October, with minimal crop residue remaining on the ground. The average slope of the site is 2.47. However, the complete slope exhibits non-uniform characteristics, displaying a concave profile overall, with notably steep inclines in both the uphill and downhill regions. The central section of the slope features gentle terrain, giving rise to localized depressions. Moreover, the micro-topography of the entire slope demonstrates pronounced spatial heterogeneity, attributed in part to continuous cultivation practices. Snowmelt erosion is severer because the slope exhibits considerable length, coupled with an extensive catchment area, leading to a substantial accumulation of runoff on the slope with enough snow cover (Figure 1).

Figure 1. The investigated catchment before snowmelt erosion, with the outlet selected at the head of a gully.

2.2. Snowmelt Erosion Process Observation

2.2.1. Snowmelt Runoff and Soil Loss

The observation was carried out from middle to late March 2018. There were mainly sunny days. The temperature gradually increased in the daytime, with a minimum of above 0 °C and a maximum of nearly 20 °C. Snow began to smelt, and then erosion occurred as a result of snowmelt runoff and freeze–thaw action. Samples were collected from 23 to 26 March, with a simple device and collecting buckets at the designated outlet, located at the head of a gully as shown in Figure 1.

The sample interval was 30 min. Sampling times and runoff amounts were recorded on-site. The collected runoff in buckets was evenly stirred, 500 mL of which was transferred into sampling bottles. Then, samples were brought to the experiment room and settled for more than 24 h and measured by using a drying method in order to obtain the sediment concentration. The samples were placed in a constant temperature drying oven at 105 °C for 8 h to make the water evaporate completely. Then, the remaining dry sediment was obtained, and sediment concentrations were accordingly calculated.

2.2.2. Factors and Measurements

The temperature and thaw depths of the surface soil were simultaneously measured or recorded simultaneously during the snowmelt period in order to analyze their influence on snowmelt erosion. Temperatures were recorded hourly using a thermometer in situ. Steel pines were used to measure thaw depths of surface soil at two locations, namely the top of the slope and a nearby place close to the outlet, at 11:00 and 15:00, separately. Steel pines were inserted into the soil until blocked by a frozen layer, and the lengths of penetration into the soil were recorded as thaw depths with a ruler. Meanwhile, the rills and micro-topography affected by contour ridge culture across the slope were also investigated through photography.

In this region, seasonal snowmelt erosion predominantly transpires within a condensed timeframe of approximately one week. Despite its significance, comprehensive observations on the fundamental aspects of snowmelt erosion worldwide remain relatively in need. Given these circumstances, the observation was planned meticulously, striving to capture intricate details and augment the sampling frequency in order to provide sufficient evidence to elucidate the distinctive characteristics of snowmelt erosion in this specific region.

2.3. Data Analysis

The snowmelt erosion process was represented by several following points: daily changes in snowmelt runoff and sediment, variations in runoff and sediment under the influence of air temperature changes at each event, and the discharge–sediment hysteresis relationship using the method proposed by Williams [26] to reveal the dynamic feature of sediment and discharge and potential sediment sources through differences between the two parameters. Subsequently, the roles of the following composite factors in the snowmelt erosion process were examined, including air temperature, discharge, freeze–thaw action or thaw depths of the surface soil, and topographic features. Snowmelt erosion has obvious spatial heterogeneity, so erosional appearance, sediment sources, and the status of the contour ridge system were also taken into consideration.

3. Results and Discussion

3.1. Daily Average Snowmelt Runoff and Soil Loss

Snowmelt erosion in this investigation took place intensively from 23 to 26 March 2018. The daily average snowmelt erosion process is illustrated in Figure 2, presenting the average snowmelt runoff and sediment concentration.

- On 23 March only 0.36% of the total snowmelt runoff amount was observed, and the low proportion was due to melted water being partially retained in the snow cover and frozen surface soil.
- On 24 March, as the air temperature continued to rise, the average snowmelt runoff exceeded that of the previous day, accounting for 4.58% of the total soil loss amount (sediment concentration multiplied by total runoff). The result could be attributed to a reduction in the water storage capacity of snow and soil. So, more melt water was released in the form of snowmelt discharge.
- On 25 March, because the air temperature increased rapidly, the snowmelt process was accelerated synchronously. Consequently, 94.49% of the total runoff amount was obtained, including both melted snow and water from thawed soil.

- On 26 March, there was no obvious snow cover. However, 0.57% of total runoff was still collected, which was mainly from the accumulation and thawing of liquid water in the surface soil.

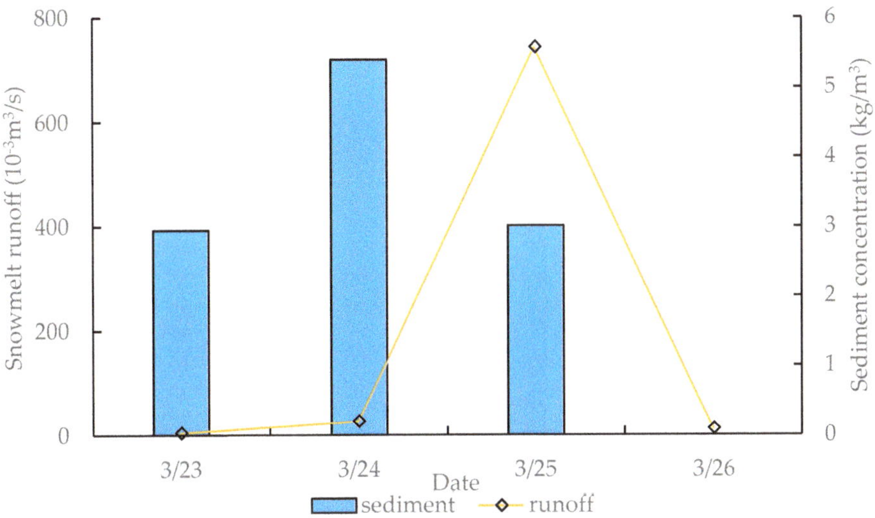

Figure 2. Daily average snowmelt runoff and sediment.

Throughout the 4 days, the air temperature remained below 0 °C at night but rose above 0 °C during the daytime (even the lowest temperature). Thus, the primary driving factor for snowmelt was the air temperature, which was consistent with temperature-driven snowmelt in an Alpine catchment [27].

- On 23 March, as the air temperature increased slowly, snow ablation was relatively obvious on the upper slope, resulting in some parts of bare soil. However, most of the site was covered by snow, with frozen surface soil. Meanwhile, snowmelt did not make a large runoff as discussed above. As a result, there was very little soil loss, accounting for only 0.2% of the total amount.
- On 24 March, much more water was released from the melting snow cover and thawed soil, which enabled the snowmelt runoff to reach the outlet, by infiltration, passing along the ridge or flowing through the furrow. Simultaneously, substantial portions of the topsoil began to thaw, making sediment particles much more susceptible to being eroded, especially in several places with broken ridge furrows. Sediment concentration increased rapidly, reaching a maximum of 17.32 kg/m^3 at 11:00, and accounted for 7.5% of the total soil loss amount.
- On 25 March, the rates of snowmelt and soil thawing further intensified, leading to the majority of the runoff and soil loss (92.3% of the total amount). Micro-topography also played a significant role in promoting snowmelt erosion [16]. Water stored in ridges and furrows would infiltrate, or run along/through them, leading to much more soil failure in forms of scouring or collapse. Local and whole distinct flow routes in situ were evident, and the centralized snowmelt discharge produced severe erosion.
- No soil loss had been observed on the 26 March.

Snow accumulation and snowmelt are critical factors in snowmelt hydrology and erosion [4]. Notably, melted water could be held by snow cover, the contour ridge system, or small dams created by straw and ice [15]. Additionally, soil particles undergoing freeze and thaw action are more susceptible to erosion [28], although during early period soil erosion, they are restricted by unthawed surface soil. These complex and composite components are crucial for analyzing the snowmelt erosion process.

3.2. Air Temperature

Figure 3 shows the air temperature variation throughout March. Snowmelt discharge was collected on 23 March, when the minimum air temperature increased and rapidly approached 0 °C. The subsequent 4 days had similar minimum air temperatures, but the maximum temperature increased rapidly. The occurrence of snowmelt erosion requires both sufficient snow accumulation and a rapid rise in air temperature.

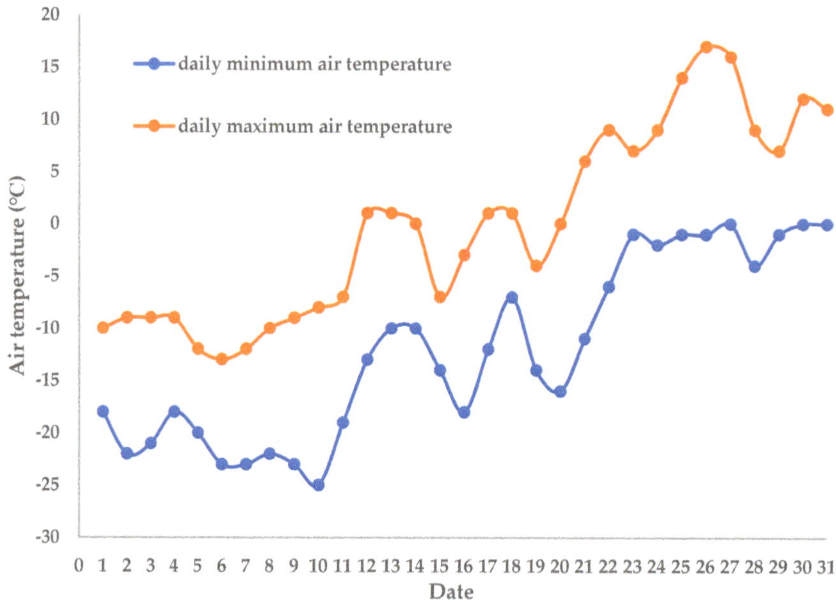

Figure 3. Daily minimum and maximum air temperatures in March.

Variations in air temperatures, snowmelt runoff, and the sediment concentration of each event are presented in Figure 4.

- On 23 March, snowmelt runoff began around 11:00, reaching its peak at 13:00, with a duration of 3.5 h. The air temperature ranged between 12 and 13 °C from 11:30 to 14:00. The sediment concentration showed a similar trend, reaching its maximum value at 12:00, 1 h earlier than the peak discharge. Despite a relatively high sediment concentration occurring momentarily, the first day of snowmelt erosion had limited snowmelt runoff and soil loss.
- On the morning of the 24 March, the sediment concentration gradually increased with the rise in snowmelt runoff, peaking at 13.72 kg/m^3 at 11:00, which was later than the peak of snowmelt runoff by 30 min. In the afternoon, as the weather changed from cloudy to sunny, and under the influence of rising temperatures and favorable radiation, the declining runoff showed a small rebound. However, due to insufficient surface soil thawing, particularly in snow-covered regions, the sediment yield rapidly decreased after 14:00. The duration of the snowmelt event was approximately 9 h, during which melted water stored in the snow and frozen surface soil was released to a significant extent, causing the breakage of some ridge furrows and the formation of an obvious water flow path near the outlet with a notable width and depth. While snowmelt water could be retained in the snow and furrows or infiltrate into thawed soil, limiting the runoff amount, the shallow soil thaw depths prevented the easy erosion of soil particles by the snowmelt discharge. Consequently, the sediment concentration was primarily influenced by the thawed surface soil particles in the

early stage, while snowmelt erosion became the limiting factor in the later stage due to the slow rate of soil thawing.

- On 25 March, the temperature and radiation improved compared to earlier days, but the weather became cloudy and the temperature slightly declined. The maximum runoff was observed at 13:30, with two peaks observed for both runoff and sediment concentration. The first peak occurred synchronously for both parameters, while the second peak exhibited a 0.5 h delay in sediment concentration compared to the runoff. Based on these observations, it can be inferred that snowmelt runoff is influenced by multiple factors, including snowmelt rate, water from thawed soil, and their relative positions. At the end of the day, most of the snow had melted, and the previously existing melted water in furrows was released, contributing to the complexity of runoff sources and potentially leading to the occurrence of multiple peaks.
- On 26 March, due to a lack of runoff source, only a slight amount of discharge was collected for less than 3 h, and no sediment yield was investigated. The main runoff sources included water from thawing surface soil and a small amount of snow and ice near the outlet.

(a)

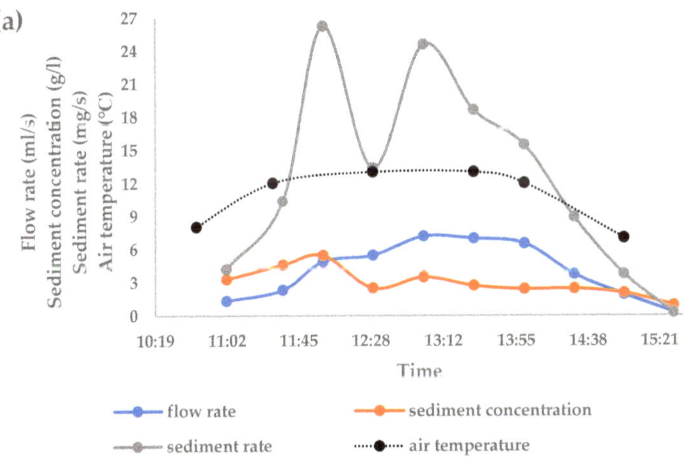

(b)

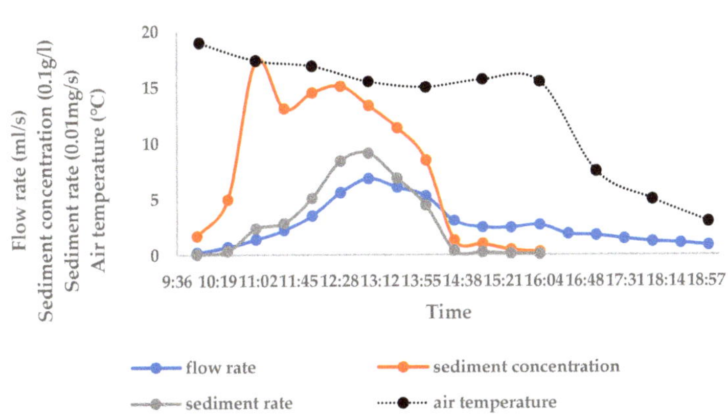

Figure 4. *Cont.*

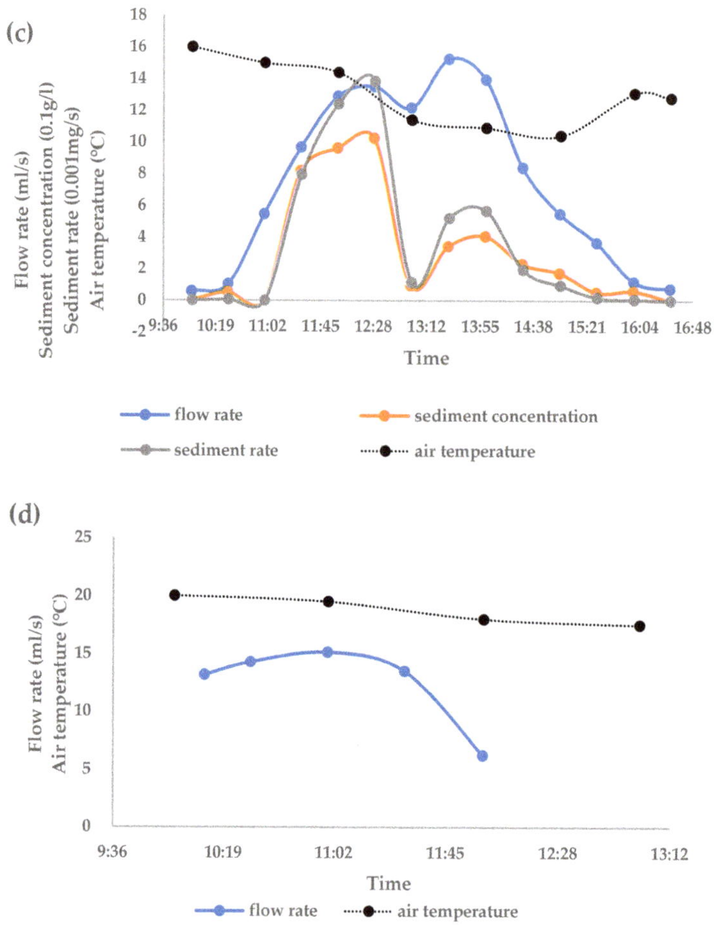

Figure 4. Response of snowmelt erosion process to variations in air temperature at each event ((**a–d**) are processes on 23 to 26, repreently).

3.3. Thaw Depths of Surface Soil

Thaw depths of surface soil at the upper slope and lower slope are indicated in Figure 5. Overall, the soil-thawing depths increased each day on the whole.

- On 24 March, because the snowmelt soil-thawing depths were shallow, even though there was sufficient snowmelt runoff, soil loss was still limited [29]. Runoff lasted from 11:00 to 19:00, while sediment yield lasted until 15:00 with an early peak occurring 1.5 h earlier compared with the runoff.
- On 25 March, there were deep soil-thawing depths throughout the entire day, coupled with an initial snowmelt runoff with a substantial early influx. Although the sediment concentration was marginally lower than the preceding day, the dominant sediment yield resulted from the significant proportion of the total runoff volume.

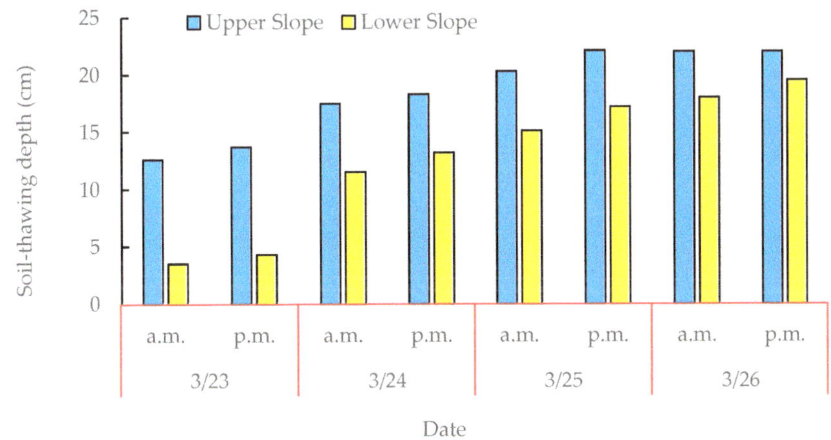

Figure 5. Variations in thaw depths of surface soil at upper slope and lower slope.

Freeze–thaw action is a remarkable characteristic of snowmelt erosion compared to rainfall-driven erosion, and the soil-thawing depth is a crucial driving factor for soil structure and the snowmelt erosion process [10,12]. The soil-thawing depth is related to sediment concentration, and the total sediment yield results from both sediment concentration and runoff volume [30]. At the beginning of the snowmelt erosion period, most of the site was covered with snow cover. Snowmelt and soil thawing predominantly occurred on the upper slope. The snowmelt gradually moved to a lower position, causing much more bare soil. While snowmelt occurred mainly in the upper region, a significant amount of melt water was retained by the snow cover. Therefore, the soil-thawing depths of the upper slope were deeper than those of the lower slope. The limited soil loss monitored can be attributed to the shallow thawed soil and a large proportion of the surface covered by snow.

In the upper and middle parts of the slope, a considerable number of furrow platforms were broken through by snowmelt runoff, leading to an obvious flow path towards the outlet. The increasing soil-thawing depth and the generation of the flow route promoted sediment yield. A large volume of snowmelt runoff carried more eroded sediment particles during the middle to late stages of the process, which is different from the previous day when the limited sediment was obvious due to the shallow soil-thawing depth. Additionally, melt water could not infiltrate into the incomplete thawing layer, making the thawed sediment particles much more susceptible to being eroded [12]. Compared with the results of other days, it is evident that soil loss increased with the deepening of the soil-thawing depth. Snowmelt erosion was sensitive to variation in soil-thawing depth. A larger depth resulted in earlier and larger initial runoff periods. Therefore, it could be concluded that snowmelt erosion is significantly affected by both snowmelt runoff and soil-thawing depth [10].

3.4. Sediment–Discharge Hysteresis Relationship

Snowmelt runoff and sediment concentration exhibited distinct variations at each event. The complexity of driving factors and material sources lead to the non-synchronous process of runoff and sediment. In order to reveal this phenomenon, the sediment–discharge hysteresis relationship method [26] was used to analyze the daily variation in snowmelt runoff and sediment, excluding 26 March, as shown in Figure 6. There was a clockwise hysteresis relationship between runoff and sediment for the 2 previous days, which is consistent with observations of snowmelt erosion of an agricultural watershed in Finland [31]. However, for 25 March, a compound style was obtained, with an initial short counter-clockwise loop followed by a clockwise loop, and another counter-clockwise loop again. Clockwise hysteresis exists when the sediment concentration peaks arrive earlier than the runoff [26]. If sediment sources were relatively limited in a basin, and taken away quickly by discharge, the reduction in sediment concentration would be more rapid than the runoff, leading to a clockwise hysteresis [32,33].

- On 23 March, snowmelt runoff was small, and topsoil had just begun to thaw, so runoff with weak scouring ability had only one peak. Consequently, only sediment near the collecting outlet could be eroded, forming a clockwise hysteresis.
- On 24 March, as the topsoil further thawed, a certain amount of sediment moved towards the lower slope but could not reach the outlet. This produced a high sediment concentration in the early stage under the influence of snowmelt runoff. However, a clockwise hysteresis showed that in the late stage, sediment turned limited again, because incompletely thawed soil was not easily scoured or flushed. The sediment sources on this day included thawed soil particles near the outlet, as well as sediment accumulated and transported along the slope, particularly from the existing flow route.
- The runoff–sediment relationship on 25 March was relatively complex and exhibited a compound hysteresis containing several parts, i.e., a clockwise hysteresis for the second part and a counter-clockwise for the other two parts. The counter-clockwise hysteresis reflects that the sediment source is farther away from the outlet [33], so the sediment path has a longer distance than that of the flow path. The slope had a length of about 330 m, and an obvious flow route originated from the middle position. Therefore, the long distance caused the repeated process of erosion and sedimentation. Additionally, the melted water also changed at different times and positions. The sediment source was abundant at the first and last stages but became limited after being flushed within the middle stage. These variations led to a changeable hysteresis pattern. Compared to the previous days, on this day, thaw depths of the lower slope were larger than ever, so soil erodibility became larger gradually, and, eventually, the main sediment source became close to the outlet again. Thus, the runoff–sediment hysteresis relationship was significantly influenced by sediment sources affected by incompletely thawed soil and flow route patterns.

3.5. Contour Ridge Culture and Micro-Topography

According to the aforementioned observation, snowmelt erosion had various erosion patterns at different sites along the slope. Generally, the upper slope had a main form of surface erosion including rill erosion, while the lower slope erosion experienced a collapse of ridge platforms and the overlying ridge by snowmelt, especially along the prominent flow route.

Most rills were observed between 60 and 90 m down-slope from the top. Snow ablation occurred first and faster at the top with shallow snow cover, and the hold capacity of snow cover prevented melted water from converting into runoff. So, rills did not form until enough runoff amount accumulated. The region mentioned above was favorable for rill generation (Figure 7). Rills occurred either along the furrows (Figure 8) or throughout the entire platform (Figure 9) when there was adequate snowmelt discharge and energy. However, in the middle positions, the topography gradually transitioned from flat to inward, leading to the convergence of the snowmelt towards the middle low-lying positions instead of flushing perpendicular to the ridge belt (Figure 10). Consequently, rills were not observed in these positions. These results showed that during the snowmelt period, rills could not generate across the whole slope but only in upper positions, indicating that snowmelt runoff in each event had a limited migration distance, likely due to a small discharge or being disturbed by local topography (Figure 11).

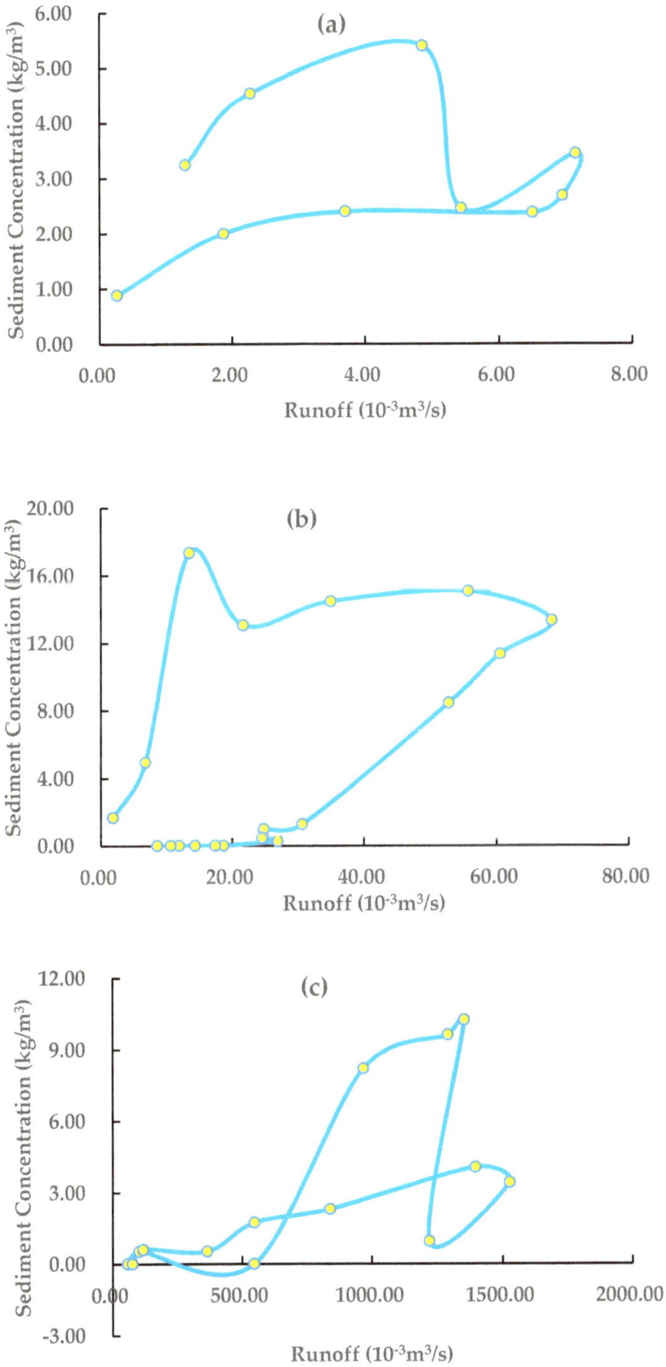

Figure 6. Hysteresis relationship between suspended sediment and snowmelt runoff at each event ((**a–c**) are processes on 23 to 25, repreently).

Figure 7. Snowmelt erosion with rills at the upper slope. (The red arrows were traces of flows).

Figure 8. A rill along a furrow.

Figure 9. A rill through a ridge.

Figure 10. Snowmelt erosion at middle position of the catchment with slightly topographic depression. (The red arrows were flow directions).

Figure 11. Snow, ice, melt water retained in the ridge, and incompletely thawed surface soil, which reduced infiltration.

In the downhill section, snowmelt runoff was gradually concentrated in a low-lying position, with the potential to flush down or overflow the ridge platform, when it exceeded the storage capacity of the ridge and furrow system. As the snowmelt process continued, the soil-thawing depth increased, and the accumulated melted water saturated the surface soil, which made the ridge platform much more prone to failure, leading to an increased sediment source and significantly promoting snowmelt erosion across the entire catchment. Consequently, the height of the ridge platform and surface soil thickness would be reduced, while sedimentation would increase in the furrows. This alteration of the local micro-topography would in turn affect the snowmelt erosion process. Therefore, the interaction of these factors contributes to enhanced soil erosion.

The field investigation revealed that different positions had various forms of broken ridges. The reason for this variation was that in some regions, where there was no efficient snowmelt runoff, the ridge platform was not stable. The first phenomenon started 220 m from the top and displayed no horizontal pattern along with the ridge belt. If there was no efficient snowmelt runoff, it could only destroy the unstable position of the ridge platform, such as relatively lower or completely thawed sites (Figure 8), resulting in an uncertain location of broken ridges. On the down-slope, increased snowmelt runoff with high energy had the ability to break ridge platforms with deeper thawing depths. When the ridge and furrow system could not withstand the force, snowmelt runoff would either overflow or break through the platform, accompanied by a collapse somewhere. Consequently, an obvious flow route near the outlet was generated by scouring and flushing, and the scope of ridge collapse was further expanded (Figure 12). Furthermore, the soil along both sides of the flow path within the ridge belt was eroded and carried away by runoff, resulting in a visible cutting-down phenomenon. The erosion near the outlet represented a typical transition from rill erosion to ephemeral gully erosion.

Figure 12. Snowmelt erosion with an obvious flow route in form of ephemeral gully at the lower slope.

The spatial distribution patterns of snowmelt are critical issues in the prevention and control of different types of soil erosion. Suitable tillage methods have been proven effective in mitigating soil and total phosphorus loss during sheet erosion and rill erosion [22]. Gully erosion would bring much more harm to land [17], while some soil conservation measurements, such as grassed waterways and check dams, have proved to be useful in gully control. Snowmelt erosion exhibits distinct characteristics when compared to rainfall-induced erosion. Therefore, it is important to conduct further observations of snowmelt erosion at various spatial and temporal scales to gain deeper insights into its underlying mechanisms. Furthermore, additional investigations should focus on evaluating the effectiveness of established traditional soil conservation measures and exploring enhanced techniques to augment their efficiency. Such endeavors are imperative for advancing our understanding and management of snowmelt-induced erosion effectively.

4. Conclusions

Seasonal snowmelt erosion is a complex composite erosion process that requires the fulfillment of several critical conditions. The occurrence of snowmelt erosion relies on the presence of a large amount of accumulated snow cover, providing a sufficient water amount source. Additionally, the rapid rise in air temperature above the melting point of ice and snow leads to the potential generation of adequate discharge. Thaw of the surface soil occurs when there is sufficient discharge to flush the soil. To minimize infiltration loss in the early stage, the soil freeze depth must reach an adequate level. However, even if snowmelt runoff is observed, the occurrence of snowmelt erosion becomes challenging without meeting these critical conditions. Moreover, snowmelt erosion exhibits high sensitivity to variations in the surrounding environment, such as fluctuations in air temperature and the freeze–thaw action caused by day and night alternation. The original topography also plays a significant role in influencing the erosion process by affecting discharge generation, influx, and sediment sources. Furthermore, the micro-topography reshaped by contour ridge cultivation determines the flow routes and erosion patterns, including the formation and development of rills.

Author Contributions: Conceptualization, H.F. and X.X.; methodology, H.F. and X.X.; software, H.F. and X.X.; validation, H.F., Y.H. and X.X.; formal analysis, X.X.; investigation, Y.H.; resources, H.F. and Y.H.; data curation, Y.H.; writing—original draft preparation, H.F. and Y.H.; writing—review and editing, H.F., X.X., C.M. and H.S.; visualization, X.X.; supervision, H.F.; project administration, H.F.; funding acquisition, H.F. All authors have read and agreed to the published version of the manuscript.

Funding: This research was funded by the National Key Research and Development Program of China (grant number 2021YFD1500701) and the Natural Science Foundation of China (grant numbers 41371272 and 41807062).

Institutional Review Board Statement: Not applicable.

Informed Consent Statement: Not applicable.

Data Availability Statement: Data can be obtained through the email of the corresponding author.

Acknowledgments: We gratefully thank Dichen Wang, Xinyu Guo, Bo Liu, Hongxi Liu, Jingyi Chi, and Tongyao Chen for their help with the observation.

Conflicts of Interest: The authors declare no conflict of interest.

References

1. Lana-Renault, N.; Alvera, B.; García-Ruiz, J.M. Runoff and Sediment Transport during the Snowmelt Period in a Mediterranean High-Mountain Catchment. *Arct. Antarct. Alp. Res.* **2018**, *43*, 213–222. [CrossRef]
2. Gusarov, A.V. The impact of contemporary changes in climate and land use/cover on tendencies in water flow, suspended sediment yield and erosion intensity in the northeastern part of the Don River basin, SW European Russia. *Environ. Res.* **2019**, *175*, 468–488. [CrossRef]
3. Zhang, L.; Ren, F.P.; Li, H.; Cheng, D.B.; Sun, B.Y. The Influence Mechanism of Freeze-Thaw on Soil Erosion: A Review. *Water* **2021**, *13*, 1010. [CrossRef]

4. Zeinivand, H.; De Smedt, F. Hydrological Modeling of Snow Accumulation and Melting on River Basin Scale. *Water Resour. Manag.* **2009**, *23*, 2271–2287. [CrossRef]
5. Panuska, J.C.; Karthikeyan, K.G. Phosphorus and organic matter enrichment in snowmelt and rainfall-runoff from three corn management systems. *Geoderma* **2010**, *154*, 253–260. [CrossRef]
6. Yakutina, O.P.; Nechaeva, T.V.; Smirnovi, N.V. Consequences of snowmelt erosion: Soil fertility, productivity and quality of wheat on Greyzemic Phaeozem in the south of West Siberia. *Agr. Ecosyst. Environ.* **2015**, *200*, 88–93. [CrossRef]
7. Liu, X.; Zhang, Y.; Zhang, L.; Fang, X.; Deng, W.; Liu, Y. Aggregate-associated soil organic carbon fractions in subtropical soil undergoing vegetative restoration. *Land Degrad. Dev.* **2023**. [CrossRef]
8. Lu, Y.F.; Liu, C.; Ge, Y.; Hu, Y.L.; Wen, Q.; Fu, Z.L.; Wang, S.B.; Liu, Y. Spatiotemporal Characteristics of Freeze-Thawing Erosion in the Source Regions of the Chin-Sha, Ya-Lung and Lantsang Rivers on the Basis of GIS. *Remote Sens.* **2021**, *13*, 309. [CrossRef]
9. Fan, H.M.; Liu, Y.J.; Xu, X.Q.; Wu, M.; Zhou, L.L. Simulation of rill erosion in black soil and albic soil during the snowmelt period. *Acta Agric. Scand. Sect. B Soil Plant Sci.* **2017**, *67*, 510–517. [CrossRef]
10. Komissarov, M.A.; Gabbasova, I.M. Snowmelt-induced soil erosion on gentle slopes in the southern Cis-Ural region. *Eurasian. Soil Sci.* **2014**, *47*, 598–607. [CrossRef]
11. Zhai, J.B.; Zhang, Z.; Melnikov, A.; Zhang, M.Y.; Yang, L.Z.; Jin, D.D. Experimental Study on the Effect of Freeze-Thaw Cycles on the Mineral Particle Fragmentation and Aggregation with Different Soil Types. *Minerals* **2021**, *11*, 913. [CrossRef]
12. Ulén, B.; Bechmann, M.; Øygarden, L.; Kyllmar, K. Soil erosion in Nordic countries—Future challenges and research needs. *Acta Agric. Scand. Sect. B Soil Plant Sci.* **2012**, *62*, 176–184. [CrossRef]
13. Shi, X.N.; Zhang, F.; Wang, L.; Jagirani, M.D.; Zeng, C.; Xiao, X.; Wang, G.X. Experimental study on the effects of multiple factors on spring meltwater erosion on an alpine meadow slope. *Int. Soil. Water Conse* **2020**, *8*, 116–123. [CrossRef]
14. Xiao, X.; Liu, Z.; Liu, K.S.; Wang, J.Q. Temporal Variation of Suspended Sediment and Solute Fluxes in a Permafrost-Underlain Headwater Catchment on the Tibetan Plateau. *Water* **2022**, *14*, 2782. [CrossRef]
15. Starkloff, T.; Stolte, J.; Hessel, R.; Ritsema, C.; Jetten, V. Integrated, spatial distributed modelling of surface runoff and soil erosion during winter and spring. *Catena* **2018**, *166*, 147–157. [CrossRef]
16. Barneveld, R.J.; van der Zee, S.E.A.T.M.; Stolte, J. Quantifying the dynamics of microtopography during a snowmelt event. *Earth Surf. Proc. Land* **2019**, *44*, 2544–2556. [CrossRef]
17. Xu, J.Z.; Li, H.; Liu, X.B.; Hu, W.; Yang, Q.N.; Hao, Y.F.; Zhen, H.C.; Zhang, X.Y. Gully Erosion Induced by Snowmelt in Northeast China: A Case Study. *Sustainability* **2019**, *11*, 2088. [CrossRef]
18. Du, P.F.; Huang, D.H.; Liu, B.; Qin, W. Using Source Fingerprinting Techniques to Investigate Sediment Sources during Snowmelt and Rainfall Erosion Events in a Small Catchment in the Black Soil Region of Northeast China. *Land* **2023**, *12*, 512. [CrossRef]
19. Zhou, Y.; Xu, Y.; Xiao, W.; Wang, J.; Huang, Y.; Yang, H. Climate Change Impacts on Flow and Suspended Sediment Yield in Headwaters of High-Latitude Regions—A Case Study in China's Far Northeast. *Water* **2017**, *9*, 66. [CrossRef]
20. Fu, Q.Y.; Meng, F.X.; Zhang, Y.; Wang, Z.L.; Li, T.X.; Hou, R.J. Ameliorating Effects of Soil Aggregate Promoter on the Physicochemical Properties of Solonetzes in the Songnen Plain of Northeast China. *Sustainability* **2022**, *14*, 5747. [CrossRef]
21. Zhu, M.; Yu, X.; Li, Z.; Xu, X.; Ye, Z. Quantifying and interpreting the hysteresis patterns of monthly sediment concentration and water discharge in karst watersheds. *J. Hydrol.* **2023**, *618*, 129179. [CrossRef]
22. Skøien, S.E.; Børresen, T.; Bechmann, M. Effect of tillage methods on soil erosion in Norway. *Acta Agric. Scand. Sect. B Soil Plant Sci.* **2012**, *62*, 191–198. [CrossRef]
23. Yu, P.F.; Li, T.X.; Fu, Q.; Liu, D.; Hou, R.J.; Zhao, H. Effect of Biochar on Soil and Water Loss on Sloping Farmland in the Black Soil Region of Northeast China during the Spring Thawing Period. *Sustainability* **2021**, *13*, 1460. [CrossRef]
24. IA Ahmed, A.; MEldoma, I.; AHElaagip, E.E.; Hou, F. Effects of Indigenous Cultivation Practices on Soil Conservation in the Hilly Semiarid Areas of Western Sudan. *Water* **2020**, *12*, 1554. [CrossRef]
25. Li, H.L.; Shen, H.O.; Wang, Y.; Wang, Y.; Gao, Q. Effects of Ridge Tillage and Straw Returning on Runoff and Soil Loss under Simulated Rainfall in the Mollisol Region of Northeast China. *Sustainability* **2021**, *13*, 10614. [CrossRef]
26. Williams, G.P. Sediment concentration versus water discharge during single hydrologic events in rivers. *J. Hydrol.* **1989**, *111*, 89–106. [CrossRef]
27. Costa, A.; Molnar, P.; Stutenbecker, L.; Bakker, M.; Silva, T.A.; Schlunegger, F.; Lane, S.N.; Loizeau, J.L.; Girardclos, S. Temperature signal in suspended sediment export from an Alpine catchment. *Hydrol. Earth Syst. Sci.* **2018**, *22*, 509–528. [CrossRef]
28. Wang, T.; Li, P.; Liu, Y.; Hou, J.M.; Li, Z.B.; Ren, Z.P.; Cheng, S.D.; Zhao, J.H.; Hinkelmann, R. Experimental investigation of freeze-thaw meltwater compound erosion and runoff energy consumption on loessal slopes. *Catena* **2020**, *185*, 104310. [CrossRef]
29. Wu, Y.Y.; Ouyang, W.; Hao, Z.C.; Yang, B.W.; Wang, L. Snowmelt water drives higher soil erosion than rainfall water in a mid-high latitude upland watershed. *J. Hydrol.* **2018**, *556*, 438–448. [CrossRef]
30. Hua, W.; Fan, H.; Xiuquan, X.U.; Jia, Y.; Liu, Y.; Tan, J.; Zhang, N. Observation on the Spring Snowmelt Erosion of Sloping Farmland in Northeast China. *J. Soil Water Conserv.* **2017**, *31*, 92–96. [CrossRef]
31. Gonzales-Inca, C.; Valkama, P.; Lill, J.O.; Slotte, J.; Hietaharju, E.; Uusitalo, R. Spatial modeling of sediment transfer and identification of sediment sources during snowmelt in an agricultural watershed in boreal climate. *Sci. Total Environ.* **2018**, *612*, 303–312. [CrossRef]

32. Vale, S.S.; Dymond, J.R. Interpreting nested storm event suspended sediment-discharge hysteresis relationships at large catchment scales. *Hydrol. Process.* **2019**, *34*, 420–440. [CrossRef]
33. Hamshaw, S.D.; Dewoolkar, M.M.; Schroth, A.W.; Wemple, B.C.; Rizzo, D.M. A New Machine-Learning Approach for Classifying Hysteresis in Suspended-Sediment Discharge Relationships Using High-Frequency Monitoring Data. *Water Resour. Res.* **2018**, *54*, 4040–4058. [CrossRef]

water

MDPI

Article

Analysis of the Coupling Relationship between Water Quality and Economic Development in Hongjiannao Basin, China

Xiaoping Liu [1,2,3], Shengdong Cheng [1,2], Ziyao Miao [3], Zhanbin Li [1,2,*], Peng Li [1,2,*], Tong Liu [1], Hegang Zhi [4], Shen Zhang [5,6], Yifan Wang [1] and Xing Zheng [1,2,3]

1 State Key Laboratory of Eco-Hydraulics in Northwest Arid Region of China, Xi'an University of Technology, Xi'an 710048, China; liuxiaoping0111@163.com (X.L.); chengsd@xaut.edu.cn (S.C.); tongliu123@163.com (T.L.); yfanw23@163.com (Y.W.); xingzheng@xaut.edu.cn (X.Z.)
2 Key Laboratory of National Forestry Administration on Ecological Hydrology and Disaster Prevention in Arid Regions, Xi'an University of Technology, Xi'an 710048, China
3 Yulin High-Tech Zone Yuheng No. 1 Industrial Sewage Treatment Company, Yulin Coal Chemical Waste Resource Utilization and Low Carbon Environmental Protection Engineering Technology Research Center, Yulin 719000, China; t.wang@xaut.edu.cn
4 Department of Civil and Environmental Engineering, Cornell University, Ithaca, NY 14853, USA; hz629@cornell.edu
5 Department of Management, Cornell University, Ithaca, NY 14853, USA; sz493@cornell.edu
6 Department of Sage Hall, Cornell University, Ithaca, NY 14853, USA
* Correspondence: zbli@xaut.edu.cn (Z.L.); lipeng74@163.com (P.L.)

Abstract: Hongjiannao is the largest inland lake in China's deserts. In recent years, the water quality and area of the Hongjiannao Lake have continued to decline, which is closely associated with the economic development in the Hongjiannao basin. To explore the coupling relationship between the water quality and economic development in the Hongjiannao basin, the water quality and economic development index of the basin has been analyzed in terms of the monthly water quality and socio-economic development from 2013 to 2020. The coupling relationship and interaction mechanism between water quality and regional economic development has been studied by coupling coordination degree model. The results show that the water pollution increased and then decreased with the seasons, while the water quality was the worst in the summer. The coordinated degree between the water quality and economic development in Hongjiannao shows an upward trend from 2013 to 2020, which has transformed from the process of lagging economic development to the process of primary coordination, finally to the process of lagging water environment. The coupling relationship between water quality and economic development changed from a state of nearly un-coordination to primary coordination from 2013 to 2016, with economic development lagging behind. The coupling relationship between the two systems changed from barely coordinated to the primary coordinated from 2017 to 2018, with the rapid development of economy and slight decline in water quality. After 2018, those two systems gradually stepped into a virtuous cycle during 2019–2020, but the phenomenon of lagging water quality still existed. Therefore, in order to maintain the stable economic development of resource-based cities, it is necessary to keep improving the current situation of water environment and water shortage in Hongjiannao, which will promote the coordinated and sustainable development of water environment and economy.

Keywords: Hongjiannao; water environment quality; economic development; coordinated development

1. Introduction

Lakes play an important role in supplying freshwater, conserving water, improving climatic conditions, and maintaining regional ecological health [1,2]. However, since the 1940s, the rapid development of economy began to threaten the lake environment [1,3,4], such as shrinking lake area [5], eutrophication [6], accumulation of organic matter [7], and discharge of heavy metals [8]. The coordinative development of the economy and the lake environment has drawn more and more remarkable attention [9–12].

To alleviate the deterioration of basin environment and promote the coordinated development of water quality and social economy, the domestic and foreign scholars had put forward the environmental resource constraint theory, sustainable development theory, and coordination degree theory [13]. The theory of coordination degree has been widely used to study the relationship between environmental pollution and economic development [14], and the theory of coordination degree points out that environmental quality will show an obvious downward trend along with economic development in the early stage of economic development; environmental quality will be gradually improved and a virtuous cycle of development will be gradually realized with the economic development reaching a higher level [15–17]. The development of the regional economy was the key factor in coordinating and stabilizing the development between water quality of the lakes and social development. A previous study had quantitatively evaluated the environmental and economic coordination of Liaoning Province since 1990, and Liaoning Province had experienced a transformation from dissonance to primary coordination [18]. Many foreign research studies focus on quantitative analysis of sustainable development of resources, environment, society, and economy [19–21], Jones et al. studied the effects of climate and land-use changes on the lake environment in Southern Australia from 1840 to 1990, and Tamm T. et al. studied the effects of climate and hydrology on dissolved organic carbon load in Lake Vortsjarv (Estonia). Several researchers applied the coupling coordination model to analyze the relationship between water quality and economic development of the Fuxian lake basin in Yunnan province and the Minjiang river basin in Fujian province [22–25]. However, there is little information about the relationship of coupling coordination between water quality and economic development of the lake basin in the desert area of northwest China, and there is a lack of research on the internal relationship between pollutants (which could reflect the water quality, such as organic matter, total nitrogen, and total phosphorus) and economic development.

Hongjiannao, which is located in Shenmu County, Shaanxi Province, is the largest freshwater lake in western China [26], which is located in the border area between the Mu Us Desert and the Loess Plateau, and plays an important role in maintaining ecosystem stability. Shenmu County is a typical resource-based city near the Hongjiannao basin and has a population density of about 76 people/km^2 and a per capita GDP of about 46,100 dollars; its economic level ranks the first among 108 counties in Shaanxi Province and the 9th among 1866 counties in China; its secondary industry accounts for more than 76% and its coal reserves are about 50 billion tons. At the present stage, most studies focused on the changes of water quality, water area, and their influencing factors in Shenmu County under the rapid development of industrial economy [27–30]. Few studies have described the coupling and coordination relationship between the change of water environmental quality and economic development.

In this study, based on water quality and economic development in the Hongjiannao basin from 2013 to 2020, the temporal evolution characteristics of the water environment in Hongjiannao were analyzed via SPSS one-way variance analysis and Daniel and Mann–Kendall trend test. Then, the coupling coordination relationship between water quality and regional socio-economic development was explored in terms of the comprehensive index method and coupling coordination degree model. Those findings will compensate the data of the coupling and coordination relationship between water quality and economic development in the Hongjianao basin. Meanwhile, this investigation aims to find the node where the water environment quality could be enhanced with the economic development kept stably increasing in order to keep sustainable development of the regional economy.

2. Methods and Materials

2.1. Study Area

Hongjiannao is located in Shenmu City, Shaanxi Province, China (Figure 1). The geographical coordinates are 39°04′–39°08′ in the north latitude and 109°49′–109°56′ in the east longitude. The current area of Hongjiannao Lake is around 41.8 square kilometers, the altitude of Hongjiannao is around 1100 m, the climate property of Hongjiannao is the continental monsoon, the average of annual rainfall is around 346.8 mm, and the average of water depth is 8.2 m, and there are many kinds of aquatic life (*Nitraria tangutorum*, *Salix cheilophila*, etc.) [27]. With the rapid development of industry and tourism economy, the underground water supply of Hongjiannao has been damaged due to coal mining, while the water quality of Hongjiannao has been affected by the discharge of domestic sewage and mineralized mine wastewater [18].

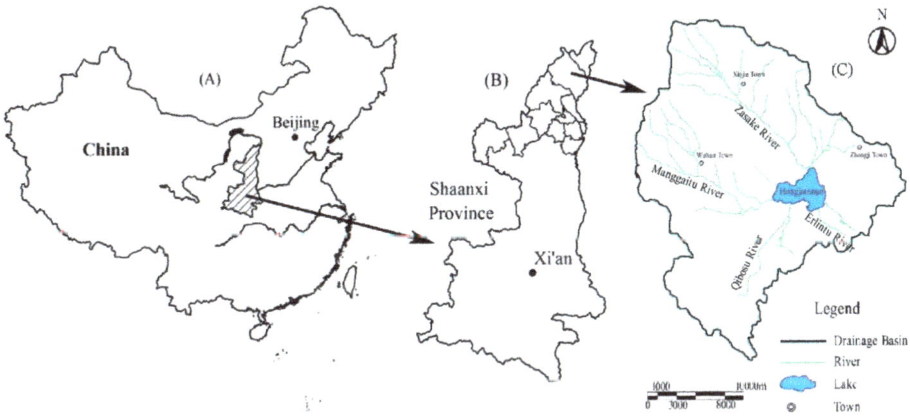

Figure 1. The location map of the Hongjiannao basin ((**A**) is for China, (**B**) is for Shaanxi, (**C**) is for Yulin).

2.2. Data Description

The water quality monitoring station of Hongjiannao Lake in Shenmu County was built in August of 2012 (http://www.cma.gov.cn/2011xzt/2013zhuant/20131126/2013112606/201311/t20131126_232560.html, accessed on 3 January 2021). The water quality of Hongjianao Lake had been detected from 2013 to 2020. The water quality cannot be effectively monitored from December to February of the next year due to the long ice sealing period in the winter. There are 10 indexes that have been used to evaluate the water quality (http://hb.yl.gov.cn/, accessed on 15 January 2021), including five-day biochemical oxygen demand (BOD_5), dissolved oxygen (DO), permanganate index (COD_{Mn}), total phosphorus (TP), total nitrogen (TN), ammonia nitrogen (NH_4^+-N), volatile phenol, chemical oxygen demand (COD_{Cr}), petroleum, and anionic surfactant. All those indexes were detected according to the Surface Water Environmental Quality Standards (GB 3838-2002) [31].

Shenmu County has entered a stable period of economic development since 2012. There are 12 economic indicators that have been selected to evaluate the economic structure and economic benefits, and the date of economy was acquired from *the Yulin Statistical Yearbook (2013–2020)* and *the Shenmu National Economic and Social Development Statistics Bulletin (2013–2020)* (http://tjj.yl.gov.cn/, accessed on 10 February 2021).

2.3. Characteristics of Water Quality Evolution

2.3.1. Seasonal Differences of Water Quality

The least significant difference (LSD) method in single factor variance (one-way ANOVA) was used to present the characteristics of water quality with seasons, and the variance analysis was applied to compare the mean values of two or more samples for evaluating the water quality [32].

2.3.2. Interannual Variation Trend of Water Quality

The annual average value of each water quality index was calculated from the monthly data, and the Daniel trend test and Mann–Kendall trend test were used to present the changeable tendency of Hongjiannao [33]. The Daniel trend test is one of the most commonly methods to measure the environmental pollution; due to its low requirement for samples number, it is commonly used in trend analysis of time series. The Mann–Kendall trend test is a nonparametric statistical method, which only requires independent data and is not susceptible to outliers, but it has the disadvantage of not being able to detect specific patterns of trends. These two methods are selected to compare and verify each other, and the water quality change trend is more reasonable. [34].

The City Water Quality Index (CWQI) has been used to evaluate water environmental quality. The smaller the index is, the better the environmental quality is. Firstly, the arithmetic mean of the concentration of each individual index $C(i)$ is calculated. Then, the water quality index for individual indicators is calculated $CWQI(i)$. Finally, the lake water quality index is calculated comprehensively $CWQI_{Lake}$ (https://www.mee.gov.cn/gkml/hbb/bgth/201606/t20160627_356096.htm, accessed on 8 November 2021).

$$CWQI(i) = \frac{C(i)}{C_s(i)} \tag{1}$$

$$CWQI_{Lake} = \sum_{i=1}^{n} CWQI(i) \tag{2}$$

where $CWQI(i)$ is the water quality index of a single index, $C(i)$ is the monitoring value of the i water quality index, $C_s(i)$ is the standard limit of Class III of the i water quality index, $CWQI_{Lake}$ is the lake water quality index, and n is the number of water quality indicators.

2.4. Coupling Model

2.4.1. Construction of Index System

The economic development is reflected by the increase of economic aggregate, the promotion of economic quality, and the improvement and optimization of economic structure [22,35]. There are 12 indicators (as shown in Table 1) that are used to present the social-economic development system of Shenmu County.

Table 1. The evaluation indexes for the economic development and water quality in Hongjiannao.

System	Sub-System	Indexes	Weight of Coefficient
Social-economic development	Economic structure	X_1	0.0294
		X_2	0.0084
		X_3	0.0237
		X_4	0.0020
	Economic benefit	X_5	0.0350
		X_6	0.0431
		X_7	0.0403
		X_8	0.0591
		X_9	0.0709
		X_{10}	0.0581
		X_{11}	0.0216
		X_{12}	0.0241

Table 1. *Cont.*

System	Sub-System	Indexes	Weight of Coefficient
Water environment	Water indexes	Y_1	0.0132
		Y_2	0.0423
		Y_3	0.1058
		Y_4	0.1086
		Y_5	0.0782
		Y_6	0.1856
		Y_7	0.1906
		Y_8	0.0666
		Y_9	0.0848
		Y_{10}	0.1243

Note: X_1 denotes share of primary industry/%, X_2 denotes share of secondary industry/%, X_3 denotes share of tertiary industry/%, X_4 denotes urbanization rate/%, X_5 denotes (GDP)/billion RMB, X_6 denotes local financial revenue/billion RMB, X_7 denotes GDP of industries/billion RMB, X_8 denotes GDP of agriculture and forestry/billion RMB, X_9 denotes increasing rate of GDP/%, X_{10} denotes total retail sales of consumer goods/thousand RMB, X_{11} denotes urban per capita disposable income/RMB, X_{12} denotes rural per capita disposable income/RMB, Y_1 denotes DO/m, Y_2 denotes COD_{Mn} /mg·L^{-1}, Y_3 denotes BOD_5/ mg·L^{-1}, Y_4 denotes NH_3-N/mg·L^{-1}, Y_5 denotes TN / mg·L^{-1}, Y_6 denotes TP/mg·L^{-1}, Y_7 denotes volatile phenol/mg·L^{-1}, Y_8 denotes COD_{Cr}/ mg·L^{-1}, Y_9 denotes petroleum/mg·L^{-1}, Y_{10} denotes anionic surfactant/mg·L^{-1}.

2.4.2. Water Environment and Economic Development Index

To eliminate the influence of different indexes, the original data are standardized. The corresponding formulas are shown as follows:

The standardized formula for positive index is:

$$r_{ij} = (x_{ij} - \min\{x_j\})/(\max\{x_j\} - \min\{x_j\}) \tag{3}$$

The standardized formula for negative indicators is:

$$r_{ij} = (\max\{x_j\} - x_{ij})/(\max\{x_j\} - \min\{x_j\}) \tag{4}$$

In this study, the relative importance of each index was reflected by the sum-variation coefficient [32], which multiplies the standardized value to obtain the water environment index ($f_{(x)}$) and the economic development index ($g_{(x)}$). x_i (i = 1, 2, 3... m) represents the standardized water quality index, x_j (j = 1, 2, 3... n) is on behalf of the standardized economic development index, and w_i and w_j are the corresponding weights of x_i and x_j, respectively. $f(x)$ and $g(x)$ are obtained by the following formulas:

$$f(x) = \sum_{i=1}^{m} w_i x_i \tag{5}$$

$$g(x) = \sum_{j=1}^{n} w_j x_j \tag{6}$$

2.4.3. Coupling Coordination

Coupling degree is a physical concept, which could reflect the mutual influence caused by multiple interactions of two or more systems [33]. In this study, the coupling degree was applied to study the interrelation between water quality and economic development in the Hongjiannao basin, which is indicated by the formula below:

$$C = \left\{(f(x)g(x))/(f(x) + g(x))^2\right\}^{1/2} \tag{7}$$

The coupling coordination degree model was also used to characterize the coordinated development between the water quality and economic development. A good coupling and coordination state is not only a benign correlation, but also a guarantee for the stable and sustainable development of each system [35]. Economic development was calculated via the following formulas:

$$T = \alpha f(x) + \beta g(x) \quad (8)$$

$$D[f(x), g(x)] = \sqrt{CT} \quad (9)$$

In the formula, $f(x)$ is the water environment index, while $g(x)$ is the economic development index. C, T, and D represent the coupling degree, comprehensive harmonic index, and coupling coordination degree, respectively. α and β represent the contribution of water environment and economic development, separately. In this study, the water environment and economic development were assumed at the same status for the social development, $\alpha = \beta = 1/2$. The classification of coupling-coordination types was shown in Table 2.

Table 2. The types of coupling coordination between water environment and economy.

<table>
<tr><th colspan="2">Types</th><th colspan="2">Sub-Types</th><th>Form</th><th>State</th></tr>
<tr><td rowspan="3">Coordination period</td><td rowspan="3">0.6 < D ≤ 1</td><td>High coordination</td><td>0.8 < D ≤ 1</td><td>$f(x) - g(x) > 0.1$
$g(x) - f(x) > 0.1$
$0 \leq |f(x) - g(x)| \leq 0.1$</td><td>lagging economic development
Lagging water environment</td></tr>
<tr><td>Intermediate coordination</td><td>0.7 < D ≤ 0.8</td><td>$f(x) - g(x) > 0.1$
$g(x) - f(x) > 0.1$
$0 \leq |f(x) - g(x)| \leq 0.1$</td><td>lagging economic development
Lagging water environment</td></tr>
<tr><td>Primary coordination</td><td>0.6 < D ≤ 0.7</td><td>$f(x) - g(x) > 0.1$
$g(x) - f(x) > 0.1$
$0 \leq |f(x) - g(x)| \leq 0.1$</td><td>lagging economic development
Lagging water environment</td></tr>
<tr><td rowspan="2">Transformation period</td><td rowspan="2">0.4 < D ≤ 0.6</td><td>Barely coordination</td><td>0.5 < D ≤ 0.6</td><td>$f(x) - g(x) > 0.1$
$g(x) - f(x) > 0.1$
$0 \leq |f(x) - g(x)| \leq 0.1$</td><td>lagging economic development
Lagging water environment</td></tr>
<tr><td>Nearly un-coordination</td><td>0.4 < D ≤ 0.5</td><td>$f(x) - g(x) > 0.1$
$g(x) - f(x) > 0.1$
$0 \leq |f(x) - g(x)| \leq 0.1$</td><td>lagging economic development
Lagging water environment</td></tr>
<tr><td rowspan="3">Un-coordination</td><td rowspan="3">0 <D ≤ 0.4</td><td>Slight un-coordination</td><td>0.3 < D ≤ 0.4</td><td>$f(x) - g(x) > 0.1$
$g(x) - f(x) > 0.1$
$0 \leq |f(x) - g(x)| \leq 0.1$</td><td>lagging economic development
Lagging water environment</td></tr>
<tr><td>Intermediate un-coordination</td><td>0.2 < D ≤ 0.3</td><td>$f(x) - g(x) > 0.1$
$g(x) - f(x) > 0.1$
$0 \leq |f(x) - g(x)| \leq 0.1$</td><td>lagging economic development
Lagging water environment</td></tr>
<tr><td>Serious un-coordination</td><td>0 < D ≤ 0.2</td><td>$f(x) - g(x) > 0.1$
$g(x) - f(x) > 0.1$
$0 \leq |f(x) - g(x)| \leq 0.1$</td><td>lagging economic development
Lagging water environment</td></tr>
</table>

3. Results and Analysis

3.1. Variation of Water Quality

3.1.1. The Seasonal Variations in Water Quality

The water quality indexes of Hongjiannao from 2013 to 2020 are listed in Table 3. As shown, the content of DO was the lowest in the summer, while the dosage of BOD_5, ammonia nitrogen, total nitrogen, total phosphorus, volatile phenol, chemical oxygen demand, petroleum, and anionic surfactant were the highest in the summer, which means that the water quality of Hongjiannao in the summer was the worst throughout the year. Meanwhile, there was no significant variation for the other indexes of water quality with the seasons ($p < 0.05$). Hence, the water quality of Hongjiannao in the summer was significantly worse than that in the other three seasons.

Table 3. The variation of water indexes in Hongjiannao with seasons.

Season	DO/ mg·L^{-1}	COD$_{Mn}$/ mg·L^{-1}	BOD$_5$/ mg·L^{-1}	NH$_3$-N/ mg·L^{-1}	TN/ mg·L^{-1}	TP/mg·L^{-1}	Volatile Phenol/ mg·L^{-1}	COD$_{Cr}$/ mg·L^{-1}	Petroleum/ mg·L^{-1}	Anionic Surfactant/ mg·L^{-1}
Spring	8.317 ^{a}	12.429 ^{a}	2.586 ^{a}	0.351 ^{a}	3.381 ^{a}	0.089 ^{a}	0.009 ^{a}	43.143 ^{a}	0.038 ^{a}	0.050 ^{a}
Summer	7.424 ^{b}	12.664 ^{a}	3.275 ^{a}	0.375 ^{a}	3.468 ^{a}	0.091 ^{a}	0.011 ^{a}	45.084 ^{a}	0.043 ^{a}	0.065 ^{a}
Autumn	8.384 ^{c}	12.844 ^{a}	1.718 ^{b}	0.268 ^{a}	2.740 ^{a}	0.085 ^{a}	0.009 ^{a}	44.118 ^{a}	0.038 ^{a}	0.059 ^{a}

Note: $^{a–c}$ represents significant differences among seasons.

3.1.2. The Annual Variation of Water Quality

The annual variation of water quality in Hongjiannao was shown in Figure 2 in terms of DO, BOD$_5$, TN, TP, and others. From 2013 to 2020, the DO, BOD$_5$, and TP had a fluctuating downward tendency; the changeable tendency of DO was proportional with that of BOD$_5$. From 2013 to 2016, the content of TN and TP behaved a significant increasing tendency, and the maximum of TN and TP reached 4.18 and 0.24 mg/L, respectively. Meanwhile, from 2014 to 2016, the level of BOD$_5$ and DO increased first and then decreased. Thus, the water quality in Hongjiannao Lake gradually deteriorated with the acceleration of eutrophication before 2016. After 2016, the content of DO, BOD$_5$, TN, and TP had a significant decreasing tendency over years, signifying the improvement of the water quality. This phenomenon was closely related to the regulation of the government for the water environment since 2017. Meanwhile, the decline of the coal economy of Shenmu city was another reason for the improvement in water quality of Hongjiannao Lake since 2015 Moreover, the content of DO kept declining and the dosage of TP became gradually stable since 2020, while the level of BOD$_5$ began to increase again. Thus, the pollution extent of Hongjiannao Lake is still serious.

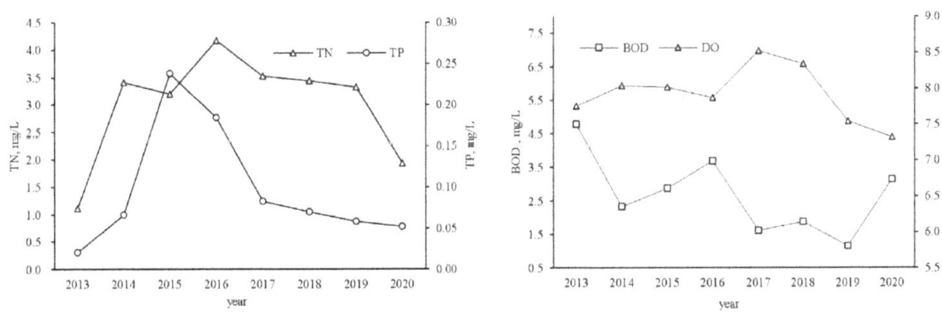

Figure 2. The variation of TN, TP, BOD$_5$, and DO in Hongjiannao Lake from 2013 to 2020.

The analytical results of the Daniel trend test and the Mann–Kendall trend test were shown in Table 4 in terms of 10 indexes of water quality, and the variation of those indexes had a similar tendency. From 2013 to 2020, the content of TN and COD kept increasing, whereas the other indexes continued to decrease to different extents, especially for the ammonia nitrogen and anionic surfactant. The pollution of water quality in Hongjiannao increased first, and then decreased with the years (Figure 2).

Table 4. The results of Daniel test and Mann–Kendall test.

	DO /mg·L^{-1}	COD$_{Mn}$ /mg·L^{-1}	BOD$_5$ /mg·L^{-1}	NH$_3$-N /mg·L^{-1}	TN /mg·L^{-1}	TP /mg·L^{-1}	Volatile Phenol /mg·L^{-1}	COD$_{Cr}$ /mg·L^{-1}	Petroleum Ether/mg·L^{-1}	Anionic Surfactant/mg·L^{-1}
Value of Daniel test	−0.167	−0.667	−0.595	−0.809 **	0.048	−0.095	−0.643 *	0.214	−0.155	−1.107 **
Value of Mann–Kendall test	−0.371	−1.608	−1.361	−2.103 *	0.000	−0.619	−1.361	0.124	−0.742	−2.969 **
Changeable tendency	Down	Down	Down	Down	Up	Down	Down	Up	Down	Down

Note: * means the correlation value was higher than 0.05; ** means the correlation value was higher than 0.01.

The $CWQI_{Lake}$ value of Hongjiannao can be obtained via Formula (1) (Figure 3). From 2013 to 2020, the water quality showed an increasing trend first, and then decreased. Combined with Figure 2, the contents of total nitrogen and total phosphorus in 2015~2016 were 3.4~6.9 times of the Class III standard limit in the "Surface Water Environmental Quality Standard" (GB3838-2002) [31], which is also the main reason for the high CWQI level. Shenmu County began to comprehensively implement the river chief system (http://www.sxsm.gov.cn/zwgk/jcxxgk/zfwj/szbf/12370.htm, accessed on 12 February 2022) after 2016, the local government closed and integrated small coal mines with less than 450,000 tons in the Hongjiannao Valley, which effectively improved the water quality. In 2020, the content of total nitrogen and total phosphorus was 1.1–1.8 times of the Class III standard limit, indicating the water environment quality was significantly improved.

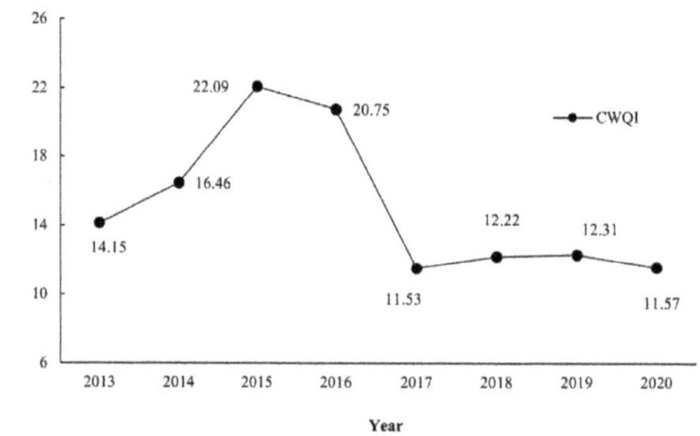

Figure 3. The variation of CWQI index in Hongjiannao Lake from 2013 to 2020.

3.2. Coupling Coordination Analysis

3.2.1. Comprehensive Index Analysis

Figure 4 presents the dynamic change of the water quality and economic development in terms of $f(x)$, $g(x)$, and the level of coupling coordination. In detail, the water environment index showed a downward trend from 2013 to 2014; it increased rapidly to a stable level from 2014 to 2016; then, the water quality declined sharply from 2016 to 2019; and finally, it began to grow again in 2020. The changeable tendency of $f(x)$ was consistent with the water quality over years. The economical level, scale, and structure kept rapidly increasing before 2015, and then they were on a downward trajectory due to the decline of the coal economy. Thus, the value of $g(x)$ rose steadily with the years. Moreover the level of $f(x)$ was lower than that of the $g(x)$ during the early stage, and then gradually exceeded that of the $g(x)$ during the later stage, which indicated that the economic development led to a significant impact on the water quality in Hongjiannao basin.

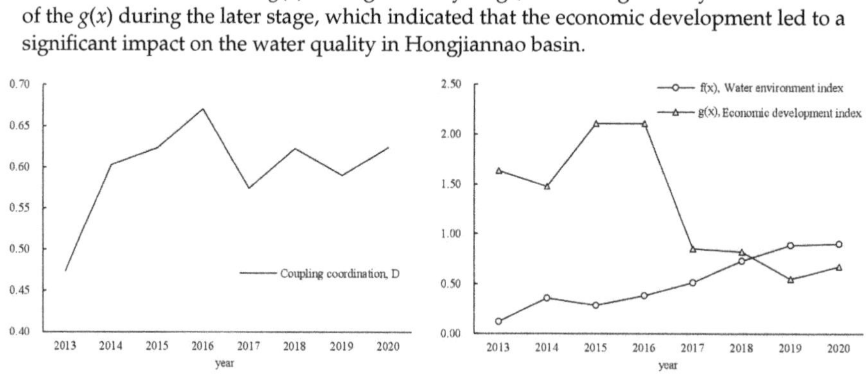

Figure 4. Tha variation of coupling coordination, $g(x)$, and $f(x)$ in Hongjiannao basin from 2013 to 2020.

3.2.2. Change in Coupling Coordination Degree and Classification of Coupling Coordination Types

The coupling coordination index got a yielding point from 2016 to 2019 (Figure 4). Before the yielding point, the $g_{(x)}$ and $f_{(x)}$ behaved in a synchronous increasing trend from 2013 to 2016; thus, the degree of coupling coordination continued to increase during this period. After the yielding point, the level of $g(x)$ kept declining from 2016 to 2019, while the content of $f(x)$ continued to increase; therefore, the degree of coupling coordination began to decline. However, the level of coupling coordination has increased again since 2019.

As presented in Table 5, the degree of coupling coordination between the water quality and economic development showed an increasing trend with time. In 2013, the type of coupling coordination was at a transition period with the degree of coupling coordination at 0.473, which was ascribed to the lagging of the economic development. The average of coupling coordination was at 0.632 from 2014 to 2016, which was in the primary coordination stage. Being affected by the domestic economic downturn and falling coal prices, the economic development index declined with the enhancement in the water quality. From 2016 to 2019, with the acceleration of industrialization and urbanization, the socio-economic development level of the basin has been growing continuously. During this period, the discharge of industrial sewage and domestic sewage increased, the water environment was seriously polluted, and the water quality began to decline. In 2019, the water environmental quality index lagged behind, which began to restrict the coordinated and stable development of those two subsystems. The coupling coordination type decreased from the coordination period to the transition period, and the coupling coordination degree decreased from 0.671 to 0.591. From 2019 to 2020, the water environment index and economic development index began to increase synchronously, the coupling coordination degree increased to 0.624, and the coordination type entered the coordination period again.

Table 5. The types of coupling coordination between water environment and economy in Hongjiannao basin.

Year	Coupling Coordination	Types	
2013	0.473	Transformation period	Nearly un-coordination—lagging economic development
2014	0.603	Coordination period	Primary coordination—lagging economic development
2015	0.623	Coordination period	Primary coordination—lagging economic development
2016	0.671	Coordination period	Primary coordination—lagging economic development
2017	0.575	Transformation period	Barely coordination—lagging economic development
2018	0.623	Coordination period	Primary coupling
2019	0.591	Transformation period	Barely coordination—lagging water environment
2020	0.624	Coordination period	Primary coupling—lagging water environment

4. Discussion

4.1. Analysis of Seasonal Variation of Water Quality in Hongjiannao Lake

According to the data of water quality with seasons in Hongjiannao from 2013 to 2020, the water quality of Hongjiannao Lake in the summer was the worst, which behaved significantly different with that in the spring; the content of TN, TP, and COD was significantly higher than the discharge standard. The reasons for the degraded water quality in the summer are as follows: (1) The Zhasak River and Manggaitu River are the main water replenishments for the Hongjiannao Lake, which accounts for more than 50% of the total water resources. The pollutants contained in the Zhasak River and Manggaitu River were seriously brought into Hongjiannao Lake in the summer; (2) the pesticides and fertilizers are widely used in the upstream and surrounding crops in the summer, especially for nitrogen fertilizer, those pollutants will enter the Hongjiannao Lake via the surface runoff. (3) Especially in recent years, the phenomenon of extreme high temperature in the summer occurs frequently, and microorganisms and organic matter proliferate in the lake, causing serious pollution to the water body. (4) With the economic development of Shenmu City, the urbanization construction did not keep the synchronous development with the environmental protection governance. The previous investigation also confirmed this phenomenon [30].

4.2. Interannual Variation Analysis of Hongjiannao Water Quality

In recent years, the area of Hongjiannao Lake continued to shrink. The pH of lake kept increasing, and even reached 9.5. The content of TN, TP, and other indexes were below the IV standard. The high level of pollution in Hongjiannao from 2013 to 2015 was mainly attributed to the discharge of chemical substances from the coal factory [18]. And then, the water quality of Hongjiannao continued to increase from 2016 to 2020. During this period, the improvement of water quality is closely related with the government regulation, which was confirmed by the previous study [11].

4.3. Coupling Coordination Analysis of Hongjiannao Water Environment and Socio-Economic Development

The degree of coupling coordination between water quality and economic development increased from 0.473 to 0.624 (from 2013 to 2020), indicating that the coupling coordination state changed from the nearly un-coordination state of the transition period to the primary coordination state of the coordination period. The coordinated development pattern between $f_{(x)}$ and $g_{(x)}$ is becoming more and more perfect. According to the development process of coupling coordination, Hongjiannao Lake has experienced three main steps: the lagging economic development, the primary coordination, and the lagging water environment.

In the early stage (2013–2016), $f(x)$ and $g(x)$ increased synchronously, while the coupling coordination degree increased from 0.473 to 0.671. At this stage, being affected by the international coal import and coal price decline, Shenmu County's exhibition economy developed slowly, while the water quality increased. However, due to the low sewage treatment rate of urban sewage treatment plants, the contents of total nitrogen, total phosphorus, and $CWQI_{Lake}$ in Hongjiannao Lake became higher.

In the interim phase (2017–2018), the economy of Shenmu County developed rapidly, the water quality declined to some extent, $CWQI_{Lake}$ increased slightly, and the coupling coordination degree increased from 0.575 to 0.623. It is worth noting that the lagging state of economic development in this stage has changed (compared with the previous stage). In 2018, $f_{(x)}$ and $g_{(x)}$ basically reached the same level, meaning those two subsystems tended to be balanced. This is mainly related to the optimization of industrial structure and the improvement of sewage treatment rate in Shenmu County. The proportion of the secondary industry in Shenmu County increased from 67% to 71%, and the industrialization level changed from the intermediate stage to the advanced stage [18,36]. In addition, according to the Chinese government's implementation of the river chief system (http://www.gov.cn/xinwen/2016--12/11/content_5146628.htm, accessed on 12 February 2022), Shenmu County raised the sewage discharge standard for coal mines, increased the recycling and utilization of mine drainage water, and reduced the discharge of highly mineralized industrial wastewater. The sewage treatment rate has increased from 82.9% to 88.7% (from 2015 to 2018).

During 2019–2020, the economic development of Shenmu County slowed down, and the water quality improved again, $CWQI_{Lake}$ decreased, and the coupling coordination degree increased from 0.591 to 0.624. The industrial structure and sewage treatment rate of the basin were further improved, the high-end coal chemical industry replaced the original high-polluting enterprises, and the proportion of the secondary industry and sewage treatment rate has been increased to 77% and 93.4%, respectively. The economic development with low emission and high efficiency gradually has been achieved preliminarily. However, there are still some problems restricting economic development, such as water shortage [37], the low level of coupling coordination degree, and the gap between the normal situation and the advanced coordination state. Therefore, with the rapid economic development, we need to strengthen the control of water environmental pollution and promote the coordinated and sustainable development of the water environment and economy in the river basin.

5. Conclusions

This paper analyzes the changeable trend of water quality in Hongjiannao basin and investigates the coupling and coordination development status between water quality and economic development, which provides a theoretical basis for the coordinated development of water pollution treatment and economy. The main conclusions were shown as follows:

(1) The water quality of Hongjiannao in the summer was worse than that in the spring and autumn. And, the water quality of Hongjiannao behaved with an increasing tendency from 2013 to 2020.

(2) The degree of coupling coordination in Hongjiannao Lake was at the middle level, and its level increased first, and then decreased, and finally raised again. From 2013 to 2020, the development of coupling coordination in Hongjiannao basin expired three steps: the lagging economic development, the primary coordination, and the lagging water environment.

(3) It can be seen that under the influence of urbanization, industrialization, and extreme hot weather, lake water levels have fallen, pollution is serious, and water resources are stressed. Therefore, governments at all levels should increase investment and management of pollution control, reduce the discharge of untreated sewage into water bodies, and adopt membrane treatment and other technologies to improve the efficiency of recycled water recovery and realize the resource utilization of sewage, especially in China and other developing countries in the world. Economic development promotes the improvement of water quality, and a good water environment also provides a guarantee for sustainable economic development.

Author Contributions: Conceptualization, Z.L. and S.C.; methodology, P.L.; software, T.L.; formal analysis, H.Z.; investigation, S.Z.; writing—review and editing, X.L.; supervision, Y.W.; project administration, X.Z.; funding acquisition, Z.M. All authors have read and agreed to the published version of the manuscript.

Funding: This study was financially supported by [the National Natural Science Foundation of China] grant number [No. 52170053, 52179043, 51779204] and [the National Forestry and Grassland Administration independent research and development program of China] grant number [No. LC-6-06] and [Yulin City Science and technology plan project] grant number [YF-2022-197] and [Yulin High-tech Zone Science and technology plan project] grant number [ZD-2021-08, YGXKG-2022-107]. The research was conducted by the "Water Saving and Reuse Innovation Team" which is supported by the Educational Department of the Shaanxi Provincial Government under The Youth Innovation Team of Shaanxi Universities.

Data Availability Statement: The datasets used and/or analysed during the current study are available from the corresponding author on reasonable request.

Conflicts of Interest: The authors declare no conflict of interest.

References

1. Yang, G.; Ma, R.; Zhang, L.; Yao, S.; Zhang, M.; Zeng, H. Lake status, major problems and protection strategy in China. *J. Lake Sci.* **2010**, *22*, 799–810.
2. Sellinger, C.E.; Stow, C.A.; Lamon, E.C.; Qian, S.S. Recent Water Level Declines in the Lake Michigan−Huron System. *Environ. Sci. Technol.* **2008**, *42*, 367. [CrossRef]
3. Song, J.X.; Tang, B.; Zhang, J.L.; Dou, X.Y.; Liu, Q.; Shen, W.B. System dynamics simulation for optimal stream flow regulations under consideration of coordinated development of ecology and socio-economy in the Weihe River Basin, China. *Ecol. Eng.* **2018**, *124*, 51–68. [CrossRef]
4. Cohen, J.E. Population growth and earth's human carrying capacity. *Science* **1995**, *269*, 341–346. [CrossRef] [PubMed]
5. Cao, H.; Han, L.; Liu, Z.; Li, L. Monitoring and driving force analysis of spatial and temporal change of water area of Hongjiannao Lake from 1973 to 2019. *Ecol. Inform.* **2021**, *61*, 101230. [CrossRef]

6. García-Nieto, P.; García-Gonzalo, E.; Alonso-Fernández, J.; Díaz-Muñiz, C. Using evolutionary multivariate adaptive regression splines approach to evaluate the eutrophication in the Pozón de la Dolores lake (Northern Spain). *Ecol. Eng.* **2016**, *94*, 136. [CrossRef]
7. Chang, J.; Zhang, E.; Liu, E.; Liu, H.; Yang, X. A 60-year historical record of polycyclic aromatic hydrocarbons (PAHs) pollution in lake sediment from Guangxi Province, Southern China. *Anthropocene* **2018**, *24*, 51–60. [CrossRef]
8. Ioannides, K.; Stamoulis, K.; Papachristodoulou, C.; Tziamou, E.; Markantonaki, C.; Tsodoulos, I. Distribution of heavy metals in sediment cores of Lake Pamvotis (Greece): A pollution and potential risk assessment. *Environ. Monit. Assess.* **2014**, *187*, 4209. [CrossRef]
9. Zhang, K.M.; Wen, Z.G. Review and challenges of policies of environmental protection and sustainable development in China. *J. Environ. Manag.* **2008**, *88*, 1249. [CrossRef] [PubMed]
10. Joshi, A.; Seth, G. Hydrochemical profile for assessing the groundwater quality of Sambhar lake City and its Adjoining area. *Environ. Monit. Assess.* **2011**, *174*, 547. [CrossRef]
11. Zhang, N.; Wu, T.; Wang, B.; Dong, L.; Ren, J.Z. Sustainable water resource and endogenous economic growth. *Technol. Forecast. Soc. Chang.* **2016**, *112*, 237–244. [CrossRef]
12. Gao, W.; Chen, Y.; Xu, M.; Guo, H.C.; Xie, Y.C. Trend and driving factors of water quality change in Lake Fuxian (1980–2011). *J. Lake Sci.* **2013**, *25*, 635–642.
13. Haken, H. Synergetics are cooperative phenomena governed by universal principles. *Naturwissenschaften* **1980**, *67*, 121. [CrossRef]
14. Liu, N.N.; Liu, C.Z.; Xia, Y.F.; Da, B. Examining the coordination between urbanization and eco-environment using coupling and spatial analyses: A case study in China. *Ecol. Indic.* **2018**, *93*, 1163–1175. [CrossRef]
15. Ouyang, X.; Shao, Q.; Zhu, X.; He, Q.; Xiang, C.; Wei, G. Environmental regulation, economic growth and air pollution: Panel threshold analysis for OECD countries. *Sci. Total Environ.* **2019**, *657*, 234–241. [CrossRef]
16. Xie, Q.; Xu, X.; Liu, X. Is there an EKC between economic growth and smog pollution in China? New evidence from semiparametric spatial autoregressive models. *J. Clean. Prod.* **2019**, *220*, 873–883. [CrossRef]
17. Isik, C.; Ongan, S.; Özdemir, D. The economic growth development and environmental degradation: Evidence from the US state-level EKC hypothesis. *Environ. Sci. Pollut. Res.* **2019**, *26*, 30772–30781. [CrossRef]
18. Liu, L.; Qiu, X.; Zhang, M.; Li, Y. Main Environmental Impact of Coal Exploitation on Hongjiannao Lake and Protective Measures. *Yellow River* **2019**, *41*, 59–62.
19. Jones, R.N.; McMahon, T.A.; Bowler, J.M. Modelling historical lake levels and recent climate change at three closed lakes, Western Victoria, Australia (c.1840–1990). *J. Hydrol.* **2001**, *246*, 159–180. [CrossRef]
20. Tamm, T.; Tiina, N.; Arvo, J.; Faycal, B. Contributions of DOC from surface and groundflow into Lake Võrtsjärv (Estonia). *Hydrobiologia* **2008**, *599*, 213–220. [CrossRef]
21. Ronchi, E.; Federico, A.; Musmeci, F. A system oriented integrated indicator for sustainable development in Italy. *Ecol. Indic.* **2002**, *2*, 197–210. [CrossRef]
22. Zhang, S.; Zhang, L.; Li, G.; Li, C.; Wu, Z. Coordination analysis between water quality and economic development in Lake Qiandao basin. *J. Lake Sci.* **2014**, *26*, 948–954.
23. Rao, Q.; Lin, X.; Li, J.; Chen, Q.; Chen, W. Analysis of coupling coordination between social economy and water environment quality in river basin. *China Environ. Sci.* **2019**, *39*, 1784–1792.
24. Xu, G.C.; Cheng, Y.T.; Zhao, C.Z.; Mao, J.S.; Li, Z.B.; Jia, L.; Zhang, Y.X.; Wang, B. Effects of driving factors at multi-spatial scales on seasonal runoff and sediment changes. *Catena* **2023**, *222*, 106867. [CrossRef]
25. Liu, X.J.; Zhang, Y.; Zhang, L.; Fang, X.M.; Deng, W.P.; Liu, Y.Q. Aggregate-associated soil organic carbon fractions in sub-tropical soil undergoing vegetative restoration. *Land Degrad. Dev.* **2023**, *72*, 446–459.
26. Zhuo, J.; Zhu, Y.; Wang, J.; He, H.; Zou, J. Spatio-temporal Change of Water Area in Hongjiannao Lake and the Effectiveness of Protection Measures. *J. Desert Res.* **2019**, *39*, 195–203.
27. Zhang, G.; Yao, T.; Chen, W.; Zheng, G.; Shum, C.K.; Yang, K.; Piao, S.; Sheng, Y.; Yi, S.; Li, J.; et al. Regional differences of lake evolution across China during 1960s–2015 and its natural and anthropogenic causes. *Remote Sens. Environ.* **2019**, *221*, 386–404. [CrossRef]
28. Borsuk, M.E.; Stow, C.A.; Reckhow, K.H. A bayesian network of eutrophication modles for synthesis prediction and uncertainty analysis. *Ecol. Model.* **2004**, *173*, 219–239. [CrossRef]
29. Zhao, M.; Xu, Y.; Zhang, C.; Rong, H.; Zeng, G. New trends in removing heavy metals from wastewater. *Appl. Microbiol. Biotechnol.* **2016**, *100*, 6509–6518. [CrossRef]
30. Paer, H.W. Assessing and Managing Nutrient Enhanced Eutrophicationin Estuarine and Coastal Waters: Interactive Effects of Human and Climatic Perturbations. *Ecol. Eng.* **2006**, *26*, 40–54. [CrossRef]
31. *GB 3838-2002*; Surface Water Environmental Quality Standards, China. 2002. Available online: https://www.mee.gov.cn/ywgz/fgbz/bz/bzwb/shjbh/shjzlbz/200206/t20020601_66497.shtml (accessed on 12 February 2022).
32. Zhang, G.; Christensen, R.; Pesko, J. Parametric boostrap and objective Bayesian testing for heteroscedastic one-way ANOVA. *Stat. Probab. Lett.* **2021**, *174*, 109095. [CrossRef]
33. Das, S.; Banerjee, S. Investigation of changes in seasonal streamflow and sediment load in the Subarnarekha-Burhabalang basins using Mann-Kendall and Pettitt tests. *Arab. J. Geosci.* **2021**, *14*, 946. [CrossRef]

34. Xin, Z. Study on the Water Environment of Larus Relictus Nature Reserve in Erdos Plateau. Ph.D. Thesis, Beijing Forestry University, Beijing, China, 2015. (In Chinese).
35. Yang, H.; Zhong, Y.; Luo, L.; Tian, X.; Wang, Z. Water Quality Evaluation and Characteristics Analysis of Xinghai Lake Wetland Based on Bayesian Method. *Water Sav. Irrig.* **2018**, *4*, 92–95.
36. Wang, Q.; Su, M. The effects of urbanization and industrialization on decoupling economic growth from carbon emission—A case study of China. *Sustain. Cities Soc.* **2019**, *51*, 101758. [CrossRef]
37. Labhasetwar, P.K.; Yadav, A. *Membrane Based Point-of-Use Drinking Water Treatment Systems*; IWA Publishing: London, UK, 2023.

water

MDPI

Article

Study on the Coupling Relationship between Carbon Emission from Sewage Treatment and Economic Development in Industrial Parks

Xiaoping Liu [1,2,3], Shengdong Cheng [1,2], Zhanbin Li [1,2,*], Peng Li [1,2], Tian Wang [1,2,*], Xingyue Guo [1], Ziyao Miao [3], Naichang Zhang [4] and Yongxiang Cao [4]

1 State Key Laboratory of Eco-Hydraulics in Northwest Arid Region of China, Xi'an University of Technology, Xi'an 710048, China
2 Key Laboratory of National Forestry Administration on Ecological Hydrology and Disaster Prevention in Arid Regions, Xi'an University of Technology, Xi'an 710048, China
3 Yulin High-Tech Zone Yuheng No. 1 Industrial Sewage Treatment Company, Yulin Coal Chemical Waste Resource Utilization and Low Carbon Environmental Protection Engineering Technology Research, Yulin 719000, China
4 Northwest Engineering Corporation Limited, Xi'an 710065, China
* Correspondence: zhanbinli@126.com (Z.L.); t.wang@xaut.edu.cn (T.W.)

Abstract: Sewage treatment carbon emissions are one of the notable sources of total carbon emission in industrial parks. In order to explore the evolutionary characteristics of sewage treatment carbon emission in industrial parks and its coupling relationship with industrial economic development, based on the quarterly sewage quality monitoring data and regional economic development data of an energy and chemical industry park in Northern Shaanxi from 2016 to 2020, this paper analyzes the evolutionary characteristics of sewage treatment carbon emissions and the coupling relationship between economic development in the industrial parks by using the Intergovernmental Panel on Climate Change carbon emission accounting method and the coupling coordination degree model. The results show that the total carbon emission of sewage treatment in the industrial parks is increasing year by year, and the indirect carbon emissions occupy the dominant position. In 2020, the direct and indirect carbon emissions in the sewage treatment process accounted for 2.4% and 97.6% of the total carbon emission, respectively. It was found that the coupling and coordination relationship between sewage treatment carbon emissions and the economy has experienced the transformation process of serious imbalance—lagging economic development, lagging carbon emission and lagging economic development. In the past five years, the coordinated development degree of the two systems has increased year by year, and the benign mutual feedback mechanism between the two systems has gradually formed. However, regional economic development has lagged due to the impact of the COVID-19 epidemic, so speeding up regional economic development while protecting the environment is recommended.

Keywords: industrial park; carbon emission; economy; coupling

1. Introduction

A large amount of human excreta and pollutants generated by the industrial sector are collectively sent to wastewater treatment plants for processing [1]. The resulting greenhouse gas emissions, including carbon dioxide, during the wastewater treatment process, account for approximately 2–5% of the total carbon emissions in society [2]. This has garnered significant attention from scholars both domestically and internationally [3–8]. Not only is China one of the largest greenhouse gas emitting countries globally, but it is also one of the fastest-growing economies [9,10]. China has over 2500 industrial parks at or above the provincial level, with the economic output of these industrial zones accounting for over 50% of the national total [11]. In 2020, China made a commitment to the world to achieve carbon peak before 2030, which puts additional pressure on the industrial economy to reduce carbon emissions during its rapid development [12]. Therefore, effectively balancing the economic development of industrial parks with green and low-carbon growth is one of the pressing issues that China urgently needs to address [13].

Wastewater treatment is an important aspect in preventing water pollution caused by industrial production, and scholars both domestically and internationally have conducted in-depth research on carbon emissions in the wastewater treatment process. Ma Xin (2011) compared and analyzed the carbon emissions of different treatment processes and scales in municipal wastewater plants using the calculation methods provided in the Intergovernmental Panel on Climate Change guidelines [14]. Yan Xu et al. (2018) analyzed the spatiotemporal distribution characteristics of greenhouse gases in urban wastewater plants in China [15]. In addition, Yu Jiao et al. (2020) conducted a study on carbon emissions from the wastewater treatment system in Zhengzhou city from the perspective of "water-energy-carbon" correlation [16]. Tian Peipei (2021) explored the coupled relationship between "water-energy-carbon" in China using a multi-regional input-output model (MRIO) [10]. Liu Yi (2019) discussed the coordinated development relationship between water environment quality and socio-economic factors in the Nansihu Basin in Shandong Province using the Environmental Kuznets Curve (EKC) model [17]. It can be observed that most scholars have conducted research on the carbon emissions in wastewater treatment processes, and some have analyzed the coordination between socio-economic development and water environment quality, while few have studied the coupled relationship between carbon emissions from wastewater treatment and regional economic development.

The Northwestern inland region possesses abundant coal resources, and while developing the coal chemical industry, the carbon emissions generated from wastewater treatment processes are also continuously increasing. With the implementation of the national dual-carbon policy, there exists a mutually dependent and restrictive relationship between regional carbon emissions and economic development. To accurately assess the level of coordination between regional carbon emissions and economic development and promote the coordinated development of water pollution control and the economy, this paper focuses on a key energy and chemical industrial park in the northern part of Shaanxi Province. Based on quarterly sewage water quality monitoring data and annual economic development statistics from 2016 to 2020, the paper uses the Intergovernmental Panel on Climate Change carbon emission calculation method [18] to analyze the dynamic evolutionary characteristics of carbon emissions from industrial wastewater treatment in the industrial park. By adopting the composite index method, the paper constructs the sewage treatment carbon emission index and economic development index, and for the first time proposes an economic development index for unit carbon emissions from wastewater treatment. Finally, this paper utilizes a coupling coordination degree model [19] to analyze and establish the coupling coordination relationship between carbon emissions from industrial wastewater treatment and regional economic development in the industrial park, aiming to evaluate the coordination between carbon emissions and economic development in the industrial park and provide reference for its low-carbon sustainable development.

2. Overview of the Study Area

The Shaanxi Northern Energy and Chemical Industrial Park is situated in the transitional area between the Loess Plateau and Inner Mongolia Plateau, at the intersection of the Mu Us Desert and the Ordos Basin. It experiences a continental monsoon climate characterized by significant temperature variations. The industrial zone boasts abundant coal resources, with a coal reserve of approximately 34 billion tons. The area has established a 1.15 million tons/year indirect coal liquefaction project, a 600,000 tons/year coal-to-olefins project, a 1.8 million tons/year coal-to-methanol project and a 10 million tons/year coal-to-electricity project, making it a pivotal national energy and chemical base. The industrial park is equipped with one wastewater treatment plant, with a processing capacity of 30,000 tons/day. The treatment process employs a "pre-treatment + membrane concentration + evaporation crystallization" technique, with all treated reclaimed water being reused, and the resulting salt mud being transported to a landfill. Figure 1 illustrates the process flow of the wastewater treatment plant in the park and the carbon emission diagram.

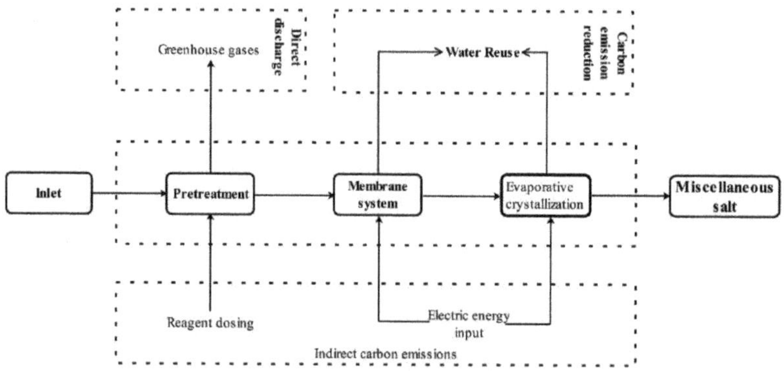

Figure 1. Schematic diagram of process flow and carbon emissions of the sewage treatment plant.

3. Research Method

3.1. Data Description

The temporal evolution of carbon emissions from wastewater treatment in the industrial park spans the years 2016 to 2020. The carbon emission indicators encompass class 3 direct carbon equivalent emissions, namely CO_2, CH_4 and N_2O. Additionally, class 4 indirect carbon emissions indicators are considered, including electricity consumption, lime usage, hydrochloric acid usage, PAM (polyacrylamide) usage and disinfectant usage (Hangzhou Shangtuo Environmental Technology Co., Ltd, Hangzhou, China). The carbon emission data used in this study are derived from the production statistics of the wastewater treatment plant for the years from 2016 to 2020. Economic development data are sourced from the statistical records of the park's administrative committee for the same time frame (2016–2020). The carbon emission and the economic development data are shown in Tables 1 and 2:

Table 1. Basic data of carbon emission from sewage treatment in the industrial zone.

Year	Quarter	BOD Emission Reduction/kg	TN Emission/kg	HCO^{3-} Emission/kg	Power Consumption × 10^4 kw/h
2016	1	0.76	0.72	532.78	337.11
	2	0.86	0.95	459.79	324.11
	3	0.82	1.20	308.37	327.50
	4	0.83	1.03	413.55	353.12
2017	1	1.15	1.66	517.67	389.63
	2	0.88	2.73	506.06	394.19
	3	1.04	2.40	704.65	403.54
	4	1.16	2.74	658.44	469.37
2018	1	0.86	3.77	711.37	438.67
	2	1.00	2.81	617.29	478.96
	3	0.92	1.11	635.31	511.97
	4	1.17	1.61	646.66	647.91
2019	1	0.85	1.77	610.62	543.54
	2	0.82	2.16	501.69	473.80
	3	0.81	1.40	652.22	498.77
	4	0.97	1.74	571.86	602.52
2020	1	0.93	3.61	532.84	568.74
	2	0.89	1.56	502.24	558.96
	3	0.89	2.22	619.04	517.50
	4	0.78	1.11	458.58	596.74

Table 2. Statistics of economic development in the industrial zones (unit: 100 million yuan).

Year	Total Operating Income	Regional Production Total Value	Total Industrial Output Value	Total Fiscal Revenue	Local Finance Income	Fixed Assets Investment
2016	630	290	280	19.0	3.5	26
2017	730	340	430	27.0	4.0	50
2018	650	260	340	36.0	5.9	75
2019	760	314	370	49.2	14.3	160
2020	870	357	446	39.1	7.3	175

3.2. *Accounting of Carbon Emission from Sewage Treatment*

3.2.1. Calculation of Direct Carbon Emissions

Direct carbon emissions in the wastewater treatment process refer to the direct release of CO_2, CH_4 and N_2O into the atmosphere during the water treatment processes [20]. These emissions are primarily determined by multiplying the reduction in pollutants such as BOD, COD, total nitrogen and alkalinity between the influent and effluent of the wastewater treatment plant by their respective emission factors. The calculation formula is as follows [21]:

$$T_{Direct} = Q(BOD_{ex} - BOD)E_{CH_4 \cdot BOD} \times \beta_{CH_4} \\ +Q(TN_{ex} - TN)E_{N_2O \cdot TN} \times \beta_{N_2O} + Q(JD_{ex} - JD)E_{CO_2 \cdot JD} \quad (1)$$

where, T_{Direct} represents the direct carbon emission, with the unit as $kgCO_2e$; Q represents the sewage treatment water volume, with the unit t; BOD_{ex}, JD_{ex}, JD_{ex} and BOD, TN, JD represent the influent and effluent BOD, COD and the concentration of TN, respectively, with the unit kg/t; $E_{CH_4 \cdot BOD}$, $E_{N_2O \cdot TN}$, $E_{CO_2 \cdot TN}$ represent CH_4 and N_2O emission factors (0.086, 0.035 and 0.721, respectively) for COD, BOD, TN and HCO^3-, emissions [18] and conversion factors for CH_4 and N_2O emissions converted to carbon equivalent emissions, at 25 and 298, respectively [18].

3.2.2. Calculation of Indirect Carbon Emissions

Indirect carbon emissions are generated by electricity and reagents consumption during sewage treatment, which includes biochemical process aeration, water pump lifting, sludge pressure filtration, air compressor, heat pump, electrical equipment and the consumption of all kinds of reagents. The calculation formula is as follows [21]:

$$T_{Indirect} = M_{CO_2 \cdot E} \times EF_{CO_2 \cdot E} + \sum_{i=1}^{n} M_{CO_2 \cdot Yi} \times EF_{CO_2 \cdot Yi} \quad (2)$$

In the formula, $T_{Indirect}$ represents the emission of CO_2 in indirect carbon emissions, measured in the unit of $KgCO_2e$; $M_{CO_2 \cdot E}$ represents the electric energy consumption of the sewage treatment plant, and its unit is KW/h; $EF_{CO_2 \cdot E}$ represents the CO_2 emission factor of electric energy consumption (0.997), measured in Kg (CO_2)/KW/h; $EF_{CO_2 \cdot Yi}$ represents the CO_2 emission factor consumed by class i chemicals (1.74 for lime, 25 for PAM, 1.4 for bactericide, and 1.6 for hydrochloric acid); and $M_{CO_2 \cdot Yi}$ represents the consumption of class i agents (Kg) [18].

3.3. *Coupling Model*

3.3.1. Construction of Index System

In order to accurately assess the coupled and coordinated relationship between carbon emissions from wastewater treatment and economic development in the industrial park, a comprehensive evaluation framework is constructed, guided by principles of scientific rigor, systematicity, representativeness and hierarchy, building upon relevant studies [22,23]. The carbon emission assessment system integrates seven indicators encompassing both direct and indirect carbon emissions from the wastewater treatment plant in the industrial park.

These indicators include the direct emissions of CO_2, CH_4 and N_2O from various stages of the wastewater treatment process, as well as carbon emissions resulting from energy and chemical consumption such as electricity, lime, hydrochloric acid, PAM and disinfectants. This framework effectively captures the carbon emission profile of the industrial park's wastewater treatment activities.

Regarding economic development, the evaluation framework primarily incorporates six indicators sourced from the industrial park's statistical bureau. These indicators encompass operational revenue, regional gross domestic product (GDP), total industrial output value, overall fiscal revenue, local fiscal revenue and fixed asset investment. Together, these indicators reflect the changes in both the scale and efficiency of the industrial park's economic development, as depicted in Table 3.

Table 3. Evaluation index system of carbon emission from sewage treatment and economic development.

System Layer	Subsystem Layer	Indicator Layer	Weight of Coefficient of Variation
Socio-economic development	Economies of scale	X_1 Operating income/100 million yuan	0.1319
		X_2 Gross regional product/100 million yuan	0.1241
		X_3 Gross industrial output value/100 million yuan	0.1814
	Economic benefits	X_4 Total fiscal revenue/100 million yuan	0.3396
		X_5 Local fiscal revenue/100 million yuan	0.6224
		X_6 Fixed assets investment/100 million yuan	0.6844
Carbon emission from sewage treatment	Direct carbon emissions	Y_1 CO_2 emission/t	0.1520
		Y_2 CH_4 carbon emission equivalent/t	0.1414
		Y_3 N_2O carbon emission equivalent/t	0.3018
	Indirect carbon emissions	Y_4 Carbon emission of electric energy consumption/t	0.1993
		Y_5 Carbon emission of hydrochloric acid consumption/t	0.1862
		Y_6 Lime consumption and carbon emission/t	0.2844
		Y_7 Carbon emission of PAM consumption/t	0.1066
		Y_8 Carbon emission of fungicide consumption/t	0.3578

3.3.2. Sewage Treatment Carbon Emission and Economic Development Index

Before conducting the calculation of a comprehensive index, it is necessary to eliminate the influence of different data dimensions by applying a range standardization to the base data.

The calculation formula for standardizing positive indicators is as follows:

$$r_{ij} = (x_{ij} - \min\{x_j\})/(\max\{x_j\} - \min\{x_j\}) \tag{3}$$

The calculation formula for standardizing negative indicators is as follows:

$$r_{ij} = (\max\{x_j\} - x_{ij})/(\max\{x_j\} - \min\{x_j\}) \tag{4}$$

The carbon emission of sewage treatment and the economic development system of the industrial zone are both complex systems composed of multiple factors. In this paper, the variation coefficient method [21] is used to objectively reflect the relative importance of each index. The carbon emission index $f(x)$ and the economic development index $g(x)$ are obtained by multiplying the standardized value and weighted summation. x_i (i = 1, 2, 3...m) represents the standardized carbon emission index, x_j (j = 1, 2, 3...n) represents the standardized economic development index, w_i and w_j are the corresponding weights of x_i and x_j, respectively. $f(x)$ and $g(x)$ are obtained by the following formulas:

$$f(x) = \sum_{i=1}^{m} w_i x_i \; g(x) = \sum_{j=1}^{n} w_j x_j \tag{5}$$

3.3.3. Economic Development Index of Unit Sewage Treatment Carbon Emission

To further elucidate the relationship between carbon emissions from wastewater treatment and the economic development index, this paper calculates the Economic Development Index per unit of sewage treatment carbon emission (*CE*) by computing the ratio between the two indices. The CE is obtained by dividing the Economic Development Index by the Sewage Treatment Carbon Emission Index.

$$CE = \frac{g(x)}{f(x)} \tag{6}$$

The level of $CE > 1$ indicates a higher level of regional economic development and green low-carbon development. When $CE = 1$, it signifies a balance between regional economic development and carbon emissions from wastewater treatment. On the other hand, when $CE < 1$, it suggests a lower level of regional economic development and green low-carbon development.

3.3.4. Coupling Coordinated Degree

Coupling degree is a physical concept that refers to the phenomenon of mutual influence caused by multiple interactions between two or more systems. Due to the existence of similarities in the coupling relationships between systems, this paper applies the concept of coupling degree to the study of the interaction between carbon emissions from wastewater treatment and economic development in industrial areas, aiming to provide references for improving the status of carbon emissions in industrial areas and achieving sustainable economic development. The formula for calculating the coupling degree is:

$$C = \left\{(f(x)g(x))/(f(x)+g(x))^2\right\}^{1/2} \tag{7}$$

Because the coupling degree reflects the strength of interaction between the two parties [23], in order to further characterize the level of coordinated development between carbon emissions from wastewater treatment and economic development, a coupling coordination degree model is introduced:

$$T = \alpha f(x) + \beta g(x) \tag{8}$$

$$D[f(x), g(x)] = \sqrt{CT} \tag{9}$$

In this formula, $f(x)$ is the carbon emission index, $g(x)$ is the economic development index, and C, T and D represent the coupling degree, comprehensive harmony index and coupling coordination degree of carbon emission and economic development, respectively; α and β represent the contribution share of carbon emission and economic development, respectively [24]. Because the two systems of carbon emission and economic development are in the same position in this study, $\alpha = \beta = 0.5$ is determined here. See Table 4 for the classification of coupling coordination types of carbon emission and economic development in industrial zones.

Table 4. Classification of coupling coordination types of carbon emission from sewage treatment and economic development.

Type		Subtype		Subtype	Status
Coordination period	$0.6 < D \leq 1$	High-level coordination	$0.8 < D \leq 1$	$f(x) - g(x) > 0.1$	High-level coordination-lagging economic development
				$g(x) - f(x) > 0.1$	Advanced Harmonization—Carbon Lagging
				$0 \leq \vert f(x) - g(x) \vert \leq 0.1$	High-level coordination
		Intermediate coordination	$0.7 < D \leq 0.8$	$f(x) - g(x) > 0.1$	Intermediate coordination-lagging economic development
				$g(x) - f(x) > 0.1$	Intermediate coordination—Carbon Lagging
				$0 \leq \vert f(x) - g(x) \vert \leq 0.1$	Intermediate coordination
		Primary coordination	$0.6 < D \leq 0.7$	$f(x) - g(x) > 0.1$	Primary coordination-lagging economic development
				$g(x) - f(x) > 0.1$	Primary Coordination—Carbon Lagging
				$0 \leq \vert f(x) - g(x) \vert \leq 0.1$	Primary coordination

Table 4. *Cont.*

Type		Subtype		Subtype	Status
Transition period	$0.4 < D \leq 0.6$	Barely coordinated	$0.5 < D \leq 0.6$	$f(x) - g(x) > 0.1$	Reluctant coordination-economic development lags behind
				$g(x) - f(x) > 0.1$	Barely Coordinated—Carbon Lagging
				$0 \leq \lvert f(x) - g(x) \rvert \leq 0.1$	Barely coordinated
		On the verge of imbalance	$0.4 < D \leq 0.5$	$f(x) - g(x) > 0.1$	On the verge of imbalance-lagging economic development
				$g(x) - f(x) > 0.1$	On the Verge of imbalance– Carbon Lagging
				$0 \leq \lvert f(x) - g(x) \rvert \leq 0.1$	On the verge of imbalance
Uncoordinated period	$0 < D \leq 0.4$	Mild imbalance	$0.3 < D \leq 0.4$	$f(x) - g(x) > 0.1$	Mild imbalance-lagging economic development
				$g(x) - f(x) > 0.1$	Mild imbalance—Carbon Lagging
				$0 \leq \lvert f(x) - g(x) \rvert \leq 0.1$	Mild imbalance
		Moderate imbalance	$0.2 < D \leq 0.3$	$f(x) - g(x) > 0.1$	Moderate imbalance-lagging economic development
				$g(x) - f(x) > 0.1$	Moderate imbalance—Carbon Lagging
				$0 \leq \lvert f(x) - g(x) \rvert \leq 0.1$	Moderate imbalance
		Severe imbalance	$0 < D \leq 0.2$	$f(x) - g(x) > 0.1$	Serious imbalance-economic development lags
				$g(x) - f(x) > 0.1$	Serious imbalance—carbon emissions lag
				$0 \leq \lvert f(x) - g(x) \rvert \leq 0.1$	Severe imbalance

4. Results and Analysis

4.1. Carbon Emission Analysis

4.1.1. Analysis of Direct Carbon Emission Indicators

Based on the monthly average BOD, TN and HCO^{3-} data of influent and effluent of the wastewater treatment plant in the park from 2016 to 2020 and the statistical data of annual pollutant reduction, the accounting results of direct carbon emission indicators from 2016 to 2020 can be obtained through the calculation of Formula (1), as shown in Figure 2.

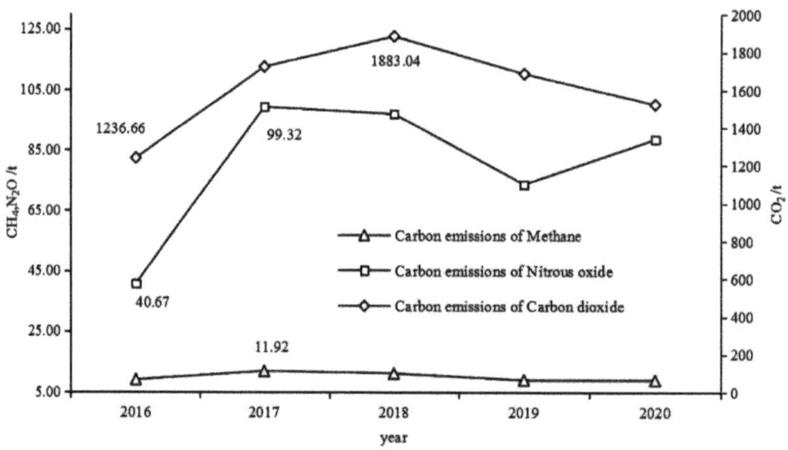

Figure 2. Annual change trend of direct carbon emission index in 2016–2020.

From the perspective of changes in key indicators such as CO_2, CH_4 and N_2O emissions at various stages of wastewater treatment, the primary contributor to direct carbon emissions is the CO_2 emission produced by the reaction between HCO^{3-} in wastewater and acid-neutralization. Following this, the carbon emission equivalent of N_2O plays a secondary role, while the carbon emission equivalent of CH_4 is minimal. In terms of annual variations, the carbon emission equivalents of CO_2, CH_4 and N_2O all exhibit an initial increase followed by a subsequent decrease trend. Among these, the peak emission of CO_2 occurred in 2018, reaching approximately 1883.04 metric tons. On the other hand, the peak emission of CH_4 and N_2O carbon equivalents both occurred in 2017, with values of 99.32 metric tons and 11.92 metric tons, respectively.

4.1.2. Coupling Coordinated Degree

Figure 3 illustrates the structural composition of carbon emissions from indirect sources, including electricity consumption and chemical reagent usage. It is evident from the graph that the primary contributor to indirect carbon emissions is the carbon emissions resulting from lime consumption, followed by electricity usage and hydrochloric acid consumption.

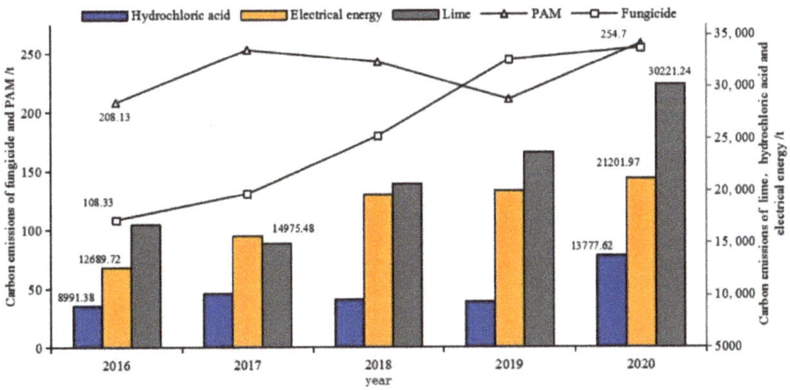

Figure 3. Annual change trend of indirect carbon emission index in 2016–2020.

Examining the annual variations in the indicators of indirect carbon emissions, a noticeable upward trend is observed in lime consumption, electricity usage, and disinfectant consumption. In contrast, the growth trends for PAM (polyacrylamide) and hydrochloric acid are relatively gradual. In the year 2020, carbon emissions from lime consumption, electricity usage and disinfectant consumption reached their peak values at 30,221.24 metric tons, 21,201.97 metric tons and 254.7 metric tons, respectively. These figures represent an increase of 80.32%, 67.08% and 135.11%, respectively, compared to the values in 2016. Notably, the most significant growth rate was observed in disinfectant consumption, followed by lime consumption. Furthermore, starting from 2018, the growth rate of carbon emissions generated by electricity consumption began to slow down, while the disparity between carbon emissions from lime and hydrochloric acid consumption became increasingly pronounced.

By considering both Figures 2 and 3, it can be deduced that the carbon emissions resulting from lime consumption exhibit an inverse relationship with CO_2 emissions. In other words, as the alkalinity represented by HCO^{3-} emissions from industrial enterprises within the park decreases, leading to reduced CO_2 emissions, the quantity of lime added to the wastewater treatment plant increases, consequently contributing to higher levels of indirect carbon emissions.

4.1.3. Analysis of Carbon Emission Results

The annual variations in direct carbon emissions and indirect carbon emissions from wastewater treatment are depicted in Figure 4. The contrasting yearly trends between direct and indirect carbon emissions are primarily influenced by fluctuations in the wastewater quality and quantity discharged by the industrial enterprises within the park. Direct carbon emissions exhibit a pattern of initial increase followed by a decrease, while indirect carbon emissions display a consistent upward trend. In the year 2020, the peak value of indirect carbon emissions was reached at 65,713.64 metric tons, approximately 1.7 times higher than the carbon emissions in 2016. The most substantial increase occurred between 2019 and 2020, primarily attributed to a significant rise in lime consumption.

From a general perspective, carbon emissions from industrial wastewater treatment have been steadily increasing year by year. Indirect emissions contribute significantly to overall carbon emissions, accounting for 97.6% of the total carbon emissions in 2020, while direct carbon emissions constitute only 2.4%. Considering Figure 3, it is evident that in 2020, the largest share of indirect carbon emissions is attributed to lime consumption, accounting for 46.0%, followed by electricity consumption (32.3%) and disinfectant usage contributing the smallest share at only 3.9%.

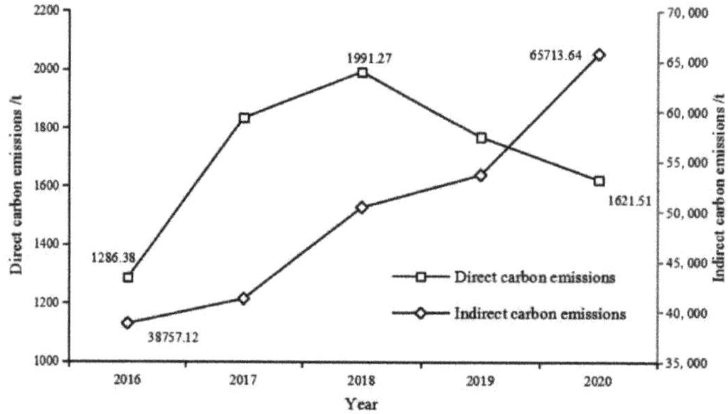

Figure 4. Annual change in direct and indirect carbon emissions from 2016 to 2020.

4.2. Coupling Coordination Analysis

4.2.1. System Index Analysis

The carbon emission index of industrial area water treatment, the economic development index and the economic development index per unit of carbon emission from water treatment can be calculated by Formulas (5) and (6), as shown in Figure 5. From 2016 to 2018, the Carbon Emission Index demonstrates a continuous upward trend. In the years 2019 to 2020, the index first experiences a decline followed by an increase, aligning closely with the annual variations observed in the major carbon emission indicators. This pattern reflects the comprehensive status of carbon emissions from industrial wastewater treatment.

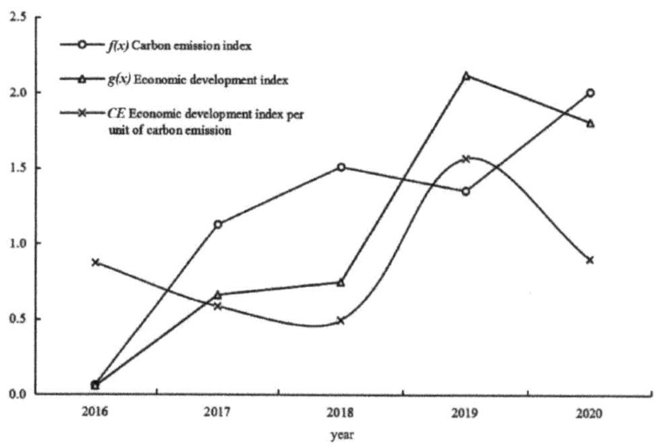

Figure 5. Annual changes of sewage treatment carbon emission index, economic development index, economic development index of carbon emission per unit of sewage treatment.

Within the economic development system, both economic efficiency and economic scale show rapid growth from 2016 to 2019. Notably, the growth rate in 2018 is relatively modest due to the impact of declining resource-based industries, such as coal. Additionally, the Economic Development Index experiences a decrease in 2020 compared to 2019, likely influenced by the disruption caused by the COVID-19 pandemic on the Chinese economy. This signifies a trend of an initially rising and then falling Economic Development Index.

Examining the Economic Development Index per unit of wastewater treatment carbon emissions, this index displays a trend of initial decline, followed by an increase, and then another decline. The lowest value for this index is observed in 2018 (0.49), while the highest value is recorded in 2019 (1.57).

Overall, both industrial wastewater treatment carbon emissions and the economic development index exhibit a consistent upward trend. In the early stages, the economic development index was lower than the carbon emission index. However, following a recovery in the coal industry in 2019, the economic development index surpassed the wastewater treatment carbon emission index. In 2020, the economic development index once again fell below the carbon emission index. This pattern indicates that the economic development of the industrial area has exerted a certain influence on wastewater treatment carbon emissions. It highlights the role of economic growth in promoting green and low-carbon development within the park, while favorable low-carbon and environmental conditions have also provided a foundation for the sustainable development of the regional industrial economy.

4.2.2. Change in Coupling Coordination Degree and Classification of Coupling Coordination Type

The coupling coordination degree between the sewage treatment carbon emission and the economic development system in the industrial area can be calculated by Formulas (8) and (9), as shown in Figure 6. The coupling coordination degree between the two systems shows an upward trend year by year. In 2016–2017, the carbon emission index and economic development index of sewage treatment in industrial zones showed a rapid growth trend, so the coupling coordination degree of the two systems increased the most; In 2017–2018, with the slowdown of the growth rate of carbon emissions and economic development index, the growth rate of the coupling coordination degree of the two also began to decrease; in 2018–2020, the economic development index first increased and then decreased, while the carbon emissions index first decreased and then increased, and the asynchrony between the two changes led to a further slowdown of the coupling coordination degree.

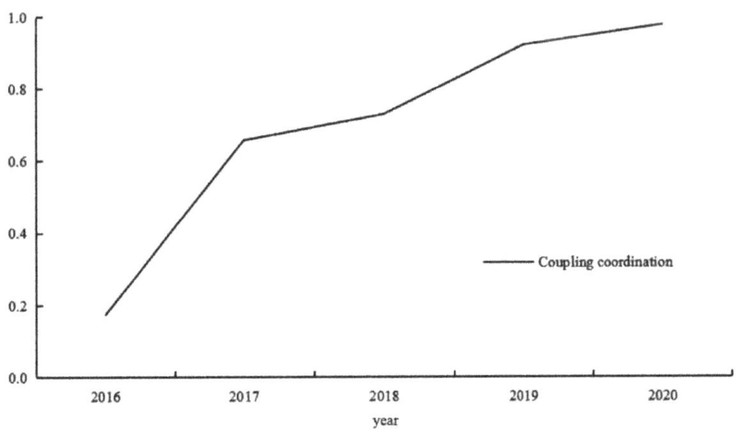

Figure 6. Coupled co scheduling of sewage treatment carbon emission and economic development.

According to Table 5, it is evident that in 2016, the coupling and coordination type between industrial wastewater treatment carbon emissions and economic development in the region was in a transitional phase, with a coupling coordination degree of 0.173, indicating a state of severe imbalance. From 2017 to 2018, the average coupling coordination degree for carbon emissions and economic development was 0.693, indicating a stage of primary and intermediate coordination within the coordination period. Economic development lagged behind carbon emission capacity during this time. In the years 2019 to 2020, the coupling and coordination type reached the advanced coordination stage within the coordination period, transitioning from a state of carbon emission lag to economic development lag.

Table 5. Types of coupling coordination degree between sewage treatment carbon emission and economic.

Year	Degree of Coordination	Type	
2016	0.173	Type incompatibility period	Severe imbalance
2017	0.657	Coordination period	Primary coordination-lagging economic development
2018	0.728	Coordination period	Primary coordination-lagging economic development
2019	0.920	Coordination period	Advanced coordination—Carbon Lagging
2020	0.976	Coordination period	High-level coordination—economic development lags

Overall, with the continuous changes in industrial wastewater treatment carbon emissions and regional economic development, the two systems remain in a dynamically coupled state. In 2016, the industrial area's economic development was just beginning, with enterprises starting production adjustments. Wastewater treatment carbon emissions were low, resulting in a lack of coordination between industrial carbon emissions and the economic system. From 2017 to 2018, although industrial carbon emissions were relatively low, the economic development level within the industrial area was even lower, leading to economic development lag. In 2019, industrial carbon emissions slowed down, with a decrease in the carbon emission index. However, the economic development index surged significantly due to the recovery of the coal economy, resulting in carbon emissions lagging economic development changes. In 2020, the impact of the COVID-19 pandemic led to a decline in industrial economic indicators. Paradoxically, carbon emission indicators began to rise noticeably, causing economic development to lag carbon emissions during this period.

5. Discussion

5.1. The Dynamic Evolutionary Characteristics of Industrial Wastewater Treatment Carbon Emissions

Combining Figures 2 and 4, it can be observed that the annual variations in direct carbon emissions from industrial wastewater treatment align with the trends in CO_2 emissions from the wastewater plant. In 2018, CO_2 emissions reached their peak (1883.04 metric tons), accounting for approximately 94.6% of the total direct carbon emissions that year. In contrast, the carbon emission equivalents of CH_4 and N_2O constituted only 5.4%. This pattern contrasts with the high proportion of organic carbon emissions in urban domestic wastewater treatment.

When considering Figures 3 and 4, the growth trend of indirect carbon emissions from industrial wastewater treatment corresponds to the trends in carbon emissions resulting from electricity consumption and lime consumption. The slower growth in indirect carbon emissions in 2019 compared to 2018 was mainly influenced by lower electricity consumption during that period. However, in 2020, there was a significant increase in indirect carbon emissions compared to 2019. This increase can be attributed to a decrease in the HCO^{3-} content in wastewater discharged by park enterprises, along with a substantial increase in the usage of lime and hydrochloric acid in the wastewater treatment plant.

Kyung et al. reported that chemical use is the major source of indirect carbon emissions, the proportion was 58.8% [25]. In this study, the use of chemical agents accounted for 65.3% of indirect carbon emissions in 2020.

5.2. Analysis on the Change in the Economic Development Index of Carbon Emission from Unit Sewage Treatment

As shown in Figure 5, the economic development index of carbon emission per unit of sewage treatment in 2016–2020 was 0.87, 0.58, 0.49, 1.57 and 0.90, of which the index shows a decreasing trend in 2016–2018, indicating that the economic and low-carbon development level of the industrial zone has been decreasing, which is mainly related to the low level of wastewater treatment at the early stage of the development of the park and the decline of the coal economy. In 2019, the index was greater than 1, which was mainly related to the reduction of carbon emissions from the sewage treatment in the park and the rebound of the coal economy. In 2020, the index was lower than in 2019, but higher than the value in 2016–2018, which was related to the economic downturn due to the New Crown Epidemic on the one hand, and the growth of carbon emissions from wastewater treatment due to stricter environmental requirements on the other.

5.3. Coupling Coordination Analysis of Sewage Treatment Carbon Emission and Social and Economic Development in Industrial Area

The coupling and coordination level between industrial wastewater treatment carbon emissions and regional economic development in the industrial area is generally high. Except for the year 2016, during which the carbon emission system and the economic development system were in a state of imbalance, the years 2017 to 2020 all fall within a coordinated period (with an average coupling coordination degree of 0.82), and both 2019 and 2020 reached an advanced coordination stage. The changes in the composite indices of the two systems reflect their continuous dynamic coupling state. In the early stages, when the carbon emission system and the economic system composite indices grow synchronously, the coupling coordination degree exhibits an upward trend. In the middle stage, as the carbon emission index starts to decline, the asynchronous changes in the two systems lead to a slowdown in the increase in the coupling coordination degree. In the later stage, as the carbon emission index begins to rise and the economic development index begins to decline, the increase in the coupling coordination degree further slows down.

From a temporal perspective, the coupling and coordination status of the two systems in the industrial area can be divided into four stages: severe imbalance (1 year)—economic development lag (2 years)—carbon emission lag (1 year)—economic development lag (1 year). This illustrates an evident influence of economic development changes on wastewater treatment carbon emissions in the industrial area, gradually forming a positive feedback mechanism between the two systems.

In addition, it is also necessary to continue to increase environmental protection investment through the recovery of biogas and biomass in the sewage and other effective measures to carry out carbon emission reduction in sewage treatment [24], and further promote the low-carbon sustainable development of the regional industrial economy.

6. Conclusions

This paper analyzes the trend of carbon emission of sewage treatment in industrial zone, discusses the coupling coordination and interaction between carbon emission of sewage treatment and economic development, and provides a theoretical basis for the coordinated development model of regional environment and economy.

(1). The carbon emissions from industrial wastewater treatment in the industrial area have shown a consistent upward trend in tandem with economic growth. Among these emissions, indirect carbon emissions were the primary contributors, accounting for 97.6% of the total carbon emissions in 2020, while direct carbon emissions constituted only 2.4%. Within the indirect carbon emissions in 2020, the largest contributor is carbon emissions from lime consumption, accounting for approximately 46.0%, followed by electricity consumption (32.3%). On the other hand, the primary contributors to direct carbon emissions are CO_2 emissions produced from the reaction of HCO^{3-} in industrial wastewater, followed by the carbon emission equivalents of N_2O.

(2). The coupling and coordination level between industrial wastewater treatment carbon emissions and the economic development system in the industrial area is generally high. This overall level falls within the advanced coordination stage in the past four years. This suggests that regional economic growth contributes to enhancing the green and low-carbon development level of the industrial park. Moreover, favorable environmental conditions provide a foundation for the sustainable development of regional economies. In the future, we should accelerate regional economic development and change the status quo of lagging economic development while strengthening environmental pollution control.

Author Contributions: Conceptualization, Data curation, Investigation, Methodology, Formal analysis, Writing—original draft: X.L.; Conceptualization, Methodology, Formal analysis, Writing—review and editing, Supervision, Funding acquisition: S.C.; Validation, Writing—review and editing: Z.L.; Project administration, Funding acquisition. P.L.; Supervision, Funding acquisition: T.W.; Validation, Writing—review and editing: X.G., Z.M., N.Z. and Y.C. All authors have read and agreed to the published version of the manuscript.

Funding: This study was financially supported by [the National Natural Science Foundation of China] grant number [No. 52170053,52179043,51779204] and [the National Forestry and Grassland Administra-tion independent research and development program of China] grant number [No. LC-6-06] and [Yulin City Science and technology plan project] grant number [YF-2022-197] and [Yulin High-tech Zone Science and technology plan project] grant number [ZD-2021-08, YGXKG-2022-107].

Data Availability Statement: The carbon emission data used in this study are derived from production statistics of the wastewater treatment plant for the years 2016 to 2020. Economic development data are sourced from the statistical records of the park's administrative committee for the same time frame (2016–2020).

Conflicts of Interest: The authors declare no conflict of interest.

References

1. Chen, W.H.; Yang, J.H.; Yuan, C.S.; Yang, Y.-H. Toward Better Understanding and Feasibility of Controlling Greenhouse Gas Emissions from Treatment of Industrial Wastewater with Activated Sludge. *Environ. Sci. Pollut. Res.* **2016**, *23*, 20449–20461. [CrossRef] [PubMed]
2. Senatore, V.; Zarra, T.; Pahunang, R.R.; Oliva, G.; Belgiorno, V.; Ballesteros, F.C., Jr.; Naddeo, V. Sustainable Odour and Greenhouse Gas Emissions Control in Wastewater Treatment Plant by Advanced Biotechnology Based System. *Chem. Eng. Trans.* **2021**, *85*, 25–30. [CrossRef]
3. Zhou, Z.; Zhuang, G.; Chen, Y. Evaluation of Low-Carbon City Construction: Theoretical Basis, Analytical Framework and Policy Implications. *Chn. Popu. Res. Envi.* **2018**, *28*, 160–169.
4. Zhai, M.; Shao, Y.; Xu, F.J. Countermeasures on greenhouse gas emission redution for the wastewater treatment plants of Xi'an. *Environ. Eng.* **2016**, *34*, 23–26.
5. Singh, P.; Kansal, A. Energy and GHG Accounting for Wastewater in Frastructure. *Resour. Conserv. Recy.* **2018**, *128*, 499–507. [CrossRef]
6. Singh, P.; Kansal, A.; Carliell-Marquet, C. Energy and Carbon Footprints of Sewage Treatment Methods. *J. Environ. Manag.* **2016**, *165*, 22–30. [CrossRef] [PubMed]
7. Xie, Q.J.; Han, P.F.; Chen, S.L.; Wang, L.; Zhen, M.X.; Du, C.J. The Characteristics of CO_2 Emissions in Sewage Treatment Plant with CAST Process. *Environ. Sci. Technol.* **2019**, *32*, 39–44.
8. Zawartka, P.; Burchart-Korol, D.; Blaut, A. Model of Carbon Footprint Assessment for the Life Cycle of the System of Wastewater Collection, Transport and Treatment. *Sci. Rep.* **2020**, *10*, 5799. [CrossRef] [PubMed]
9. Liu, X.; Dai, H.; Wada, Y.; Kahil, T.; Ni, J.; Chen, B.; Chen, Y.; Guo, C.; Pan, C.; Liu, X. Achieving Carbon Neutrality Enables China to Attain Its Industrial Water-Use Target. *One Earth* **2022**, *5*, 188–200. [CrossRef]
10. Tian, P.P. Water-Energy-Carbon Nexus in China's Trade Network: An Analysis with Multi-Regional Input-Output Paradigm. Ph.D. Thesis, North China Electric Power University, Beijing, China, 2021. [CrossRef]
11. Guo, Y.; Tian, J.; Chen, L. Managing Energy Infrastructure to Decarbonize Industrial Parks in China. *Nat. Commun.* **2020**, *11*, 981. [CrossRef] [PubMed]
12. Notice of the State Council on Issuing the Action Plan for Carbon Peak before 2030. Available online: http://www.gov.cn/zhengce/content/2021-10/26/content_5644984.htm (accessed on 18 August 2023).
13. Zhu, E.Y. Study on the Spatialtemporal Pattern of Carbon Emission and Its Response to Urbanization in Zhejiang Province. Ph.D. Thesis, North China Electric Power University, Hangzhou, China, 2020. [CrossRef]
14. Ma, X. Greenhouse Gas Emission Analysis from Municipal Sewage Treatment Plants of China. Master's Thesis, Beijing Forestry University, Beijing, China, 2011.

15. Yan, X.; Qiu, D.Z.; Guo, D.L.; Qi, X.H.; Zheng, S.K.; Cheng, K.; Sun, J.H.; Liu, J.W. Emission Inventory of Greenhouse Gas from Urban Wastewater Treatment Plants and Its Temporal and Spatial Distribution in China. *Environ. Sci.* **2018**, *39*, 1256–1263.
16. Yu, J.; Zhao, R.Q.; Xiao, L.G.; Zhang, L.J.; Wang, S.; Chuai, X.W.; Han, Y.C.; Jiao, S.X. Emission Inventory of GreenhouseGas from Urban Wastewater TreatmentPlants and Its Temporal and Spatial Distribution in China. *Resour. Sci.* **2020**, *42*, 1052–1062.
17. Wang, D. Carbon Emissions, Economic Development and Environmental Protection Coupling in China. Master's Thesis, Northwest Normal University, Lanzhou, China, 2020. [CrossRef]
18. Eggleston, H.S.; Buendia, L.; Miwa, K.; Ngara, T.; Tanabe, K. *IPCC Guidelines for National Greenhouse Gas Inventories, Cambridge*; Cambridge University Press: Cambridge, UK, 2006.
19. Rao, X.H.; Lin, X.Z.; Li, J.B.; Chen, Q.; Chen, W.H. Analysis of coupling coordination between social economy and water environment quality in river basin. *China Environ. Sci.* **2019**, *39*, 1784–1792.
20. Malila, R.; Lehtoranta, S.; Viskari, E.-L. The Role of Source Separation in Nutrient Recovery—Comparison of Alternative Wastewater Treatment Systems. *J. Clean. Prod.* **2019**, *219*, 350–358. [CrossRef]
21. Zhai, G.H.; Zuo, X.F. Spatial Differences and Influencing Factors of CarbonEmissions from Urban Wastewater Treatment in China. *J. Shijiazhuang Tiedao Univ.* **2020**, *14*, 1–10+24.
22. Zhang, S.S.; Zhang, L.C.; Li, G.; Li, C.F.; Wu, Z.X. Coordination analysis between water quality and economic development in Lake Qiandao basin. *J. Shijiazhuang Tiedao Univ.* **2014**, *26*, 948–954.
23. Liu, Y. Study on the Coordinated Development between Water Environmental Quality and Social Economy in Nansi Lake Catchment. Master's Thesis, Jinan University, Jinan, China, 2020. [CrossRef]
24. Almomani, F.; Al Ketife, A.; Judd, S.; Shurair, M.; Bhosale, R.R.; Znad, H.; Tawalbeh, M. Impact of CO_2 Concentration and Ambient Conditions on Microalgal Growth and Nutrient Removal from Wastewater by a Photobioreactor. *Sci. Total Environ.* **2019**, *662*, 662–671. [CrossRef] [PubMed]
25. Kyung, D.; Kim, M.; Chang, J.; Lee, W. Estimation of greenhouse gas emissions from a hybrid wastewater treatment plant. *J. Clean. Prod.* **2015**, *95*, 117–123. [CrossRef]

water

MDPI

Article

The Lag Effect of Riverine Flow-Discharge and Sediment-Load Response to Antecedent Rainfall with Different Cumulative Durations in Red Hilly Area in China

Lixiang Zhao [1,2], Xiaofei Nie [1,3,*], Haijin Zheng [1,3], Kaitao Liao [1,3] and Jinjuan Zhang [4]

1 Jiangxi Academy of Water Science and Engineering, Nanchang 330029, China; lixiangzhao1102@163.com (L.Z.); haijinzheng@163.com (H.Z.); liaokaitao@126.com (K.L.)
2 College of Resources and Environment, Huazhong Agricultural University, Wuhan 430070, China
3 Jiangxi Provincial Key Laboratory of Soil Erosion and Prevention, Nanchang 330029, China
4 Zhejiang Institute of Hydraulic & Estuary (Zhejiang Institute of Marine Planning and Design), Hangzhou 310020, China; zjj558@126.com
* Correspondence: xfnie85@163.com

Abstract: Rainfall is an important factor that causes riverine flow and sediment transport, and extreme rainfall has a particularly significant effect on the fluctuations of riverine flow and sediment load. Based on the daily rainfall from 1990 to 2020, in the upper watershed of the Lianjiang River, which is one of the source tributaries of China's largest freshwater lake (Poyang Lake), the 95th percentile method and minimum event interval time were employed to identify extreme rainfall events. Mann–Kendall test was used to check for abrupt changes in annual rainfall, riverine flow discharge, and riverine sediment loads, and to identify abrupt-change years; thus, different periods were divided via the abrupt change years. Multiple linear regression was applied to explore the lag effect of riverine flow-discharge and sediment-load response to antecedent rainfall, with different cumulative durations for each period. The results of the study indicated that (1) the expansion of garden land in 1995 caused a significant and abrupt change in sediment load. (2) Extreme rainfall events had a greater impact on riverine flow and sediment load as compared to ordinary rainfall. These events were found to explain more variations in riverine flow and sediment load, which led to longer lag times for both riverine flow and sediment transport. (3) The expansion of garden land under extreme rainfall conditions resulted in longer lag times for riverine flow and sediment transport, and reduced the need for antecedent rainfall with a longer pre-event time. Therefore, the analysis of antecedent rainfall and the lag response of riverine flow discharge and sediment load can help in understanding the response mechanism of riverine flow discharge and sediment load for the current era of increasing extreme rainfall. This analysis is crucial for improving the accuracy of simulating riverine flow and sediment under extreme rainfall conditions. Ultimately, it can contribute to effective watershed management during extreme rainfall events.

Keywords: antecedent rainfall; lag time; rainfall event; rainfall types

1. Introduction

The riverine flow–sediment relationship is an extremely complex hydrological process, and the impacts of climate change and human activities on riverine flow–sediment changes are hot topics that have been discussed [1,2]. Numerous studies have shown that climate change has altered riverine flow discharge and sediment loads, which has been exacerbated by intense human activities [3,4]. Riverine flow discharge and sediment loads depend on a series of flow and sediment generation and transportation related processions, which spatial and temporal variations are easily affected by the temporal and spatial distribution of rainfall, evaporation, infiltration, runoff generation, and soil erosion [3]. Clarifying the study period is the key to discussing flow–sediment changes [3–7]. There is a lag effect in the influence of rainfall events on runoff and sediment, and the antecedent rainfall also affects the lag effect [8]. Antecedent rainfall and rainfall events both play an important role in the variation of flow discharge and sediment loads. There have been limited investigations into the impact of rainfall on river flow and sediment loads, specifically exploring the periods of flow–sediment change, the duration of the antecedent rainfall, and differentiating between various types of rainfall.

In the red hilly area of China, the annual rainfall is high, ranging from 1.9 to 2.8 times the national average level, and is disproportionally distributed throughout the year [9,10]. This region is particularly vulnerable to extreme rainfall from April to September, making it one of China's most affected areas by water erosion [10]. Riverine sediment load is primarily triggered by rainfall [11,12], and its fluctuation is particularly affected by extreme rainfall due to induced serious soil erosion [13]. Extreme rainfall events demonstrate more comprehensively the effects of rainfall on riverine flow discharge and sediment loads, providing valuable insights into this relationship. Previous studies have observed the impact of extreme rainfall (storms) on riverine flow and sediment loads, but event-based studies offer better insights into the nature of riverine flow and sediment loads.

Antecedent rainfall increases soil moisture and reduces soil infiltration, so the significance of antecedent rainfall in sediment-load processes cannot be understated. Studies have confirmed that the runoff coefficient can double when the soil is moistened by antecedent rainfall, due to the acceleration of runoff generation with moist soils [14,15]. Antecedent rainfall increases soil moisture, improves runoff conversion efficiency, accelerates riverine runoff discharge [13,16], enhances water erosion, and increases riverine sediment loads. The amount of riverine sediment load after rainfall is influenced by moist soil before the event [17]. Meanwhile, nearly 70% of the runoff was related to the antecedent soil water content [18]. Antecedent soil moisture correlates more strongly with riverine flow. Antecedent soil moisture and rainfall are the important factors influencing riverine flow discharge [14,19], and the combination of antecedent rainfall and intra-event rainfall is vital to generate riverine flow discharge and sediment loads [14,20]. Larger floods caused by extreme rainfall respond more strongly to antecedent rainfall [21,22]. Riverine sediment is also affected by antecedent rainfall that accumulates in the soil [23], which is an easily overlooked effect [24]. Rainfall intensity and antecedent rainfall are also important contributors to changes in riverine flow discharge and riverine sediment loads [2,25,26], and their specific roles can differ depending on the region [8,22,27,28]. Only sufficiently accumulated antecedent rainfall in the soil can trigger massive riverine sediment in response to extreme rainfall [29,30]. However, the response of riverine flow discharge and sediment loads to antecedent rainfall of different cumulative durations remains unknown. By simultaneously considering the impact of rainfall events and antecedent rainfall on riverine flow and sediment loads, we can gain a deeper understanding of the mechanisms of how rainfall influences riverine flow and sediment loads.

In this study, the daily rainfall, and riverine flow discharge and sediment loads from 1990 to 2020 in the upper watershed of the Lianjiang River were analyzed. The main objectives of this study were to (1) clarify the change characteristics of rainfall, and riverine flow discharge and sediment load; (2) check the abrupt changes in annual rainfall, riverine flow discharge, and riverine sediment loads, and to identify the abrupt-change years; (3) identify the most significant days with antecedent rainfall affecting riverine flow discharge and sediment load, and the most significant lag effecting days of riverine flow discharge and sediment load. Thus, the lag effect of riverine flow-discharge and sediment-load response to antecedent rainfall were explored. The results of this study would help to understand the underlying mechanisms that govern the response of riverine flow and sediment to extreme rainfall conditions and in improving predictive models to accurately simulate these phenomena.

2. Materials and Methods

2.1. Study Area

Lianjiang River watershed (115°11′53″~115°11′53″ E, 25°02′52″~25°21′10″ N) with a total watershed area of 2339 km^2, which is located in the southern red soil area in China. The Lianjiang River is a first-class tributary of the upper Ganjiang River. Ganjiang River is the largest inlet river of Poyang Lake (Figure 1), in China.

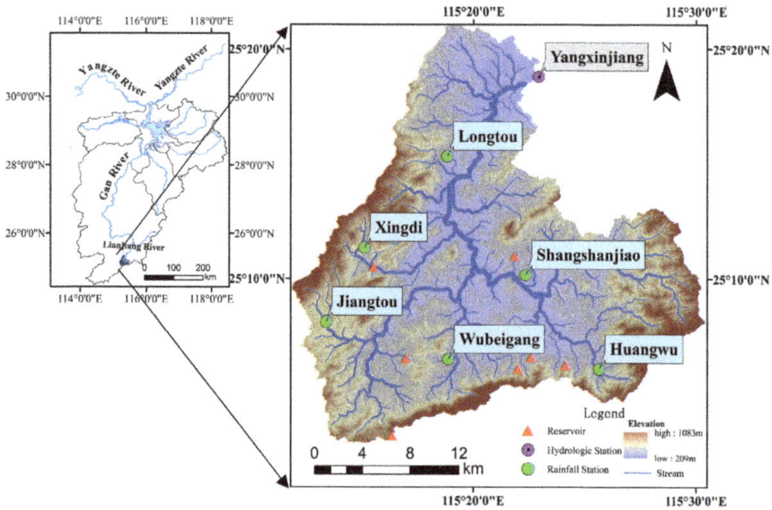

Figure 1. Location of the study area in Poyang Lake watershed and the distributions of rainfall and hydrologic stations.

The average annual temperature is 18.7 °C; the annual runoff is 1.92 billion m^3; and the annual suspended mass sediment transport is 259,000 t [9]. The upper watershed of the Lianjiang River, which was the focus of this study, is monitored by the hydrological station of Yangxinjiang. This station is responsible for managing a watershed area of 568 km^2. The region is located in Anyuan County, where there was extensive promotion of citrus cultivation in the early 20th century due to policy initiatives. As a result, there have been significant changes in the land use in this area [9].

2.2. Data Collection

The hydrological data were obtained from the Jiangxi Provincial Hydrological Monitoring Center, which is the official hydrological monitoring institution, with the standard specifications for hydrological monitoring to acquire detailed and reliable hydrological data. This paper collected daily rainfall data from six rainfall stations (Huangya, Shangshanjiao, Wubeigang, Xingdi, Jiangtou, and Longtou) and one hydrological station (Yangxinjiang hydrological station), in the upper watershed of the Lianjiang River from 1990 to 2020, as well as daily runoff and sediment data from the Yangxinjiang hydrological station for the same period.

The land-use data were obtained from the geographic monitoring cloud platform (http://www.dsac.cn/, the accessed data was 12 August 2020) with a spatial resolution of 30 m during 1990–2020, with four periods of 1990, 2000, 2010, and 2020 [31], which are some of the most authoritative land-use data in China. Land-use types are classified into 6 main categories and 25 subcategories. The classification accuracy of crop land and urban-rural, industrial, mining, and residential land is consistently above 85%, while the accuracy of other land-use types is above 75% on average. This indicates a high level of accuracy in the classification process [31]. In these data, the other wooded land refers to young afforested land, traces, nurseries, and various types of gardens (orchards, mulberry gardens, tea gardens, hot crop gardens, etc.). Referring to the corresponding garden area of Anyuan County, in the statistical yearbook of Ganzhou City from 1992 to 2020 [32], the other wood land could be identified as garden land in the study area.

2.3. Definition of Extreme Rainfall Events

The rainfall data of the Lianjiang River watershed is based on rainfall data from seven stations. The average daily rainfall is calculated using the Thiessen polygon as the following formula.

$$R = \sum_{i=1}^{n} R_i \frac{s_i}{S} \quad (1)$$

where R_i is the amount of rainfall greater than 1 mm at each rainfall station (mm); s_i is the control area of each rainfall station (km^2); S is the total area of the watershed (km^2); n is the number of rainfall stations; and, here, $n = 7$ is taken. In this paper, the daily rainfall is selected to be greater than 1 mm, and the 95th percentile value was used as the threshold value [33]. The threshold for extreme rainfall in the upper reaches of the Lianjiang River watershed was calculated based on the average daily rainfall and is determined to be 41.71 mm.

2.4. Mann–Kendall Test

The Mann–Kendall method, or M–K Test, is a non-parametric diagnostic technique that can be applied to determine whether there is an abrupt change in time series data, and if so, when it occurs [34]. The M–K test does not require a normal distribution of the data series, which is very data-friendly [35]. In this study, the M–K test was applied to analyze the yearly rainfall, and riverine flow discharge and sediment load data to detect whether there was an abrupt change, and in which year the abrupt change happened.

2.5. Pre-Processing of Hydrological Data

In this study, a single rainfall event was defined as a continuous rainfall event where every daily rainfall was greater than 1 mm. Extreme rainfall events were defined as the rainfall events in which the daily rainfall exceeded the extreme rainfall threshold (41.71 mm) on any given day, while other rainfall events were considered as ordinary rainfall events. By analyzing the flood's hydrological element extraction table of Yangxinjiang hydrological station from Jiangxi Provincial Hydrological Monitoring Center, it was found that the duration of flood events at Yangxinjiang hydrological station was almost no more than 5 days. In order to eliminate the interference of different flood events, this study only selected rainfall events with an interval of more than 7 days for analysis. During the period 1990–2020, 150 extreme rainfall events and 740 ordinary rainfall events were selected in this study.

2.6. Multiple Regression Analysis

Multiple linear regression models are an intuitive and efficient way to analyze complex problems [36]. In this study, we used the rainfall of the rainfall event as the fixed independent variable, and the riverine flow discharge or sediment load within the event as the dependent variable. We added antecedent rainfall at different accumulation times as an independent variable and riverine flow discharge or sediment load at different time durations as a dependent variable. The multiple linear regressions were fitted sequentially with the combination of every AP and RA or SA list in Table 1. The optimal model was selected based on the highest R^2, which implied the highest degree of explanation.

Table 1. Indicator labels and content.

Labels	Context
AP1	Rainfall on the last day before an extreme rainfall event
AP2	Cumulative rainfall of the last 2 days before an extreme rainfall event
AP3	Cumulative rainfall of the last 3 days before an extreme rainfall event
AP5	Cumulative rainfall of the last 5 days before an extreme rainfall event
AP7	Cumulative rainfall of the last 7 days before an extreme rainfall event
RA1	Flow discharge of the next day after an extreme rainfall event
RA2	Cumulative flow discharge of the next 2 days after an extreme rainfall event
RA3	Cumulative flow discharge of the next 3 days after an extreme rainfall event
RA5	Cumulative flow discharge of the next 5 days after an extreme rainfall event
RA7	Cumulative flow discharge of the next 7 days after an extreme rainfall event
SA1	Sediment load of the next day after an extreme rainfall event
SA2	Cumulative sediment load of the next 2 days after an extreme rainfall event
SA3	Cumulative sediment load of the next 3 days after an extreme rainfall event
SA5	Cumulative sediment load of the next 5 days after an extreme rainfall event
SA7	Cumulative sediment load of the next 7 days after an extreme rainfall event

3. Results

3.1. Change Characteristics of Rainfall, Riverine Flow Discharge, Sediment Load, and Garden Land

From 1990–2020, there were a total of 205 days with daily rainfall exceeding the given threshold and 3888 days with daily rainfall between 1 mm and the threshold. The yearly rainfall varied between 1080.9 and 2174 mm (Figure 2a); the yearly riverine flow discharge ranged from 172.39 to 1017.07 $\times$ 10^6 m^3 (Figure 2c); and the yearly riverine sediment load ranged from 0.57 to 64.30 $\times$ 10^4 t (Figure 2e) from 1990 to 2020. The M–K test showed that rainfall (Figure 2b) and riverine flow discharge (Figure 2d) did not undergo a significant mutation ($p > 0.05$) from 1990 to 2020. Only riverine sediment load underwent a significant mutation, and the year of mutation was 1995 ($p = 0.041$) (Figure 2f). Therefore, the study period could be split into two periods with the abrupt year; thus, 1990–1995 is period 1 (P1) and 1996–2020 is period 2 (P2). Subsequently, the effects of rainfall events on riverine runoff discharge and sediment load were dissected under the above two different periods.

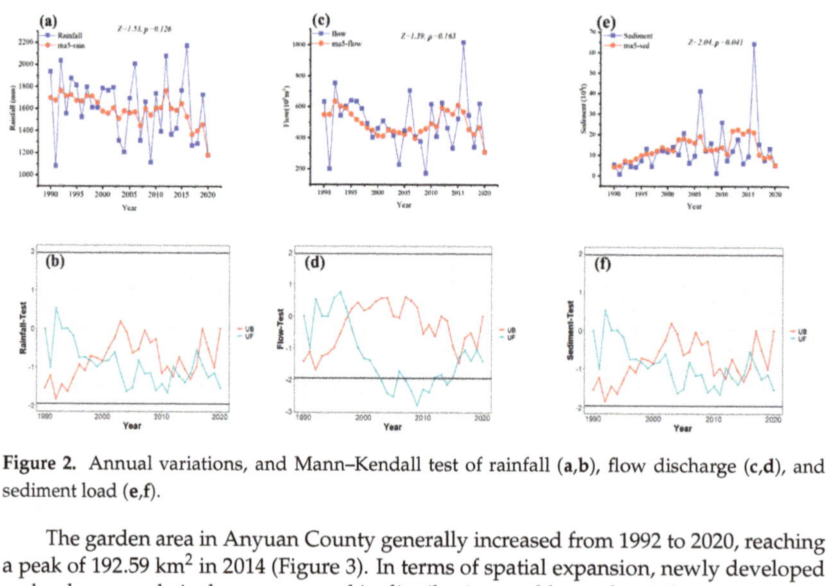

Figure 2. Annual variations, and Mann–Kendall test of rainfall (**a**,**b**), flow discharge (**c**,**d**), and sediment load (**e**,**f**).

The garden area in Anyuan County generally increased from 1992 to 2020, reaching a peak of 192.59 km^2 in 2014 (Figure 3). In terms of spatial expansion, newly developed orchards were relatively concentrated in distribution, and located mainly near the main stream and main tributaries of the Lianjiang River (Figure 3).

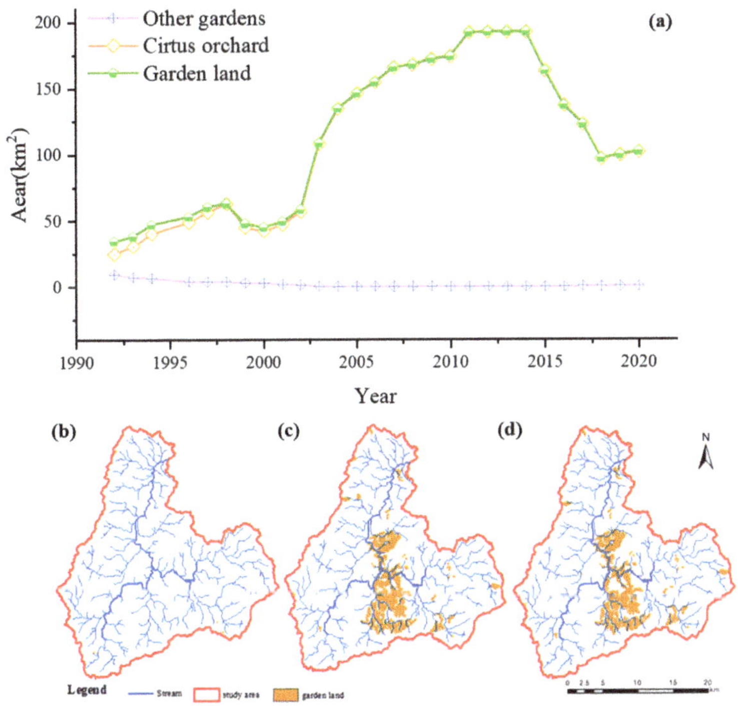

Figure 3. The area of garden land in Anyuan country, from 1992 to 2020 (**a**), Spatial distribution of garden land in the upper Lianjiang River watershed ((**b**): 1990, (**c**): 2010, and (**d**): 2020).

3.2. Statistical Characteristics of Rainfall, Riverine Flow, and Riverine Sediment Loads in Different Periods

Compared with the P1 period, the rainfall decreased significantly in the P2 period. The mean extreme rainfall, mean ordinary rainfall, and total rainfall decreased from 328.10 mm, 796.55 mm and 1716.53 mm to 278.53 mm, 726.91 mm and 1584.54 mm, respectively (Figure 4). Similarly, the mean extreme rainfall, mean ordinary rainfall and total rainfall of river flow decreased from 241.01×10^6 m^3 in P1, 146.36×10^6 m^3, and 562.85×10^6 m^3 to 222.17×10^6 m^3, 125.02×10^6 m^3, and 484.92×10^6 m^3 in P2 (Figure 4). Interestingly, despite the decrease in rainfall and riverine flow, the riverine sediment loads increased significantly during P2 (Figure 4). The mean sediment loads of extreme, ordinary, and total rainfall increased from 0.60×10^4 t, 1.50×10^4 t, and 4.72×10^4 t in P1 to 2.89×10^4 t, 4.31×10^4 t, and 14.69×10^4 t, respectively, in P2 (Figure 4). The frequency of extreme and ordinary rainfall events was relatively stable at about 5 and 24 events, respectively (Figure 5). The proportion of ordinary and extreme rainfall events was also about the same in both periods, at over 45% and over 15%, respectively (Figure 5).

It is noteworthy that the sediment-production capacity per unit of rainfall for extreme rainfall events in the P2 period was higher than that of ordinary rainfall; although, the frequency of occurrence was lower. Even small decreases in rainfall and a lower frequency of extreme events have the potential to cause an observed increase in sediment loads within river ecosystems.

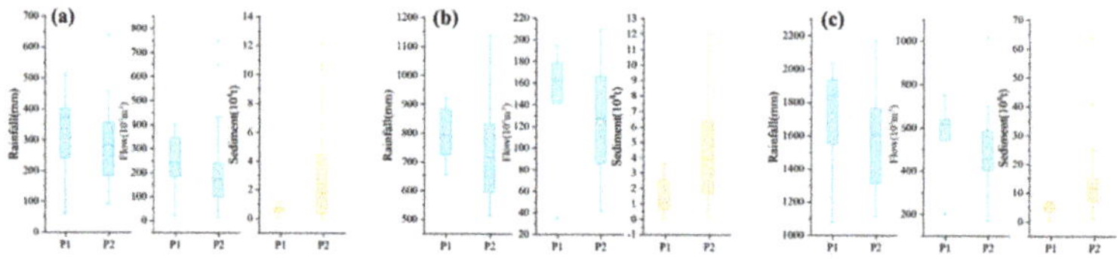

Figure 4. Distribution characteristics of extreme rainfall (**a**), ordinary rainfall (**b**), total rainfall (**c**), and their corresponding riverine flow discharge and sediment load for different periods. Solid lines and square in this figure represent the median and mean, respectively. The box boundaries represent the 75% and 25% quartiles; the whisker caps represent the 90% and 10% quartiles; and the circles represent the 95% and 5% quartiles.

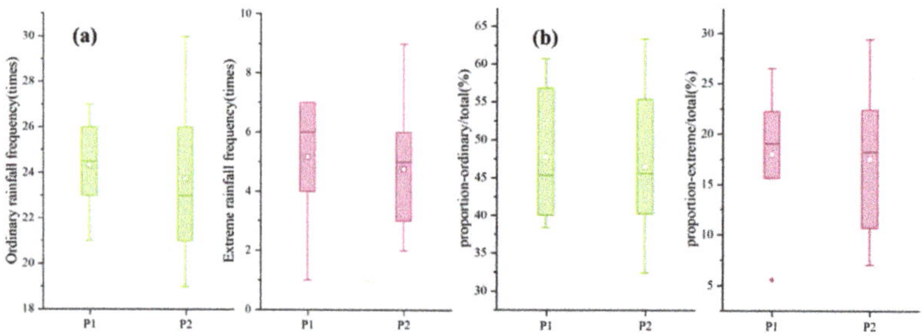

Figure 5. The frequency (**a**) and proportion (**b**) distribution of ordinary and extreme rainfall in different periods. Solid lines and square in this figure represent the median and mean, respectively. The box boundaries represent the 75% and 25% quartiles; the whisker caps represent the 90% and 10% quartiles; and the circles represent the 95% and 5% quartiles.

3.3. Impact of Antecedent Rainfall on Runoff

From the period of P1 to P2, the optimal influence factor of the preceding rainfall on runoff in the extreme rainfall scenario changed from AP5 to AP7. The affected riverine flow discharge was extended from RA1 to RA7 (Table 2), and the degree of explanation in the period of P2 declined from 0.729 to 0.470. In contrast, the optimal influencing factor of riverine flow discharge by the antecedent rainfall changed from AP7 to AP2 in the ordinary rainfall scenario, and the affected riverine flow discharge was RA1 in both periods (Table 2).

Table 2. Multiple regression modeling of antecedent rainfall and riverine flow discharge, under extreme and ordinary rainfall scenarios for different periods.

Type	Periods	Antecedent Rainfall	Flow Discharge	R^2	*n*	*p*-Value
Extreme rainfall	P1 (1990–1995)	AP5	RA1	0.729	31	***
	P2 (1996–2020)	AP7	RA7	0.470	119	***
Ordinary rainfall	P1 (1990–1995)	AP7	RA1	0.461	146	***
	P2 (1996–2020)	AP2	RA1	0.477	594	***

Note: R^2 represents the extent to which the model can explain the variation in the data; "*n*" represents the total number of samples available in the dataset; and "***" represents a high level of significance, indicating that the $p < 0.001$.

3.4. Impact of Antecedent Rainfall on Sediment

The regression model was statistically significant ($p < 0.05$) only in the extreme rainfall scenario (Table 3). During the P1 period, AP3 was the antecedent rainfall with the highest degree of explanation of riverine flow discharge, whereas the corresponding dependent variable was SA1. During the P2 period, AP7 and SA7 were the most highly explained independent and dependent variables, respectively. From P1 to P2, R^2 decreased from 0.554 to 0.245 for the extreme rainfall scenario (Table 3).

Table 3. Multiple regression modeling of antecedent rainfall and sediment under extreme and ordinary rainfall scenarios for different periods.

Type	Periods	Antecedent Rainfall	Sediment	R^2	n	p-Value
Extreme rainfall	P1 (1990–1995)	AP3	SA1	0.554	31	***
	P2 (1996–2020)	AP7	SA7	0.245	119	***
Ordinary rainfall	P1 (1990–1995)	AP7	SA2	0.059	146	0.077
	P2 (1996–2020)	AP7	SA7	0.011	594	0.816

Note: R^2 represents the extent to which the model can explain the variation in the data; "n" represents the total number of samples available in the dataset; and "***" represents a high level of significance, indicating that the $p < 0.001$.

4. Discussion

Previous researchers have broadly attributed changes in riverine flow discharge and sediment load to both climate change and human activities [6,37,38]. Rainfall changes in the upper Lianjiang River watershed were not significant, so it was inferred that the dramatic increase in sediment could be due to human activities. The expansion of the garden land in the study area was drastic, and no large- or medium-sized reservoirs were built in the study area [9]. During the P2 period, extreme rainfall explained less of the riverine flow discharge (Table 2) and sediment load (Table 3), which may be related to the enhancement of human activities (orchard expansion). Conversion of orchards from forested land reduces surface vegetation cover and, consequently, runoff losses [39,40], which has a reduced scouring and transporting effect on soil particles.

Garden expansion increases riverine sediment loads in the red hilly area in China [5], and the same phenomenon is observed in the provinces of Granada and Málaga [41]. During the P2 period, the degree of explanation (R^2) of rainfall events on riverine flow discharge decreased to 0.470 for the extreme rainfall scenario (Table 2). Rainfall and river flow decreased during P2, while river sediment loads increased significantly. Extreme and ordinary rainfall as a proportion of total rainfall did not change much between the two periods, while the corresponding average sediment loads increased by 4.8 and 2.8 times, respectively. The decrease in rainfall directly reduces river flow, while the increase in riverine sediment load is attributed to human activities [3,4]. Most of the new gardens were found near the main stem of the river (Figure 3). The conversion of land to orchards significantly impacts riverine flow discharge and sediment loads [41]. Newly reclaimed orchards often have severe soil erosion, and increased riverine sediment loads [1,11,15]. The presence of orchards near the main stem of the river reduces the distance over which runoff reaches the river, thereby increasing riverine flow discharge and sediment loads [37]. Orchard development or clean-cultivated orchards increased riverine sediment load [42–45]. In orchards, more than 80 percent of rainfall is lost through runoff, which can be reduced by 86 percent with optimal tillage practices [18]. Management practices in orchards are vital to riverine flow discharge and sediment loads [46]. The implementation of soil and water conservation measures in orchards is effective in reducing riverine sediment loads [43,45]. Extreme rainfall events and orchard-management practices are key factors influencing sediment yield in orchards [47].

The degree of explanation (R^2) for the response of riverine sediment loads to antecedent rainfall was reduced to 0.245 (Table 3) during the P2 period. Soil and water conservation measures have played a crucial role in orchards [48]. With the fully functioning of soil and water conservation measures (e.g., horizontal terraces, anti-slope terraces, grass strip, etc.) after orchard maturity, water infiltration, and water retention capacity was effectively increased, and the time for runoff to reach the river was prolonged [46]. In addition, restored vegetation can effectively obstruct runoff from saturated soils [49] and reduce sediment loads [43,47]. Vegetation restoration also lengthens the time of runoff from the slope, to the flow, and into the river. Those all explain the insignificant response of riverine sediment loads to rainfall in the ordinary rainfall scenario during P2.

Extreme rainfall is a strong driver of riverine flow discharge and sediment load changes [2,9]. In other words, extreme rainfall is an important environmental factor for the lag time of antecedent rainfall on riverine flow discharge and sediment. Extreme rainfall produces fast-flowing runoff that is less consumptive. The lag time of riverine flow discharge depends on the soil-moisture conditions before the event [50]. Rainfall intensity severely affects flow-discharge lag time, and high-intensity and long-duration rainfall shortens the response time of river runoff [51]. Ordinary rainfall, on the other hand, requires the soil to reach the moist threshold for runoff to occur [14]. Rainfall and rainfall intensity are also important factors in the lag time of riverine flow discharge [52,53].

Lag time is a significant indicator of the lag effect, which varies during different periods of flow–sediment. During the P2 period, the lag time of the riverine flow discharge response to the rainfall was shortened from AP7 to AP2, in the ordinary rainfall scenario (Table 2). Garden-land reclamation removed vegetation from the surface. The consumption of runoff is reduced, and the demand for antecedent rainfall is reduced. Under the extreme rainfall scenario, the lag time for both riverine flow discharge and sediment during P2 is 7 days, and the duration of the antecedent rainfall corresponding to the riverine flow discharge and sediment is also 7 days (Tables 2 and 3). Changes in riverine flow discharge are closely related to antecedent rainfall and land use [38]. The conversion process destroys the original ground cover [54], reducing evapotranspiration and moisture absorption from antecedent rainfall [55]. Soil-moist conditions can significantly increase riverine sediment concentration [56]. The lag effect is bound to change as more rainfall converges into the river. The lag time for riverine flow discharge is one day after the rainfall event for extreme rainfall in period P1 and for ordinary rainfall in both periods, while the lag time for riverine flow discharge is 7 days after the rainfall event for extreme rainfall in period P2. Haga et al. conducted a study and discovered that the antecedent soil-moisture conditions and the amount and intensity of rainfall played a crucial role in determining the lag time [29]. Davis conducted another study and demonstrated that implementing biological measures had a significant impact on both the peak flow and the delay in reaching the peak [57]. He et al. conducted a study and found that the lag effect in watershed was a consequence of the interaction among four factors: rainfall, watershed storage, human activities, and the lag period [52]. The increase in human activities, particularly garden development, in the Lianjiang River basin will undoubtedly alter the lag time of river runoff and sediment loading. Sultan et al. emphasized that implementing soil and water conservation measures in shrub forests and natural forests did not effectively reduce runoff [58]. However, the implementation of such measures in plantations could lead to a reduction in runoff of up to 34% [58]. On the one hand, ordinary rainfall events bring less moisture and even less runoff into the river, and the lag time of riverine flow discharge changes weakly. On the other hand, riverine flow discharge from extreme rainfall events is more responsive to orchard expansion, and a longer lag time for riverine flow discharge means more riverine flow discharge is generated. During the P1 period, there was a high amount of rainfall and frequency, and forest land remained untouched. This allowed for more interception of rainfall by the ground surface and less runoff into the river, resulting in lower sediment loads. In the P2 period, however, rainfall and frequency reduced, while garden land began to emerge, resulting in increased stripping of the ground surface and erosion. This led to more-successful conversion of rainfall to riverine flow, ultimately increasing its effect on the riverine sediment loads in the river.

5. Conclusions

In this paper, we examined the rainfall, flow discharge, and sediment load data from the upper Lianjiang River watershed, located in the source area of Chinese largest freshwater lake (Poyang Lake), with humid climate, from 1990 to 2020. We defined extreme rainfall events, and then identified the mutation years in the annual flow-discharge and sediment-load data series, which divided the study period into different periods for further analysis. We also investigated the lag effect of riverine flow-discharge and sediment-load response to antecedent rainfall, with different cumulative durations for each period and rainfall event type. The results showed the following:

(1) During the period of 1990–2020, the sediment load in the upper Lianjiang River watershed experienced a significant abrupt change in 1995, while no such change occurred for rainfall and flow discharge. The increase in garden area, particularly near the main stem of the river, was identified as a potential factor contributing to the change in riverine sediment load.

(2) Extreme rainfall events had a greater impact on riverine flow discharge and sediment load, compared to regular rainfall events, leading to a more significant effect on the lag time of riverine flow discharge and sediment. Expanding garden land increased the lag times for riverine flow discharge and sediment load, and prolonged their response time to preceding rainfall. Garden growth and development reduced soil erosion by improving riverine flow and sediment loads through increased soil stability and better water infiltration.

The results could help to understand the response mechanism of riverine flow discharge and sediment load to antecedent rainfall, and improve the accuracy of simulating riverine flow and sediment under extreme rainfall conditions, which would contribute to effective watershed management during extreme rainfall events.

Author Contributions: Conceptualization, L.Z., X.N., H.Z. and K.L.; supervision and funding acquisition, X.N., H.Z. and J.Z.; data analysis, L.Z. and X.N.; writing the manuscript, L.Z. and X.N. All authors have read and agreed to the published version of the manuscript.

Funding: This study was jointly supported by the National Natural Science Foundation of China (42267058), Jiangxi Provincial Department of Science and Technology Major Science and Technology R&D Special "Unveiling" Project (20213AAG01012), Double Thousand Plan of Jiangxi Province (No. JXSQ2023102244),the Science Foundation of Jiangxi Provincial Water Conservancy Department in China (202124ZDKT25, 202223YBKT18), and Science and Technology Plan Project of Zhejiang Provincial Water Conservancy Department (RB2023).

Data Availability Statement: Data for this work can be found within the article, and for further data, contact the first author or corresponding author.

Conflicts of Interest: The authors declare no conflict of interest.

References

1. Sriwongsitanon, N.; Taesombat, W. Effects of land cover on runoff coefficient. *J. Hydrol.* **2011**, *410*, 226–238. [CrossRef]
2. Ziadat, F.M.; Taimeh, A.Y. Effect of Rainfall Intensity, Slope, Land Use and Antecedent Soil Moisture on Soil Erosion in an Arid Environment. *Land Degrad. Dev.* **2013**, *24*, 582–590. [CrossRef]
3. Liu, Y.X.; Hou, X.L.; Qiao, J.X.; Zhang, W.C.; Fang, M.; Lin, M. Evaluation of soil erosion rates in the hilly-gully region of the Loess Plateau in China in the past 60 years using global fallout plutonium. *Catena* **2023**, *220*, 106666. [CrossRef]
4. Wei, W.; Chen, L.D.; Zhang, H.D.; Chen, J. Effect of rainfall variation and landscape change on runoff and sediment yield from a loess hilly catchment in China. *Environ. Earth Sci.* **2015**, *73*, 1005–1016. [CrossRef]
5. Xiao, L.L.; Yang, X.H.; Chen, S.X.; Cai, H.Y. An assessment of erosivity distribution and its influence on the effectiveness of land use conversion for reducing soil erosion in Jiangxi, China. *Catena* **2015**, *125*, 50–60. [CrossRef]
6. Tian, P.; Zhai, J.Q.; Zhao, G.J.; Mu, X.M. Dynamics of Runoff and Suspended Sediment Transport in a Highly Erodible Catchment on the Chinese Loess Plateau. *Land Degrad. Dev.* **2016**, *27*, 839–850. [CrossRef]
7. Guo, W.X.; Sang, Y.; Hu, J.W.; Wang, W.P.; Wang, H.X. Characteristics and attribution analysis of runoff and sediment evolution in the Wei River mainstream, China. *J. Water Clim. Chang.* **2023**, *14*, 2432–2447. [CrossRef]
8. Wasko, C.; Nathan, R. Influence of changes in rainfall and soil moisture on trends in flooding. *J. Hydrol.* **2019**, *575*, 432–441. [CrossRef]
9. Zhao, L.X.; Guo, Z.L.; Nie, X.F.; Liao, K.T.; Zheng, H.J. Effects of extreme rainfall events on runoff and sediment in the southern red soil area: A long series analysis based on the Upper Lianjiang River of Ganjiang River (1984–2020). *J. Lake Sci.* **2023**, *35*, 2133–2143. [CrossRef]
10. Liang, Y.; Li, D.C.; Lu, X.X.; Yang, X.; Pan, X.Z.; Mu, H.; Shi, D.M.; Zhang, B. Soil Erosion Changes over the Past Five Decades in the Red Soil Region of Southern China. *J. Mt. Sci.* **2010**, *7*, 92–99. [CrossRef]
11. Duan, J.; Liu, Y.J.; Yang, J.; Tang, C.J.; Shi, Z.H. Role of groundcover management in controlling soil erosion under extreme rainfall in citrus orchards of southern China. *J. Hydrol.* **2020**, *582*, 124290. [CrossRef]

12. Liu, X.; Zhang, Y.; Xiao, T.; Li, P.; Zhang, L.; Liu, Y.; Deng, W. Runoff velocity controls soil nitrogen leaching in subtropical restored forest in southern China. *For. Ecol. Manag.* **2023**, *548*, 121412. [CrossRef]
13. Bennett, B.; Leonard, M.; Deng, Y.; Westra, S. An empirical investigation into the effect of antecedent precipitation on flood volume. *J. Hydrol.* **2018**, *567*, 435–445. [CrossRef]
14. Wang, Y.J.; Gao, L.; Huang, S.S.; Peng, X.H. Combined effects of rainfall types and antecedent soil moisture on runoff generation at a hillslope of red soil region. *Eur. J. Soil Sci.* **2022**, *73*, e13274. [CrossRef]
15. Li, X.Y.; Contreras, S.; Sole-Benet, A.; Canton, Y.; Domingo, F.; Lazaro, R.; Lin, H.; Van Wesemael, B.; Puigdefabregas, J. Controls of infiltration-runoff processes in Mediterranean karst rangelands in SE Spain. *Catena* **2011**, *86*, 98–109. [CrossRef]
16. Zhang, W.J.; Zhu, X.I.; Xiong, X.; Wu, T.; Zhou, S.Y.D.; Lie, Z.; Jiang, X.J.; Liu, J.X. Changes in soil infiltration and water flow paths: Insights from subtropical forest succession sequence. *Catena* **2023**, *221*, 106748. [CrossRef]
17. Tu, A.G.; Zeng, J.L.; Liu, Z.; Zheng, H.J.; Xie, S.H. Effect of minimum inter-event time for rainfall event separation on rainfall properties and rainfall erosivity in a humid area of southern China. *Geoderma* **2023**, *431*, 116332. [CrossRef]
18. Abrisqueta, J.M.; Plana, V.; Mounzer, O.H.; Mendez, J.; Ruiz-Sanchez, M.C. Effects of soil tillage on runoff generation in a Mediterranean apricot orchard. *Agric. Water Manag.* **2007**, *93*, 11–18. [CrossRef]
19. Dugan, H.A.; Lamoureux, S.F.; Lafreniere, M.J.; Lewis, T. Hydrological and sediment yield response to summer rainfall in a small high Arctic watershed. *Hydrol. Process.* **2009**, *23*, 1514–1526. [CrossRef]
20. Kim, Y.; Rahardjo, H.; Nistor, M.M.; Satyanaga, A.; Leong, E.C.; Sham, A.W.L. Assessment of critical rainfall scenarios for slope stability analyses based on historical rainfall records in Singapore. *Environ. Earth Sci.* **2022**, *81*, 39. [CrossRef]
21. Wasko, C.; Nathan, R.; Stein, L.; O'Shea, D. Evidence of shorter more extreme rainfalls and increased flood variability under climate change. *J. Hydrol.* **2021**, *603*, 126994. [CrossRef]
22. Najibi, N.; Devineni, N. Scaling of Floods With Geomorphologic Characteristics and Precipitation Variability Across the Conterminous United States. *Water Resour. Res.* **2023**, *59*, e2022WR032815. [CrossRef]
23. Defersha, M.B.; Quraishi, S.; Melesse, A. The effect of slope steepness and antecedent moisture content on interrill erosion, runoff and sediment size distribution in the highlands of Ethiopia. *Hydrol. Earth Syst. Sci.* **2011**, *15*, 2367–2375. [CrossRef]
24. Rahimi, A.; Rahardjo, H.; Leong, E.C. Effect of Antecedent Rainfall Patterns on Rainfall-Induced Slope Failure. *J. Geotech. Geoenviron. Eng.* **2011**, *137*, 483–491. [CrossRef]
25. Wang, L.; Zheng, F.L.; Hu, W.; Zhang, X.C.J.; Shi, H.Q. Interactive effects of rainfall intensity, kinetic energy and antecedent soil moisture regime on splash erosion in the Ultisol region of South China. *Catena* **2023**, *222*, 106863. [CrossRef]
26. Jin, Z.; Guo, L.; Yu, Y.L.; Luo, D.; Fan, B.H.; Chu, G.C. Storm runoff generation in headwater catchments on the Chinese Loess Plateau after long-term vegetation rehabilitation. *Sci. Total Environ.* **2020**, *748*, 141375. [CrossRef]
27. Ran, Q.H.; Wang, J.; Chen, X.X.; Liu, L.; Li, J.Y.; Ye, S. The relative importance of antecedent soil moisture and precipitation in flood generation in the middle and lower Yangtze River basin. *Hydrol. Earth Syst. Sci.* **2022**, *26*, 4919–4931. [CrossRef]
28. He, J.; Cai, Q.; Li, G.; Wang, Z. Integrated erosion control measures and environmental effects in rocky mountainous areas in northern China. *Int. J. Sediment Res.* **2010**, *25*, 294–303. [CrossRef]
29. Haga, H.; Matsumoto, Y.; Matsutani, J.; Fujita, M.; Nishida, K.; Sakamoto, Y. Flow paths, rainfall properties, and antecedent soil moisture controlling lags to peak discharge in a granitic unchanneled catchment. *Water Resour. Res.* **2005**, *41*, W12410. [CrossRef]
30. Zou, Z.Q.; Tao, Y.; Gao, Y.H.; Liu, Z.X.; Li, W.K.; Tian, Z.C.; Lin, L.R.; He, Y.B.; Chen, J.Z. Soil moisture dynamics near a gully head in relation to the trigger of collapse in granite red soil slope in southern China. *Geomorphology* **2023**, *420*, 108493. [CrossRef]
31. Liu, J.; Kuang, W.; Zhang, Z.; Xu, X.; Qin, Y.; Ning, J.; Zhou, W.; Zhang, S.; Li, R.; Yan, C.; et al. Spatiotemporal characteristics, patterns, and causes of land-use changes in China since the late 1980s. *J. Geogr. Sci.* **2014**, *24*, 195–210. [CrossRef]
32. Ganzhou Municipal Bureau of Statistics, National Bureau of Statistics Ganzhou Survey Team. *Ganzhou Statistical Yearbook*; China Statistics Press: Beijing, China, 1993–1995, 1997–2021; pp. 0–376.
33. Fang, N.F.; Shi, Z.H.; Yue, B.J.; Wang, L. The Characteristics of Extreme Erosion Events in a Small Mountainous Watershed. *PLoS ONE* **2013**, *8*, e76610. [CrossRef] [PubMed]
34. Mann, H.B. Nonparametric Tests Against Trend. *Econometrica* **1945**, *13*, 245–259. [CrossRef]
35. Shahid, M.; Rahman, K.U. Identifying the Annual and Seasonal Trends of Hydrological and Climatic Variables in the Indus Basin Pakistan. *Asia-Pac. J. Atmos. Sci.* **2021**, *57*, 191–205. [CrossRef]
36. Li, K.; Jacob, D.J.; Liao, H.; Shen, L.; Zhang, Q.; Bates, K.H. Anthropogenic drivers of 2013-2017 trends in summer surface ozone in China. *Proc. Natl. Acad. Sci. USA* **2019**, *116*, 422–427. [CrossRef]
37. Zheng, M.G.; Yang, J.S.; Qi, D.L.; Sun, L.Y.; Cai, Q.G. Flow-sediment relationship as functions of spatial and temporal scales in hilly areas of the Chinese Loess Plateau. *Catena* **2012**, *98*, 29–40. [CrossRef]
38. Gao, G.Y.; Fu, B.J.; Zhang, J.J.; Ma, Y.; Sivapalan, M. Multiscale temporal variability of flow-sediment relationships during the 1950s-2014 in the Loess Plateau, China. *J. Hydrol.* **2018**, *563*, 609–619. [CrossRef]
39. Marchi, L.; Borga, M.; Preciso, E.; Gaume, E. Characterisation of selected extreme flash floods in Europe and implications for flood risk management. *J. Hydrol.* **2010**, *394*, 118–133. [CrossRef]
40. Liu, X.; Zhang, Y.; Zhang, L.; Fang, X.; Deng, W.; Liu, Y. Aggregate-associated soil organic carbon fractions in subtropical soil undergoing vegetative restoration. *Land Degrad. Dev.* **2023**, *34*, 4296–4306. [CrossRef]
41. Zuazo, V.H.D.; Ruiz, J.A.; Raya, A.M.; Tarifa, D.F. Impact of erosion in the taluses of subtropical orchard terraces. *Agric. Ecosyst. Environ.* **2005**, *107*, 199–210. [CrossRef]

42. Rodrigo-Comino, J.; Novara, A.; Gyasi-Agyei, Y.; Terol, E.; Cerda, A. Effects of parent material on soil erosion within Mediterranean new vineyard plantations. *Eng. Geol.* **2018**, *246*, 255–261. [CrossRef]
43. Rodrigo-Comino, J.; Taguas, E.; Seeger, M.; Ries, J.B. Quantification of soil and water losses in an extensive olive orchard catchment in Southern Spain. *J. Hydrol.* **2018**, *556*, 749–758. [CrossRef]
44. Cerd, A.; Terol, E.; Daliakopoulos, I.N. Weed cover controls soil and water losses in rainfed olive groves in Sierra de Enguera, eastern Iberian Peninsula. *J. Environ. Manag.* **2021**, *290*, 112516. [CrossRef] [PubMed]
45. Tsanis, I.K.; Seiradakis, K.D.; Sarchani, S.; Panagea, I.S.; Alexakis, D.D.; Koutroulis, A.G. The Impact of Soil-Improving Cropping Practices on Erosion Rates: A Stakeholder-Oriented Field Experiment Assessment. *Land* **2021**, *10*, 964. [CrossRef]
46. Duan, J.; Liu, Y.J.; Tang, C.J.; Shi, Z.H.; Yang, J. Efficacy of orchard terrace measures to minimize water erosion caused by extreme rainfall in the hilly region of China: Long-term continuous in situ observations. *J. Environ. Manag.* **2021**, *278*, 111537. [CrossRef]
47. Comino, J.R.; Senciales, J.M.; Ramos, M.C.; Martinez-Casasnovas, J.A.; Lasanta, T.; Brevik, E.C.; Ries, J.B.; Sinoga, J.D.R. Understanding soil erosion processes in Mediterranean sloping vineyards (Montes de Malaga, Spain). *Geoderma* **2017**, *296*, 47–59. [CrossRef]
48. Bayat, F.; Monfared, A.B.; Jahansooz, M.R.; Esparza, E.T.; Keshavarzi, A.; Morera, A.G.; Fernandez, M.P.; Cerda, A. Analyzing long-term soil erosion in a ridge-shaped persimmon plantation in eastern Spain by means of ISUM measurements. *Catena* **2019**, *183*, 104176. [CrossRef]
49. Bombino, G.; Denisi, P.; Gomez, J.A.; Zema, D.A. Water Infiltration and Surface Runoff in Steep Clayey Soils of Olive Groves under Different Management Practices. *Water* **2019**, *11*, 240. [CrossRef]
50. Li, J.Y.; Zhang, F.B.; Wang, S.W.; Yang, M.Y. Combined influences of wheat-seedling cover and antecedent soil moisture on sheet erosion in small-flumes. *Soil Tillage Res.* **2015**, *151*, 1–8. [CrossRef]
51. Deng, L.Z.; Sun, T.Y.; Fei, K.; Zhang, L.P.; Fan, X.J.; Wu, Y.H.; Ni, L. Effects of erosion degree, rainfall intensity and slope gradient on runoff and sediment yield for the bare soils from the weathered granite slopes of SE China. *Geomorphology* **2020**, *352*, 106997. [CrossRef]
52. He, Z.H.; Zhao, C.W.; Zhou, Q.; Liang, H.; Yang, Z.H. Temporal-spatial evolution of lagged response of runoff to rainfall in Karst drainage basin, Central Guizhou of China. *Theor. Appl. Climatol.* **2022**, *147*, 437–449. [CrossRef]
53. Singh, N.K.; Emanuel, R.E.; McGlynn, B.L.; Miniat, C.F. Soil Moisture Responses to Rainfall: Implications for Runoff Generation. *Water Resour. Res.* **2021**, *57*, e2020WR028827. [CrossRef]
54. Zhang, X.; Yu, G.Q.; Li, Z.B.; Li, P. Experimental Study on Slope Runoff, Erosion and Sediment under Different Vegetation Types. *Water Resour. Manag.* **2014**, *28*, 2415–2433. [CrossRef]
55. Li, X.Y.; Zou, L.; Xia, J.; Dou, M.; Li, H.W.; Song, Z.H. Untangling the effects of climate change and land use/cover change on spatiotemporal variation of evapotranspiration over China. *J. Hydrol.* **2022**, *612*, 128189. [CrossRef]
56. Napolitano, E.; Fusco, F.; Baum, R.L.; Godt, J.W.; De Vita, P. Effect of antecedent-hydrological conditions on rainfall triggering of debris flows in ash-fall pyroclastic mantled slopes of Campania (southern Italy). *Landslides* **2016**, *13*, 967–983. [CrossRef]
57. Davis, A.P. Field performance of bioretention: Hydrology impacts. *J. Hydrol. Eng.* **2008**, *13*, 90–95. [CrossRef]
58. Sultan, D.; Tsunekawa, A.; Haregeweyn, N.; Adgo, E.; Tsubo, M.; Meshesha, D.T.; Masunaga, T.; Aklog, D.; Fenta, A.A.; Ebabu, K. Impact of Soil and Water Conservation Interventions on Watershed Runoff Response in a Tropical Humid Highland of Ethiopia. *Environ. Manag.* **2018**, *61*, 860–874. [CrossRef]

water

MDPI

Article

Transpiration Water Consumption by *Salix matsudana* and *Populus simonii* and Water Use Patterns at Different Developmental Stages on Sandy Land

Qiaoting Zhai [1], Li Xu [1,*], Tiegang Zhang [2], Jianying Guo [2], Haibo Gao [3], Rui Jiao [2] and Bo Yang [3]

[1] College of Desert Science and Engineering, Inner Mongolia Agricultural University, Hohhot 010018, China; zqtwangyi2021@163.com

[2] Yinshanbeilu Grassland Eco-Hydrology National Observation and Research Station, China Institute of Water Resources and Hydropower Research, Beijing 100038, China; zhang_tiegang@163.com (T.Z.); guojianying1980@163.com (J.G.); jiaorui@iwhr.com (R.J.)

[3] Ordos City Water Conservancy Development Center, Ordos 017000, China; imaugaohb@163.com (H.G.); 13947761370@163.com (B.Y.)

* Correspondence: xulinmg@163.com; Tel.: +86-137-0478-3343

Abstract: Moisture plays a pivotal role in the establishment of vegetation in sandy areas, underscoring the need to comprehend the water utilization strategies employed by established trees for the judicious use of water resources. Despite this significance, there exists a research gap concerning the water uptake patterns and consumption disparities between the dominant trees, namely the dry willow (*Salix matsudana*) and small-leafed poplar (*Populus simonii*), in the Mu Us sandy region. Consequently, our study sought to investigate the water utilization patterns and transpiration water consumption of these two plants. This was achieved through the analysis of hydrogen and oxygen isotope compositions in xylem water, soil water, and groundwater, coupled with the assessment of stem flow rates of tree trunks. The findings reveal that both *Salix matsudana* and *Populus simonii* exhibited variations in soil water content with soil depth, characterized by an initial increase followed by a subsequent decrease. During the months of July, August, and September, both species demonstrated the ability to absorb water from multiple sources concurrently. Specifically, *Salix matsudana* and *Populus simonii* predominantly utilized middle and shallow soil water sources in July and September, respectively. However, in August, both species primarily relied on shallow soil water for absorption. Over the period from July to September, the sap flow rate of *Salix matsudana* surpassed that of *Populus simonii* by 1888.2 $mL \cdot h^{-1}$ to 2499.04 $mL \cdot h^{-1}$, representing a 1.5 to 2.2 times increase. This underscores the necessity for *Salix matsudana* to draw water from middle and deep soil layers to compensate for shallow water deficits. In summary, schemes for the establishment of vegetation in sandy areas should consider the dynamic nature of water uptake and evapotranspiration, emphasizing the importance of regulating these processes for efficient water conservation and utilization.

Keywords: Mu Us sandy land; hydrogen and oxygen isotopes; MixSIAR model; plant water uptake; transpiration water consumption

1. Introduction

In arid and semi-arid regions, the structure and function of ecosystems, as well as the physiological characteristics and spatial distribution of plants, are profoundly influenced by water availability [1,2]. Vegetation, as a vital component of the ecosystem, plays a crucial role in the regional water cycle [3]. Adaptations to the scarcity of rainfall in arid areas lead plants to employ various water use strategies for survival [4]. In response to seasonal water shortages or prolonged precipitation deficits, deep-rooted plants may absorb water from deep soil layers or groundwater to cope with or evade drought conditions [5].

However, inappropriate plant water uptake can result in soil water deficits and even disrupt the balance of the entire ecosystem. Therefore, understanding the water use patterns of artificial vegetation in arid areas, assessing the utilization rate of potential water sources by vegetation under varying water conditions, elucidating the water use and competition dynamics among different vegetation types in the same ecosystem, and comprehending the characteristics of vegetation transpiration stem flow are essential for optimizing water resource utilization and promoting rational water use.

The advancement of stable isotope technology, particularly the application of stable isotopes of hydrogen and oxygen as natural tracers, has emerged as a valuable method for examining plant water sources. This technology has been widely utilized to identify and quantify the contribution rates of different water sources to plant water absorption [6,7]. Plant roots absorb water in a mixture from various sources without isotope fractionation during the uptake process [8]. Comparative analysis of hydrogen and oxygen isotopes in plant xylem and different water sources enables the confirmation of the absorption and utilization of diverse water sources by plants [9]. Researchers such as Rothfuss and Javaux [6] and West [10], have successfully employed stable isotopes of hydrogen and oxygen for water traceability in various ecosystems and regions. Wang [11] utilized this technique to investigate water use patterns in plants in arid and semi-arid regions, determining that the contribution of soil water sources to plants can be estimated through a two-layer or three-layer mixed model. McCole and Stern [12] applied a two-layer mixed model to examine juniper water supply in different seasons in central and southern Texas, revealing that juniper primarily used groundwater in dry and hot seasons and soil water in cold and wet seasons. Other studies, such as those by Darrouzet-Nardi [13], have leveraged $\delta^{18}O$ values to discern that sage in the southern Nevada Mountains predominantly utilizes deeper water than most herbaceous plants but also acquires 10–30% of water from shallow (<30 cm) soil. Ding [14] and McCole and Stern noted that, when shallow soil water supply is insufficient, deep-rooted trees and shrubs may resort to absorbing deep soil water or groundwater, whereas shallow-rooted plants like herbs primarily rely on shallow soil water throughout the growing season. Furthermore, many plants in arid and semi-arid regions exhibit dimorphic roots, allowing them to switch between shallow and deep water sources depending on water availability [15].

Water use patterns among different plants are additionally influenced by variations in soil moisture and plant transpiration water demand. Jackson's study, for instance, discovered that the tropical rainforest in Barro Colorado adapts to dry seasons by acquiring deep soil water, while it relies on shallow soil water during the rainy season, even increasing transpiration water consumption in the dry season [16]. Similarly, Schwendenmann investigated seasonal water uptake patterns of trees in an experimental plantation in central Panama and found that trees capable of utilizing deep soil water maintained higher transpiration rates during the dry season [17].

The Mu Us sandy land represents a typical arid and semi-arid region where seasonal drought is prevalent. This phenomenon is primarily attributed to the dry climate, decreased rainfall, and the excessive planting of high-water-consuming plants. The cumulative effect of these factors results in the seasonal fluctuation of groundwater levels, progressively declining year by year. The diminishing soil water supply and water scarcity have emerged as critical constraints to the ecological restoration and sustainable development of vegetation in the Mu Us sandy land [18].

Although recent years have witnessed an increase in research on plant water sources in the Mu Us sandy land, much of this work has focused on native vegetation such as *Artemisia ordosica*, *Juniperus sabina*, and *Salix matsudana* [19]. However, there remains a notable gap in understanding how artificial vegetation utilizes soil water and groundwater, particularly the impact of transpiration water consumption in areas characterized by shallow groundwater depths. Therefore, an investigation into the root water absorption sources and transpiration water consumption patterns of typical artificial vegetation, namely *Salix matsudana* and *Populus simonii*, in the Mu Us sandy land is imperative to enhance water resource utilization efficiency and promote the sustainability of ecological restoration in this region [20].

In this study, the Mu Us sandy land served as the research focus. The stable isotope technique and stem flow method were employed to examine variations in water absorption, utilization, and transpiration water consumption by *Salix matsudana* and *Populus simonii* at different temporal intervals. The objectives of this study are threefold: (1) to quantitatively analyze the relative contribution rates of different water sources to the water absorption of *Salix matsudana* and *Populus simonii* by measuring stable isotope ratios of hydrogen (δ^2H) and oxygen (δ^{18}O) in plant xylem water, soil water, and groundwater; (2) to determine the patterns of transpiration water consumption by *Salix matsudana* and *Populus simonii* through the measurement of stem flow rates; and (3) to elucidate the differences in water sources and transpiration water consumption between *Salix matsudana* and *Populus simonii* at various time points. The outcomes of this study are expected to offer valuable insights for the development of artificial sand-fixing vegetation and the judicious allocation of groundwater resources in the Mu Us sandy land.

2. Materials and Methods

2.1. Overview of the Study Area

This study was conducted at Meilin Temple (109°22′59″ E, 38°50′44″ N) in Wushen Banner, Ordos City, Inner Mongolia Autonomous Region, situated in the heart of the Mu Us sandy land at an average altitude of 1100 m (refer to Figure 1). The region experiences a semi-arid continental monsoon climate, characterized by an annual average temperature of 6.8 °C, annual sunshine duration ranging from 2800 to 3000 h, an effective accumulated temperature between 2800 and 3000 °C, and an average annual rainfall of 358.2 mm. Precipitation is predominantly concentrated in the months of July to September, accounting for over 70% of the annual precipitation. The annual evaporation rate ranges from 2200 to 2800 mm, the average annual wind speed is 3.4 m/s, and the frost-free period spans 113 to 156 days.

The soil particle composition in the study area is relatively uniform throughout the soil profile of 0–210 cm (see Table 1). The predominant landform type consists of fixed and semi-fixed dunes, and the soil type is characteristic of aeolian sandy soil. The groundwater level in the study area is relatively shallow, typically remaining around 4 m from the surface, providing a potential water source for vegetation growth. The age of *Salix matsudana* and *Populus simonii* in the study area is 16 years. *Salix matsudana* stands at a height of 5.8 m with a diameter at breast height of 12.2 cm and a crown width of 3.2 m^2. *Populus simonii* measures 8.6 m in height, 13.1 cm in diameter at breast height, and has a crown width of 2.8 m^2.

Table 1. Basic physical and chemical properties of the studied soils.

Soil Depth (cm)	SOC (g/kg)	TN (g/kg)	TP (g/kg)	TK (g/kg)	Clay (%)	Silt (%)	Sand (%)
0–30	3.64	0.48	0.40	3.53	2.78	5.54	91.68
30–60	2.25	0.42	0.44	4.02	4.28	13.18	82.53
60–90	1.56	0.36	0.37	3.45	10.94	27.13	61.92
90–120	2.19	0.19	0.24	3.58	8.72	25.4	65.88
120–150	2.02	0.48	0.34	7.70	1.44	3.4	95.16
150–180	1.96	0.08	0.20	2.18	1.51	2.8	95.69
180–210	2.06	0.07	0.17	1.80	2.54	5.36	92.1

Notes: SOC, soil organic carbon; TN, total nitrogen; TP, total phosphorus; TK, total potassium; clay, <0.002 mm; silt, 0.002–0.02 mm; sand, 0.02–2 mm.

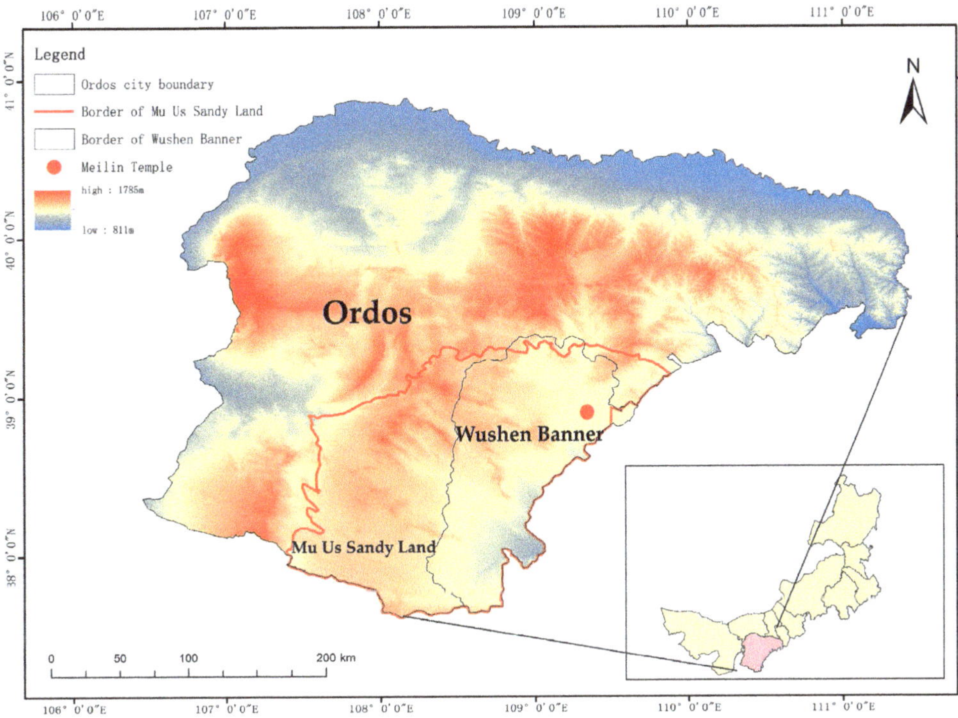

Figure 1. Study area location map.

2.2. Sample Collection and Determination

2.2.1. Experimental Sample Collection

During the peak plant growth season from July to September 2022, we conducted monthly collections of soil water, plant xylem, and groundwater samples. At each sampling date, soil samples were obtained using a spiral drill (diameter = 6 cm, height = 40 cm) positioned 1 m away from *Salix matsudana* and *Populus simonii*. The sampling depths were categorized as follows: 0–30 cm, 30–60 cm, 60–90 cm, 90–120 cm, 120–150 cm, 150–180 cm, and 180–210 cm. Three soil samples were collected from each soil layer of each tree, once a month, and a total of 126 soil samples were collected for 3 months. The collected soil samples underwent a dual partitioning process: one portion was placed in a glass bottle with a spiral cap, enveloped in polyethylene (Parafilm) sealing film, and stored in a refrigerator at −20 °C until water extraction, while the other part was loaded into an aluminum box for soil water content (SWC) determination.

Simultaneously, plant xylem samples were gathered on each sampling date. Healthy twigs from each plant, fully exposed to sunlight and situated at different positions in the canopy, were collected on sunny mornings between 9:00 and 12:00. To prevent contamination of xylem samples by isotope-rich phloem tissue, the sampled twigs underwent the removal of phloem tissue [4]. The collected twig segments (3–5 cm each) were promptly placed in glass bottles with threaded caps, sealed with polyethylene (Parafilm) film, and rapidly transferred to a refrigerator at −20 °C for subsequent isotope analysis. Three plant samples were collected from each tree every month, and a total of 18 plant samples were collected in 3 months. Groundwater samples were extracted from the groundwater observation well (diameter = 11 cm) located 5 m from the sampling point. The timing, frequency, and storage of groundwater samples were synchronized with the collection of soil and plant xylem samples. Temperature and precipitation data were recorded using a small weather station (Campbell Scientific, Utah, USA) installed in the study area.

2.2.2. Measurement of Soil Water Content

The aluminum box, employed for determining soil water content, underwent a two-step process. Initially, it was placed on an electronic balance and weighed. Subsequently, the box was transferred to a constant temperature blast drying oven and dried until reaching a constant weight at 105 °C. The calculation of soil water content utilized Formula (1):

$$SWC = \frac{Sw - Sd}{Sw} * 100 \quad (1)$$

This formula provides the percentage of water content in the soil based on the weight difference before and after drying. In the formula, SWC is soil water content (%), and Sw and Sd are soil wet weight and soil dry weight, respectively.

Based on the depth and seasonal variation characteristics of soil water content (SWC), the soil moisture profiles of *Salix matsudana* and *Populus simonii* were categorized into three levels: 0–60 cm was strongly affected by soil evaporation and precipitation, and was classed as shallow soil water. The SWC of the 60~150 cm soil layer changed greatly and the activity was strong; this was considered middle soil water. The variation of SWC in the 150–210 cm soil layer was small and the activity was stable, and this region was classed as deep soil water.

1. Shallow depth: 0~60 cm
2. Middle depth: 60~150 cm
3. Deep depth: 150~210 cm

This classification allows for a nuanced understanding of the moisture distribution at different soil depths, aiding in the analysis of the water absorption patterns and requirements of *Salix matsudana* and *Populus simonii* across various levels in the soil profile.

2.2.3. Analysis of Hydrogen and Oxygen Isotope Sample Measurement

The plant xylem water and soil water were extracted using a low-temperature vacuum condensation extraction system (L1-2100 type, LICA United Technology Limited, BeiJing, China) [1]. The plant xylem water, soil water, and groundwater were filtered through a size 0.22 μm filter, and the hydrogen and oxygen isotope values were determined by utilizing a liquid water isotope analyzer (TLWIA-921, LICA United Technology Limited, Beijing, China) [2].

$$\delta^2H(‰) = 1000 \times [(R_{sample}/R_{standard}) - 1] \quad (2)$$

$$\delta^{18}O(‰) = 1000 \times [(R_{sample}/R_{standard}) - 1] \quad (3)$$

Here, δ^2H and $\delta^{18}O$ are the hydrogen and oxygen isotope ratios of the corresponding samples, respectively. Rsample and Rstandard are the ratios of abundance of heavy and light isotopes of elements in the samples according to international common standards (such as $^{18}O/^{16}O$), respectively. The accuracy of $\delta^{18}O/^{16}O$ is better than 0.1‰, and $^2H/H$ is better than 0.3‰. The test error of the δ^2H value is less than 1‰, and the test error of the $\delta^{18}O$ value is less than 0.2‰.

The MixSIAR model was used to quantify the proportion of water absorption:

$$^{\delta}X = C1^{\delta}X1 + C2^{\delta}X2 + C3^{\delta}X3 + C4^{\delta}X4 + C5^{\delta}X5 + C6^{\delta}X6 + C7^{\delta}X7 \quad (4)$$

$$c1 + c2 + c3 + c4 + c5 + c6 + c7 = 1$$

In the formula, X is the $^2H/^{18}O$ value (‰) of the tree branch water; and X1, X2, X3, X4, X5, X6, and X7 are the $^2H/^{18}O$ values of soil water at depths of 0~30, 30~60, 60~90, 90~120, 120~150, 150~180, 180~210 cm, respectively. In the formula, C1, C2, C3, C4, C5, C6, C7 represent the absorption ratio of trees for 0~30, 30~60, 60~90, 90~120, 120~150, 150~180, 180~210 cm soil water, respectively.

2.2.4. Measurement of Transpiration Sap Flow

Stem flow density was determined using the Granier thermal diffusion probe method [21]. The trunks of *Salix matsudana* and *Populus simonii*, located at a height of 1.3 m, were meticulously polished with sandpaper. A pair of Granier thermal diffusion probes, each with a length of 2 cm and a diameter of 2 mm, were inserted into the sapwood at intervals of 10 cm and sealed with silica gel. To mitigate the impact of solar radiation and rain, aluminum foil was wrapped around the trunk at the probe installation site [22].

The emitted signals from the stem flow probe were scanned every 60 s, and the data recorder (Delta-T, Cambridge, UK) recorded the average value over 30 min intervals. It is noteworthy that, in this study, azimuthal and radial changes in stem flow within the tree were not considered, given that each sampled tree was equipped with only one pair of probes. The monitoring period aligned with the experimental sample collection time, spanning from July to September 2022. Stem flow density and stem flow were calculated using Formula (5), below. This calculation provides insight into the density of stem flow per unit area of the stem cross-section, contributing to a quantitative understanding of water transport dynamics in *Salix matsudana* and *Populus simonii* during the specified monitoring period.

$$Js = 3600 \times 0.0119 \times \left(\frac{\Delta Tm - \Delta T}{\Delta T}\right)^{1.231} \quad (5)$$

In the formula, 3600 is the unit conversion coefficient, Js is sap flow density ($mL/cm^2 \cdot h$), ΔTm is the maximum temperature difference between day and night, and ΔT is the instantaneous temperature difference.

$$Fs = Js \times SA \quad (6)$$

In the formula, Fs is stem flow ($mL \cdot h^{-1}$), and SA is the sapwood area at DBH (cm^2).

The sapwood area of the whole sample tree was calculated by using the growth cone drill to sample the xylem of the tree to determine the sapwood thickness. The calculation formula is as follows:

$$SA = \pi \left[\left(\frac{DBH}{2} - d \right)^2 - r^2 \right] \quad (7)$$

where DBH is the diameter at breast height (cm) of the tree, d is bark thickness (cm), and r is the radius of heartwood (cm).

2.2.5. Data Processing

SPSS27.0 software was used to linearly fit the plant water and soil water of *Salix matsudana* and *Populus simonii*. Then, using ArcGIS10.2 version to draw the general map of the study area. Origin2018 software was used to plot and analyze the temperature and rainfall, the soil water content of *Salix matsudana* and *Populus simonii*, the changes of hydrogen and oxygen isotopes, and the changes of stem flow rate. Finally, based on R software (v.4.1.2, R Core Team, 2021), the MixSIAR model was used to analyze the water use from each soil layer by *Salix matsudana* and *Populus simonii*.

Through the above processing results, the transpiration water consumption characteristics and water use patterns of *Salix matsudana* and *Populus simonii* were analyzed, and then the differences between the two were compared.

2.2.6. Flowchart for Methodology

According to the process of the experiment in this study, the flow chart of the method is drawn(refer to Figure 2).

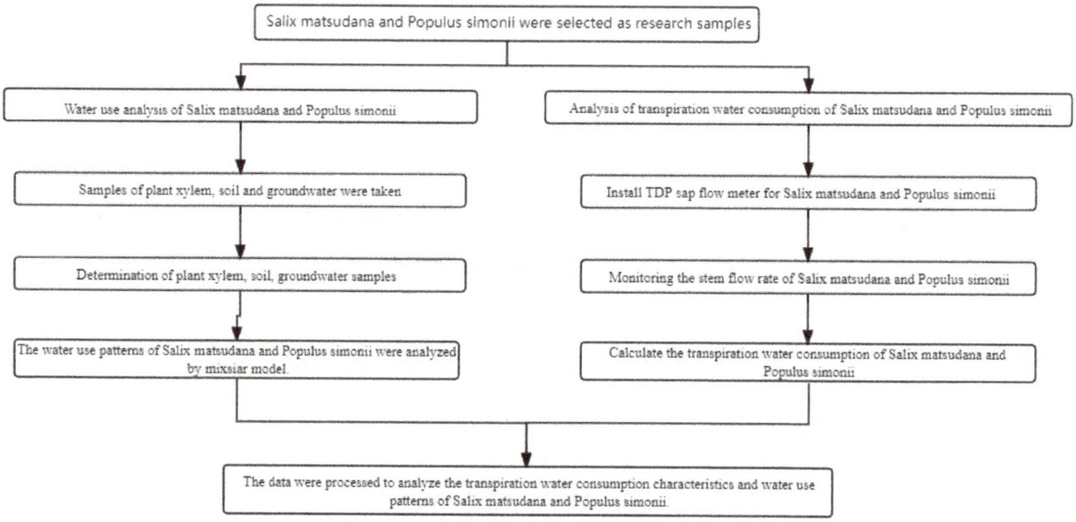

Figure 2. Experimental procedure flow chart.

3. Results

3.1. Temperature and Rainfall

The temporal variations in temperature and rainfall from July to September 2022 are summarized as follows(refer to Figure 3). The peak values for monthly temperature and precipitation both occurred in July. Temperature fluctuated above 20 °C until August 22, after which it exhibited a declining trend, settling around 15–20 °C. Cumulatively, the observed precipitation during the entire monitoring period was 271.7 mm, with a predominant concentration in July and August, accounting for 96.7% of the total precipitation. Specifically, July experienced 117.6 mm of precipitation, and August witnessed 145.4 mm. The highest daily precipitation was recorded on July 11, reaching 43.6 mm. These weather patterns and precipitation levels provide context for understanding the environmental conditions during the study period, essential for interpreting the observed dynamics in soil moisture, plant water utilization, and transpiration.

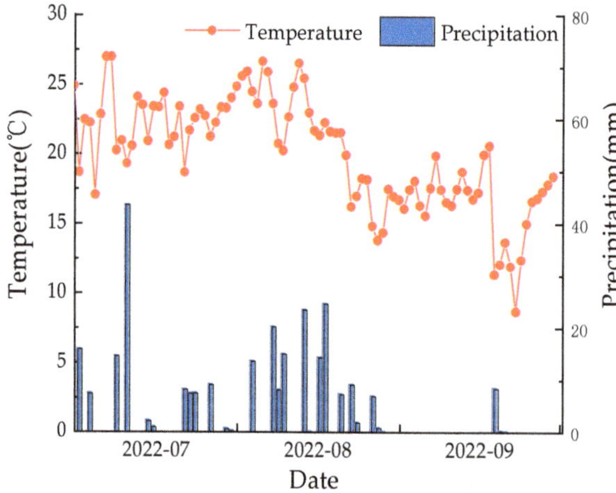

Figure 3. Variation characteristics of temperature and precipitation during monitoring period.

3.2. Changes of Soil Water Content

The monthly soil water content for *Salix matsudana* and *Populus simonii* exhibited a pattern of initial increase followed by a decrease with changes in soil depth(refer to Figure 4). Following a heavy rainfall event (P = 43.6 mm) in July, soil water content increased, along with an increase in infiltration depth. Consequently, the soil water content for *Populus simonii* and *Salix matsudana* reached its maxima at 120 and 150 cm, respectively, registering values of 5.02% and 4.86%. These figures were 1.86% and 1.70% higher than the average value of 3.16%. Subsequently, soil water content began to decrease.

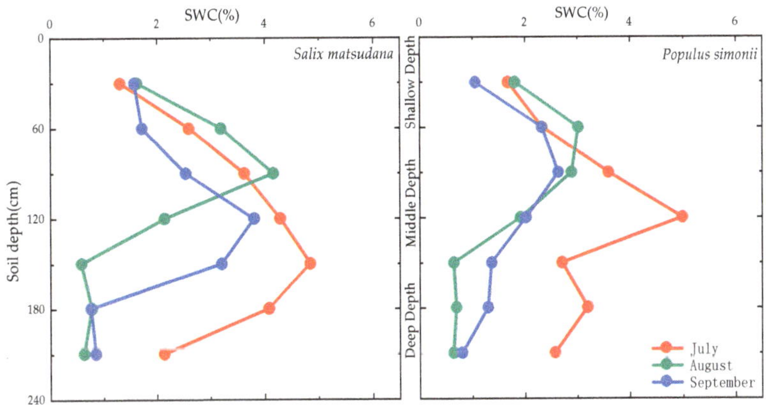

Figure 4. Variation characteristics of soil water content in *Salix matsudana* and *Populus simonii*.

In August, despite higher overall rainfall, individual rainfall events were smaller, resulting in increased soil moisture content but with a shallower infiltration depth. *Populus simonii* and *Salix matsudana* achieved their respective maxima at 60 and 90 cm, recording values of 4.18% and 3.05%. These figures were 2.4% and 1.27% higher than the average value of 1.78%. Following this period, soil moisture content decreased with increasing depth, with a significant decline observed below 150 cm.

In September, characterized by reduced rainfall, the soil water content of *Populus simonii* and *Salix matsudana* decreased overall. The maxima were recorded at 90 and 120 cm, with values of 3.82% and 2.66%, respectively. These figures represented increases of 1.95% and 0.79%, respectively, compared with the average value of 1.87%. Overall, soil moisture content exhibited minimal change in the shallow soil layer (0–60 cm) but displayed pronounced variations in the middle depth layer (60–150 cm). In the deep soil layer below 150 cm, the range of change decreased, and water content began to decrease significantly.

3.3. Composition of Xylem Water, Soil Water, and Groundwater in Salix matsudana and Populus simonii

The $\delta^{18}O$ and $\delta^{2}H$ of plant xylem water were different with regards to time(refer to Figure 5). In the months of July, August, and September, the distribution of $\delta^{18}O$ values of the xylem moisture in *Salix matsudana* ranged from −4.27‰ to 4.89‰, −4.24‰ to −3.26‰, and −3.82‰ to 3.35‰, with average values of −0.59‰, −4.04‰, and 0.95‰, respectively. The$\delta^{2}H$ values were distributed between −41.01‰ and −17.76‰, −35.20‰ and −31.50‰, −48.88‰ and −27.89‰; and the average values were −30.78‰, −33.69‰ and −40.32‰, respectively. In the months of July, August, and September, the distribution of $\delta^{18}O$ values of *Populus simonii* wood moisture ranged from −3.51‰ to 1.29‰, −3.98‰ to 5.70‰, and 3.03‰ to 13.80‰, with average values of −1.00‰, −0.22‰, and 8.33‰, respectively. The $\delta^{2}H$ values were distributed between −41.39‰ and −4.06‰, −40.76‰ and −20.02‰, and −53.89‰ and −40.91‰, with average values of −28.26‰, −32.31‰, and −48.44‰, respectively. The average value of $\delta^{18}O$ in xylem water of *Salix matsudana* followed the pattern September > July > August, the pattern for the average value of $\delta^{18}O$ in xylem water of *Populus simonii* was September > August > July, and that for the average value of δ^2H in xylem water of both was July > August > September. The $\delta^{18}O$ and δ^2H values of xylem water in *Salix matsudana* and *Populus simonii* were linearly correlated, and the slopes of the plant water line (PWL) in July, August, and September were 1.40–2.77 and 2.20–2.79, respectively.

The isotopic composition of soil moisture was also different with regards to time (Figure 3). In the months of July, August, and September, the $\delta^{18}O$ values of soil moisture in *Salix matsudana* were -8.41 ± 2.33‰, -8.57 ± 2.61‰, and -8.60 ± 3.21‰, respectively. The δ^2H values were -51.27 ± 18.36‰, -59.68 ± 13.66‰, and -64.20 ± 13.52‰, respectively. The $\delta^{18}O$ values of soil moisture for *Populus simonii* in July, August, and September were -7.53 ± 2.62‰, -8.51 ± 3.10‰, and -7.99 ± 1.44‰, respectively. The corresponding δ^2H values were -51.41 ± 18.06‰, -59.14 ± 14.84‰, and -63.58 ± 11.50‰. The changes of $\delta^{18}O$ and δ^2H values of soil water in *Salix matsudana* and *Populus simonii* were similar in July, August, and September. The soil water line (SWL) fitted by $\delta^{18}O$ and δ^2H linear regression showed that the SWL slope of *Salix matsudana* from July to September was in the range of 5.77~7.34, and the SWL slope of *Populus simonii* from July to September was in the range of 5.17~6.29.

The $\delta^{18}O$ value of groundwater from July to September was between -9.40‰ and -8.57‰, and the δ^2H value was between -57.05‰ and -55.80‰, which is relatively stable compared with plant xylem water and soil water.

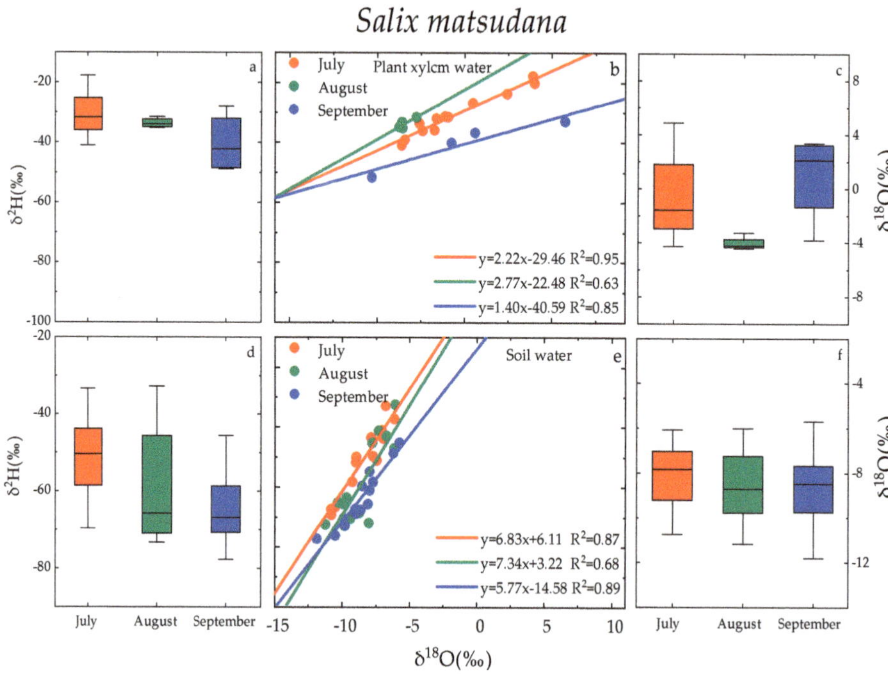

Figure 5. *Cont.*

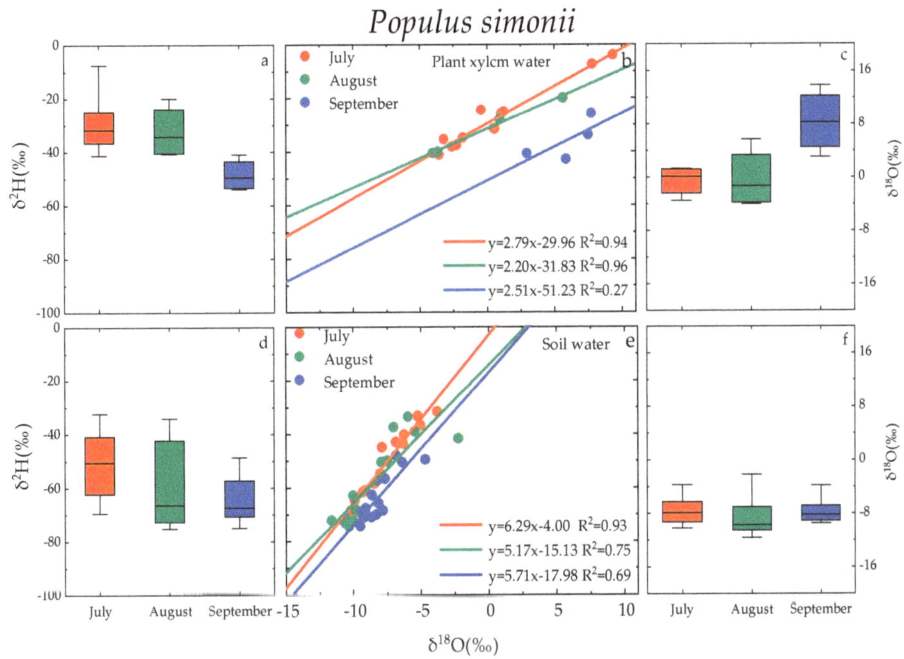

Figure 5. Variation characteristics of xylem water (**a–c**) and soil water (**d–f**) in *Salix matsudana* and *Populus simonii*. Note: The scatter plot and box plot show the isotopic composition of xylem water and soil water of *Salix matsudana* and *Populus simonii* in July, August, and September. Box plots show the median values (black line in box), interquartile range (extent of the box), and range of data (whiskers).

3.4. Seasonal Variation of Water Use Patterns of *Salix matsudana* and *Populus simonii*

During the measurement period, distinct changes in the soil profile water use patterns of *Salix matsudana* and *Populus simonii* were observed in July, August, and September (refer to Figure 6). In July, *Salix matsudana* primarily sourced water from the shallow layer (29.7%), middle layer (33.5%), deep layer (18.5%), and groundwater (18.3%). The highest water contribution rate was observed in the 60~90 cm layer, accounting for 15.7% of the total. In August, *Salix matsudana* predominantly utilized water from the shallow layer (52.8%), middle layer (19.1%), deep layer (13.6%), and groundwater (14.5%). The highest water contribution rate occurred in the 30~60 cm layer, constituting 16.7% of the total. In September, *Salix matsudana* obtained water from the shallow layer (24.3%), middle layer (34.5%), deep layer (16.4%), and groundwater (24.8%). The highest water contribution rate was observed in the 60~90 cm layer, accounting for 16.4% of the total. For *Populus simonii* in July, the major water sources were the shallow layer (40.3%), middle layer (25.4%), deep layer (18.1%), and groundwater (16.1%). The highest water contribution rate occurred in the 0~30 cm layer, constituting 14.6% of the total. In August, *Populus simonii* primarily utilized water from the shallow layer (58.3%), middle layer (15.8%), deep layer (11.3%), and groundwater (14.7%). The highest water contribution rate was observed in the 30~60 cm layer, constituting 18.4%. In September, *Populus simonii* sourced water from the shallow layer (32.7%), middle layer (22.3%), deep layer (18.7%), and groundwater (26.3%). The highest water contribution rate occurred in the 30~60 cm layer, accounting for 13.4%.

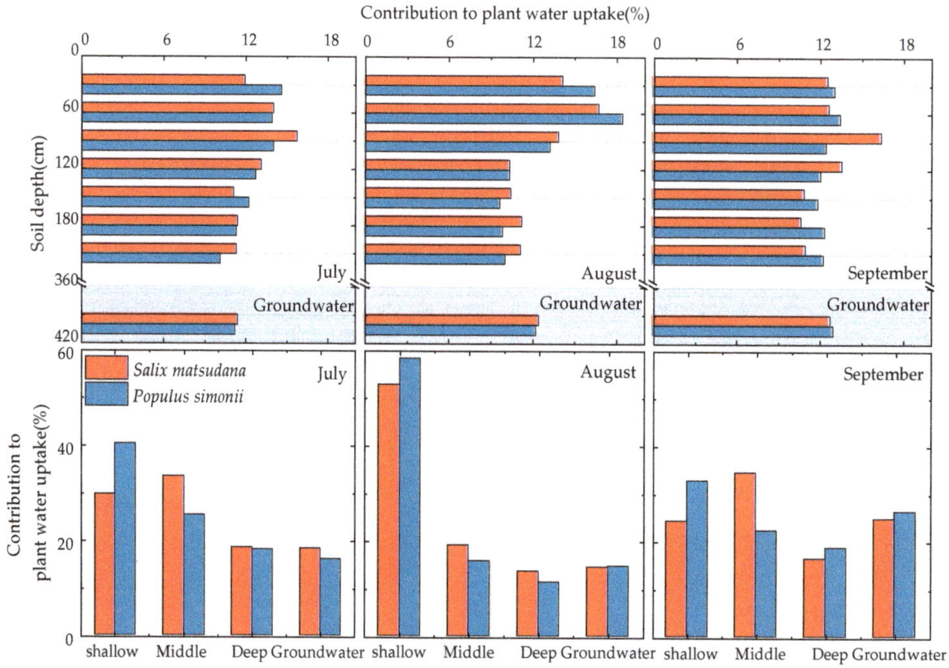

Figure 6. Contribution of potential water sources to *Salix matsudana* and *Populus simonii* during the measurement period.

When both *Salix matsudana* and *Populus simonii* have an ample supply of shallow soil water, they prioritize its utilization. For instance, during periods of increased rainfall, such as in August, the utilization of shallow soil water can account for 50–60% of the total water usage. Conversely, during drier periods, like in September, when shallow soil water becomes insufficient, both species increase their utilization of middle and lower soil layers as well as groundwater.

3.5. Changes of Sap Flow Rate of *Salix matsudana* and *Populus simonii*

During the measurement period, the diurnal variation characteristics of stem flow rate for *Salix matsudana* and *Populus simonii* were analyzed (refer to Figure 7). It was observed that both *Salix matsudana* and *Populus simonii* exhibited a single-peak curve in the diurnal variation of stem flow rate, but there were differences in the start time, peak time, and peak size of the stem flow between the two species.

For *Salix matsudana*, the sap flow rate began to rise rapidly after 7:00 in the morning, peaked around 13:00 at noon, and started to decline thereafter. The decrease in sap flow rate weakened around 21:00 in the evening, remaining relatively stable. The sap flow rates for *Salix matsudana* in July, August, and September ranged from 3573.38 mL·h^{-1} to 5810.30 mL·h^{-1}, 4211.24 mL·h^{-1} to 7889.36 mL·h^{-1}, and 3957.66 mL·h^{-1} to 7939.50 mL·h^{-1}, respectively. The daily average sap flow rates were 4551.86 mL·h^{-1}, 5760.56 mL·h^{-1}, and 5554.33 mL·h^{-1}, respectively. Notably, the sap flow rate of *Salix matsudana* was highest in August, with decreases of 1208.7 mL·h^{-1} and 206.23 mL·h^{-1} in July and September, respectively, compared with August.

For *Populus simonii*, the sap flow rate started to increase rapidly after 6:00 in the morning, with the peak time shifting from about 14:00 in July to about 13:00 in August and about 12:00 in September. The peak was followed by a decline, with a weakened decrease in sap flow rate around 21:00 in the evening, stabilizing thereafter. The sap flow rates for *Populus simonii* in July to September ranged from 634.33 mL·h^{-1} to 3633.12 mL·h^{-1}, 1773.68 mL·h^{-1} to 5876.43 mL·h^{-1}, and 1816.90 mL·h^{-1} to 5944.66 mL·h^{-1}, respectively. The daily average sap flow rates were 2052.82 mL·h^{-1}, 3620.01 mL·h^{-1}, and 3666.13 mL·h^{-1}, respectively. Notably, the sap flow rate in September increased by 1613.31 mL·h^{-1} and 46.12 mL·h^{-1} compared with July and August, reaching its maximum.

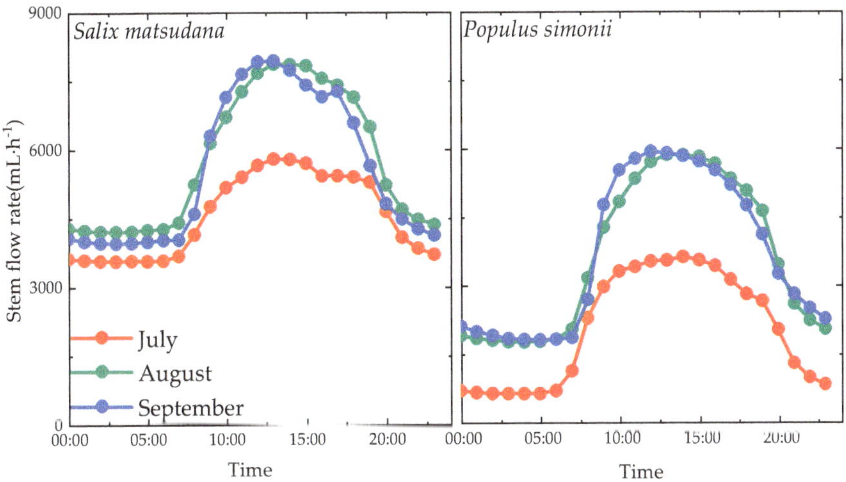

Figure 7. Diurnal variation characteristics of sap flow rate of *Salix matsudana* and *Populus simonii.*

Overall, the changes in stem flow rate for *Salix matsudana* and *Populus simonii* were quite similar from July to September. Both species showed a rapid increase in stem flow rate from 6:00 to 7:00 in the morning, reaching a peak around 12:00 to 14:00, followed by a gradual decrease. The rates stabilized around 21:00 but did not drop to zero. However, the sap flow rate of *Salix matsudana* increased by 2499.04 mL·h^{-1}, 2140.55 mL·h^{-1}, and 1888.2 mL·h^{-1} from July to September compared with *Populus simonii*, representing roughly 1.5–2.2 times that of *Populus simonii.*

4. Discussion

4.1. Isotope Composition and Seasonal Variation in Various Water Bodies

Soil moisture stands out as the primary factor influencing the growth and distribution patterns of vegetation in arid sandy regions. Plants in these areas predominantly rely on soil water and groundwater as their primary water sources, both of which are influenced by precipitation, as well as factors such as soil evaporation and movement. This leads to changes in soil water content, affecting the δ^2H and δ^{18}O values of soil water at various depths(refer to Figure 8). The relative natural abundance of δ^{18}O and δ^2H in different water bodies can serve as indicators of hydrological processes. Throughout the measurement period, the isotope composition of soil moisture in *Salix matsudana* and *Populus simonii* remained essentially similar. This similarity can be attributed to their proximity and the shared soil environment. Notably, *Populus simonii* predominantly absorbed water from the shallow layer from July to September, a period marked by stronger evaporation, resulting in enriched isotopes in the shallow soil water [23]. In contrast, *Salix matsudana* primarily utilized soil water from the middle layer in July and September, where the isotopes were relatively lighter compared with the shallow layer. Consequently, the average values of δ^{18}O and δ^2H in the xylem water of *Populus simonii* were higher than those of *Salix matsudana* [24,25].

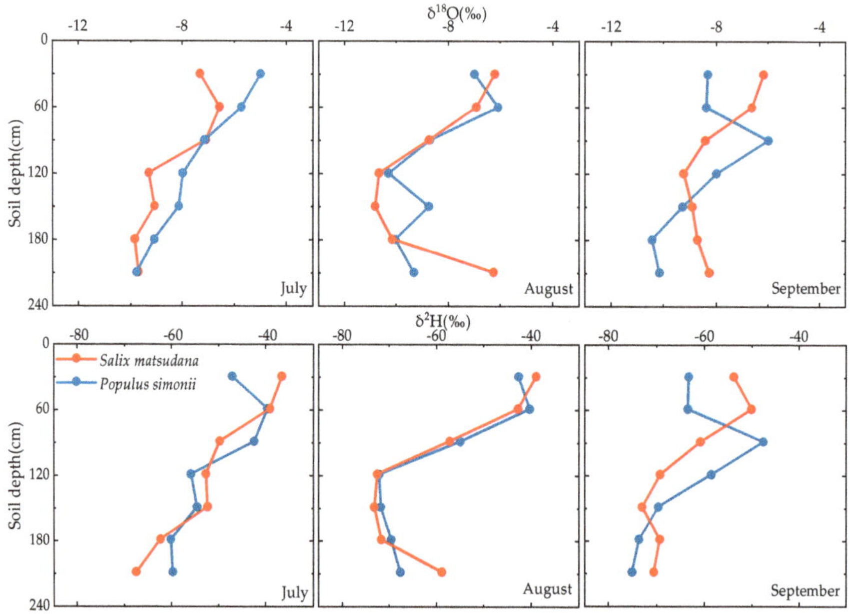

Figure 8. Vertical distribution of soil water isotopes ($\delta^{18}O$ and $\delta^{2}H$) in *Salix matsudana* and *Populus simonii* from July to September.

The isotopic composition of soil water varied significantly with depth and time due to the combined effects of precipitation and evaporation. The average values of $\delta^{18}O$ and $\delta^{2}H$ in soil water for both *Salix matsudana* and *Populus simonii* followed the sequence July > August > September. Shallow soil water, influenced by precipitation and evaporation, exhibited isotopic enrichment [26]. Large precipitation events coupled with existing soil water generated new soil water during the infiltration process, leading to changes in its stable isotope values over time and further contributing to isotopic enrichment [27]. With increasing soil depth, the isotope fractionation effect of soil water diminished, resulting in a gradual decrease in $\delta^{2}H$ and $\delta^{18}O$ values. Notably, the $\delta^{18}O$ and $\delta^{2}H$ values of groundwater from July to September remained relatively consistent, indicating that seasonal precipitation events had minimal impact on the isotope composition of groundwater.

4.2. Water Use Patterns of Salix matsudana and Populus simonii

Plants in arid regions demonstrate a remarkable sensitivity to variations in soil moisture, adjusting their water absorption strategies accordingly. Their ability to absorb and utilize soil water from stable layers contributes to their overall viability. Throughout the measurement period, the water utilization strategies of *Salix matsudana* and *Populus simonii* exhibited temporal fluctuations, with varying proportions of water absorption and utilization at different soil depths.

Trees generally prioritize easily accessible, stable, and sufficient water sources to support their growth [28]. Liu's analysis [29] of the water use characteristics of *Platycladus orientalis* and *Quercus variabilis* in Beijing revealed that *Platycladus orientalis* primarily absorbed soil water from the 0–20 cm layer during the rainy season, while Quercus variabilis predominantly utilized soil water and groundwater from the same layer. In the dry season, *Platycladus orientalis* relied heavily on deep soil water, while *Quercus variabilis* favored shallow soil water [30]. Consequently, when shallow soil water is abundant, vegetation tends to prioritize its utilization.

In degraded forests, Zhang Huan observed a shift in the water source for *Populus simonii* from deep to shallow with increasing degradation degree, suggesting an adaptive response to drought [31,32]. In the current study, *Salix matsudana* demonstrated a higher utilization rate of middle soil water in July, while *Populus simonii* exhibited the highest utilization rate of shallow soil water. In August, with increased rainfall, both species favored shallow soil water. Come September, *Salix matsudana* returned to maximizing the utilization of middle soil water, while *Populus simonii* continued to predominantly rely on shallow soil water, albeit with increased utilization of deep soil water and groundwater.

The observed variations in water uptake by *Salix matsudana* and *Populus simonii* at different times can be attributed to the following factors: (1) The two-state nature of the root system plays a crucial role. In arid and semi-arid regions, plants with shallow roots, such as herbs, generally rely on surface soil water and precipitation. In contrast, deep-rooted plants like trees or shrubs tend to utilize more stable water sources, such as groundwater and deep soil water. Although a two-layer water use model hypothesis [33] has been proposed, some studies, like Le Roux's [34], have found that regardless of root depth, plants predominantly depend on surface soil water. (2) The variation in soil water content is a key determinant. In months with higher rainfall, such as August, the content of shallow soil water is elevated and closer to the plant roots. This results in a shorter duration for water absorption, making it more accessible for *Salix matsudana* and *Populus simonii*. Conversely, during periods when the shallow soil water content is low, plants shift their water absorption towards the middle and deep soil layers and groundwater. The uniform utilization of potential water sources by both species suggests an optimal water absorption mode. When shallow soil water is insufficient, the utilization rate of middle and deep soil water and groundwater increases, allowing plants to access more stable water sources. (3) Variations in plant water requirements and transpiration rates also contribute. *Salix matsudana* exhibits a higher sap flow rate compared with *Populus simonii*, indicating a greater demand for water to meet transpiration needs. Consequently, *Salix matsudana* tends to absorb water from the middle and deep soil layers to compensate for the deficit in shallow soil water. In summary, the interplay of root system characteristics, soil water content, and plant water requirements leads to dynamic changes in the water uptake strategies of *Salix matsudana* and *Populus simonii* at different times, enabling them to adapt to varying environmental conditions.

4.3. Changes of Transpiration Sap Flow of Salix matsudana and Populus simonii

During the measurement period, the sap flow rate of both *Salix matsudana* and *Populus simonii* exhibited similar diurnal variations, following bell-shaped curves from July to September. This pattern aligns with previous studies on *Caragana microphylla* conducted by Yue [35]. The sap flow rates of both species were higher in August and September compared with July, probably due to these months being the peak seasons for transpiration. During this period, the leaves of *Salix matsudana* and *Populus simonii* are the most mature and dense, and the high temperatures and low humidity, along with strong radiation, create optimal conditions for leaf transpiration, as observed in Jia's research [36].

The diurnal variation analysis revealed that the sap flow rate increased rapidly from 6:00 to 7:00 in the morning, reaching its peak around 12:00 to 14:00, followed by a gradual decline. The sap flow rate remained relatively stable from around 21:00 but did not drop to zero. This phenomenon is attributed to the substantial water consumption by *Salix matsudana* and *Populus simonii* during the day, leading to water deficit in the plants [37]. To counteract this, weak transpiration occurs at night, and water is actively transported to the plant body under root pressure, replenishing the water consumed during the day. This helps restore the water balance in the plants and aids in adapting to the arid environment with water deficits. This finding is consistent with Yu Feng's study [38] on the transpiration water consumption pattern of artificial *Caragana korshinskii* in the arid zone of central Ningxia.

Salix matsudana exhibited a higher sap flow rate from July to September, indicating a greater demand for water to meet its transpiration needs. Despite more rainfall in August, which could satisfy its transpiration demands with shallow soil water, *Salix matsudana* still absorbed water from the middle and deep layers in July and September. In contrast, *Populus simonii* had a lower stem flow rate during this period, and the water demands for its transpiration were met by shallow soil water, making it the main water-absorbing layer from July to September. These observations highlight the dynamic nature of water absorption strategies employed by these species based on their water requirements and the availability of soil moisture.

5. Conclusions

In this study, the water use patterns and sap flow transpiration of *Salix matsudana* and *Populus simonii* in the Mu Us sandy land were investigated using stable isotope techniques and thermal diffusion sap flow methods. The findings indicate that both *Salix matsudana* and *Populus simonii* absorbed water from multiple sources simultaneously from July to September. *Populus simonii* predominantly absorbed water from the shallow soil layer (0~60 cm), with a contribution ranging from 32.7% to 58.3%. On the other hand, *Salix matsudana* exhibited a different water absorption pattern, drawing water from deeper soil layers and groundwater during dry periods. The diurnal variation of sap flow rate for both *Salix matsudana* and *Populus simonii* showed a single-peak curve. The sap flow rates of *Salix matsudana* and *Populus simonii* ranged from 3573.38 $mL \cdot h^{-1}$ to 7939.50 $mL \cdot h^{-1}$ and from 634.33 $mL \cdot h^{-1}$ to 5944.66 $mL \cdot h^{-1}$, respectively, during the period from July to September. *Salix matsudana* consistently exhibited higher sap flow rates, ranging from 1888.2 $mL \cdot h^{-1}$ to 2499.04 $mL \cdot h^{-1}$ more than *Populus simonii*. This difference indicates that *Salix matsudana* required 1.5 to 2.2 times more water for transpiration compared with *Populus simonii*. Therefore, in contrast to *Populus simonii*, which primarily absorbed water from shallow soil layers, *Salix matsudana* needed to extract water from middle and deep soil layers to compensate for the water deficit in the shallow soil. These findings contribute to a better understanding of the water use strategies and transpiration patterns of these two vegetation species in the Mu Us sandy land.

Author Contributions: Conceptualization, Q.Z. and T.Z.; methodology, L.X. and J.G; validation, T.Z. and Q.Z.; investigation, H.G.; resources, B.Y.; data curation, R.J. writing—original draft preparation, Q.Z.; review and editing, T.Z.; visualization, R.J.; supervision, L.X. and J.G.; project administration, H.G.; funding acquisition, B.Y. All authors have read and agreed to the published version of the manuscript.

Funding: This work was supported by the Inner Mongolia Natural Science Foundation (2021MS04022), Inner Mongolia Science and Technology Program (2021GG0198, 2022YFHH0063), Collaborative Innovation Center for Grassland Ecological Security (Jointly Supported by the Ministry of Education of China and Inner Mongolia Autonomous Region), the Science and Technology Program of Inner Mongolia Autonomous Region, China (KJXMEEDS-2020006).

Data Availability Statement: The data are not publicly available due to confidentiality. The data presented in this study are available on request from the corresponding author.

Acknowledgments: The authors are very grateful for the strong support provided by all the cooperative institutions involved in the project research. We would like to thank Savannah Grace at the University of Florida for her assistance with English language and grammatical editing of the manuscript.

Conflicts of Interest: The authors declare no conflict of interest.

References

1. Zhang, Z.; Wang, W.; Gong, C.; Zhao, M.; Franssen, H.-J.H.; Brunner, P. Salix psammophila afforestations can cause a decline of the water table, prevent groundwater recharge and reduce effective infiltration. *Sci. Total Environ.* **2021**, *780*, 146336. [CrossRef]
2. Brunel, J.-P.; Walker, G.R.; Kennett-Smith, A.K. Field validation of isotopic procedures for determining sources of water used by plants in a semi-arid environment. *J. Hydrol.* **1995**, *167*, 351–368. [CrossRef]
3. Yu, X.; Liu, H.; Wang, L.; Wang, Q.; Liu, X.; Wen, L.; Dong, S.; Pan, B. Analysis on Water Sources of Riverbank Vegetation in Tributary of Yellow River Based on Stable Isotope Technique. *J. Soil. Water Conserv.* **2021**, *41*, 75–82+91. (In Chinese) [CrossRef]

4. Ehleringer, J.R.; Dawson, T.E. Water uptake by plants: Perspectives from stable isotope composition. *Plant Cell Environ.* **2010**, *15*, 1073–1082. [CrossRef]
5. Wang, Y.L.; Liu, L.C.; Gao, Y.H.; Li, G.; Zhao, J.C.; Xie, M. Analysis of water sources of plants in artificial sand-fixation vegetation area based on large rainfall events. *Chin. J. Appl. Ecol.* **2016**, *27*, 1053–1060. (In Chinese) [CrossRef]
6. Youri, R.; Mathieu, J. Reviews and syntheses: Isotopic approaches to quantify root water uptake: A review and comparison of methods. *Biogeosciences* **2017**, *14*, 2199–2224. [CrossRef]
7. Penna, D.; Hopp, L.; Scandellari, F.; Allen, S.T.; Benettin, P.; Beyer, M.; Geris, J.; Klaus, J.; Marshall, J.D.; Schwendenmann, L. Ideas and perspectives: Tracing terrestrial ecosystem water fluxes using hydrogen and oxygen stable isotopes—Challenges and opportunities from an interdisciplinary perspective. *Biogeosciences* **2018**, *15*, 6399–6415. [CrossRef]
8. Montgomery, J.A.; Klimas, C.A.; Arcus, J.; Deknock, C.; Rico, K.; Rodriguez, Y.; Vollrath, K.; Webb, E.; Williams, A. Soil quality assessment is a necessary first step for designing urban green infrastructure. *J. Environ. Qual.* **2016**, *45*, 18–25. [CrossRef]
9. Midwood, A.J.; Boutton, T.W.; Archer, S.R.; Watts, S.E. Water use by woody plants on contrasting soils in a savanna parkland: Assessment with δ2H and δ18O. *Plant Soil.* **1998**, *205*, 13–24. [CrossRef]
10. West, A.G.; Dawson, T.E.; February, E.C.; Midgley, G.F.; Bond, W.J.; Aston, T.L. Diverse functional responses to drought in a Mediterranean-type shrubland in South Africa. *New Phytol.* **2012**, *195*, 396–407. [CrossRef]
11. Peng, W.; Song, X.; Han, D.; Xin, L. A study of root water uptake of crops indicated by hydrogen and oxygen stable isotopes: A case in Shanxi Province. *Agric. Manag. Water Qual.* **2010**, *97*, 475–482. (In Chinese) [CrossRef]
12. McCole, A.A.; Stern, L.A. Seasonal water use patterns of Juniperus ashei on the Edwards Plateau, Texas, based on stable isotopes in water. *J. Hydrol.* **2007**, *342*, 238–248. [CrossRef]
13. Darrouzet-Nardi, A.; D'Antonio, C.M.; Dawson, T.E. Depth of water acquisition by invading shrubs and resident herbs in a Sierra Nevada meadow. *Plant Soil.* **2006**, *285*, 31–43. [CrossRef]
14. Ding, Y.; Nie, Y.; Schwinning, S.; Chen, H.; Yang, J.; Zhang, W.; Wang, K. A novel approach for estimating groundwater use by plants in rock-dominated habitats. *J. Hydrol.* **2018**, *565*, 760–769. [CrossRef]
15. Zhao, Y.; Wang, Y.; He, M.; Tong, Y.; Zhou, J.; Guo, X.; Liu, J.; Zhang, X. Transference of Robinia pseudoacacia water-use patterns from deep to shallow soil layers during the transition period between the dry and rainy seasons in a water-limited region. *For. Ecol. Manag.* **2019**, *457*, 117727. [CrossRef]
16. Jackson, P.C.; Cavelier, J.; Goldstein, G.; Meinzer, F.C.; Holbrook, N.M. Partitioning of water resources among plants of a lowland tropical forest. *Oecologia* **1995**, *101*, 197–203. [CrossRef] [PubMed]
17. Schwendenmann, L.; Pendall, E.; Sanchez-Bragado, R.; Kunert, N.; Hulscheer, D. Tree water uptake in a tropical plantation varying in tree diversity: Interspecific differences, seasonal shifts and complementarity. *Ecohydrology* **2015**, *8*, 1–12. [CrossRef]
18. Hao, B.; Ai, N.; Jia, Y.; Zong, Q.; Liu, J.; Liu, C. Soil water characteristics of different vegetation types in the south of MU Us Sandy Land. *J. Arid Environ.* **2020**, *34*, 196–200. (In Chinese) [CrossRef]
19. Ohte, N.; Koba, K.; Yoshikawa, K.; Sugimoto, A.; Matsuo, N.; Wang, K.L. Water Utilization of Natural and Planted Trees in the Semiarid Desert of Inner Mongolia, China. *Ecol. Appl.* **2003**, *13*, 337–351. [CrossRef]
20. Li, R.; Huang, L.; Pei, Y.; Shao, M.; Zhang, Y. Water Use Source of Salix psammophila in Gechougou Small Watershedof Mu Us Sandy Land. *J. Soil Water Conserv.* **2021**, *35*, 122–130. (In Chinese) [CrossRef]
21. Granier, A. Evaluation of transpiration in a Douglas-fir stand by means of sap flow measurements. *Tree Physiol.* **1987**, *3*, 309–320. [CrossRef]
22. Yin, L.H.; Huang, J.T.; Wang, X.Y.; Dong, J.Q.; Ma, H.Y.; Zhang, J. Characteristics of night time sap flow of Salix matsudana and Populus simonii in Yulin, Shaanxi. *J. Northwest A F Univ.* **2013**, *41*, 85–90. (In Chinese) [CrossRef]
23. Gazis, C.; Feng, X.H. A stable isotope study of soil water: Evidence for mixing and preferential flow paths. *Geoderma* **2004**, *119*, 97–111. [CrossRef]
24. Allen, S.T.; Kirchner, J.W.; Braun, S.; Siegwolf, R.T.W.; Goldsmith, G.R. Seasonal origins of soil water used by trees. *Hydrol. Earth Syst. Sci.* **2019**, *23*, 1199–1210. [CrossRef]
25. Wang, J.; Fu, B.; Lu, N.; Zhang, L. Seasonal variation in water uptake patterns of three plant species based on stable isotopes in the semi-arid Loess Plateau. *Sci. Total Environ.* **2017**, *609*, 27–37. [CrossRef] [PubMed]
26. Bowen, G.J.; Putman, A.; Brooks, J.R.; Bowling, D.R.; Oerter, E.J.; Good, S.P. Inferring the source of evaporated waters using stable H and O isotopes. *Oecologia* **2017**, *187*, 1025–1039. [CrossRef]
27. Tian, C.; Meng, P.; Zhang, J.; Sun, S.; He, C.; Jia, C.; Li, J. Seasonal characteristics of 8D and 8180 in precipitation and its water vaporsources in the Xiaolangdi Reservoir area of Yellow River, China. *J. Appl. Ecol.* **2015**, *26*, 3579–3587. (In Chinese) [CrossRef]
28. Eggemeyer, K.D.; Awada, T.; Harvey, F.E.; Wedin, D.A.; Zhou, X.; Zanner, C.W. Seasonal changes in depth of water uptake for encroaching trees Juniperus virginiana and Pinus ponderosa and two dominant C4 grasses in a semiarid grassland. *Tree Physiol.* **2009**, *29*, 157–169. [CrossRef]
29. Liu, Z.; Yu, X.; Deng, W.; Jia, G.; Jia, J.; Lou, Y.; Li, H. Water sources of *Pinus tabuliformis* and *Platycladus orientalis* before and after rain in northern China. *J. Soil Water Conserv.* **2016**, *14*, 111–119. (In Chinese) [CrossRef]
30. Liu, Z.; Yu, X.; Jia, G.; Jia, J.; Lou, Y.; Zhang, K. Water Use Characteristics of *Platycladus orientalis* and Quercus variabilis in Beijing Mountain Area. *Sci. Silvae Sin.* **2016**, *52*, 22–30. (In Chinese)
31. Hasselquist, N.J.; Allen, M.F. Increasing demands on limited water resources: Consequences for two endangered plants in Amargosa Valley, USA. *Am. J. Bot.* **2009**, *96*, 620–626. [CrossRef]

32. Zhang, H.; Jia, G.; Liu, Z.; Yu, X.; Cao, J.; Wang, H.; Song, B.; Zeng, J. Water utilization characteristics of the degraded poplar shelterbelts inZhangbei, Hebei, China. *J. Appl. Ecol.* **2018**, *29*, 1381–1388. (In Chinese) [CrossRef]
33. Ward, D.; Wiegand, K.; Getzin, S. Walter's two-layer hypothesis revisited—Back to the roots! *S. Afr. J. Bot.* **2013**, *86*, 141. [CrossRef]
34. Roux, X.L.; Bariac, T.; Mariotti, A. Spatial partitioning of the soil water resource between grass and shrub components in a West African humid savanna. *Oecologia* **1995**, *104*, 147–155. [CrossRef]
35. Yue, G.Y.; Zhao, H.L.; Zhang, T.H.; Yun, J.Y.; Niu, L.; He, Y.H. Characteristics of Caragana microphylla sap flow and water consumptionunder different weather conditions on Horqin sandy land of northeast China. *J. Appl. Ecol.* **2007**, *18*, 2173–2178. (In Chinese)
36. Jia, T.; Liu, T.; Duan, L.; Zhang, Y.; Tong, X.; Bao, Y. Transpiration and water consumption of poplar trees in semi-arid dunemeadow transition zone. *Chin. J. Ecol.* **2020**, *39*, 3255–3264. (In Chinese) [CrossRef]
37. Li, L.; Nie, H.; Jing, Q.; Zhang, Y.; Lan, Y.; Zhu, L. Correlations between the Transpiration Water Consumption of Hedysarum-mongolicum and Environmental Factors in Mu Us Sandy Land. *Water SA* **2022**, *12*, 103–109+118. (In Chinese) [CrossRef]
38. Yu, F.; Chen, L.; Wang, F.; Liu, H.; Du, J.M.; Guo, Y.Z.; Li, H.X. The Study on Sap Flow and Transpiration of Artificial Caragana korshinkii in the Arid Area of Central Ningxia Hui Autonomous Region. *Soil Water Conserv.* **2011**, *18*, 143–147. (In Chinese)

water

Article

Transformative Trends in Runoff and Sediment Dynamics and Their Influential Drivers in the Wuding River Basin of the Yellow River: A Comprehensive Analysis from 1960 to 2020

Jingwei Yao [1,2], Zhanbin Li [1], Biao Zhu [3], Pan Zhang [2], Jingshu Wang [2], Weiying Sun [2], Shasha Mei [4], Yaqiang Zhang [5] and Peiqing Xiao [2,*]

1 State Key Laboratory of Eco-Hydraulics in Northwest Arid Region of China, Xi'an University of Technology, Xi'an 710048, China; solofromchina@163.com (J.Y.)
2 Key Laboratory of Soil and Water Conservation on the Loess Plateau of Ministry of Water Resources, Yellow River Institute of Hydraulic Research, Zhengzhou 450003, China
3 Hydrology Bureau of Yellow River Conservancy Commission, Zhengzhou 450004, China
4 School of Management Engineering, Zhengzhou University of Aeronautics, Zhengzhou 450046, China
5 Henan Water Conservancy Resettlement Affairs Center, Zhengzhou 450003, China
* Correspondence: 12022131378@stu.nxu.edu.cn; Tel.: +86-10-136-6385-8913

Abstract: The correlation between runoff and sediment challenges ecological preservation and sustainable development in the Yellow River Basin. An understanding of the key factors influencing variations in runoff and sediment transport in crucial river basins is essential for effective soil erosion management within the context of ecological and economic development. The Mann–Kendall test, Pettitt test, and Morlet wavelet analysis were employed in the Wuding River Basin to analyze the trends in runoff and sediment changes from 1960 to 2020. We explored the double cumulative curve method to assess the contribution rates of precipitation and human activities to the variability of runoff and sediment transport. We explored the primary factors driving the changes in runoff and sediment transport through random forest regression analysis. (1) From 1960 to 2020, annual precipitation in the Wuding River Basin increased minimally, while annual runoff and sediment transport decreased strongly with abrupt changes. Abrupt changes in annual runoff and sediment transport occurred in 1971 and 1979, respectively. (2) The relationship between runoff and sediment transport changed in approximately 1972 and 2000. The distribution of monthly runoff became more uniform during Periods II (1973–2001) and III (2002–2020) compared to that during the baseline period (1960–1972, Period I), while sediment transport became increasingly concentrated in the flood season. (3) During Period II, the contribution rates of climate and human activities to runoff and sediment transport were 11.94% and −14.5%, respectively, compared to the baseline period. During Period III, the contribution rates of climate and human activities to runoff and sediment transport were −11.9% and −17.7%, respectively. Human activities substantially reduced runoff and sediment, with greater impacts on sediment reduction. Climate weakly influenced basin sediment transport variations. (4) The normalized difference vegetation index (NDVI) and grassland area extent had the greatest impact on runoff, while the NDVI and forest area extent affected sediment transport.

Keywords: runoff–sediment correlation; sediment burden; anthropogenic interference; climatic perturbations; propelling mechanisms

1. Introduction

The dynamics of flow and sediment, an integral facet of surface material circulation, arise from the synergistic impacts of climatic factors and underlying surface characteristics, assuming a pivotal role in shaping and advancing regional water and soil resources [1,2]. Globally, climate change and anthropogenic endeavors have variably modified hydrological processes within watersheds, thereby influencing the stability of ecosystemic realms [3,4].

In response to the deteriorating ecological milieu, extensive ecological restoration initiatives have been undertaken worldwide since the 20th century, aiming to ameliorate the prevailing conditions [5–7]. Human-induced modifications on the Earth's surface have profoundly impacted the structure, functionality, and spatial arrangement of ecosystems, concurrently yielding alterations in the sediment and water processes within watersheds. In recent decades, significant changes have been observed in the flow and sedimentation of many rivers globally [8–10], driven primarily by climate variability and human activities. These changes exhibit significant spatiotemporal variations [11–13]. Consequently, enhancing our comprehension of the mechanisms and processes governing changes in watershed sediment and water conditions is of paramount significance to safeguard and comprehensively manage watershed ecosystems.

The Yellow River Basin holds significant ecological and economic importance in China, playing a crucial role in the strategic framework of social and economic development. Specifically, the middle reaches of the Yellow River exemplify an ecologically vulnerable area with severe soil erosion due to its high sediment content and coarse sand. Since 1950, the Loess Plateau has implemented large-scale measures to conserve soil and water, including afforestation, terracing, and the construction of check dams [14,15]. Notably, the Grain for Green Program, initiated in 1999, has brought about significant changes in the underlying landscape of the basin [2,16]. Human activities have exerted a profound influence on the water cycle and erosion sediment production in the Yellow River Basin. Monitoring data reveal a substantial decline of approximately 70% in the flow and sediment load of the Yellow River over the past 60 years. For instance, the sediment load at the Tongguan hydrological station on the main course of the Yellow River has dropped from an average of 16×10^8 tons per year in the 1970s to approximately 3×10^8 tons since 2000 [17]. Several studies have investigated the characteristics and driving mechanisms of water sediment evolution in the middle reaches of the Yellow River, with a consensus emerging that both flow and sediment load have significantly decreased, with human activities playing a more prominent role than climate change as the primary driver behind these declines [18].

The Wuding River, situated in the Loess Plateau, is a significant tributary in the middle reaches of the Yellow River, playing a substantial role in the hydrological and sedimentary dynamics of the Yellow River. Therefore, understanding the patterns and factors influencing water sediment changes in the Wuding River Basin is essential for comprehending runoff and sediment fluctuations in the Yellow River. The Wuding River Basin is a critical area for implementing soil and water conservation measures in the Loess Plateau and is also highly vulnerable to climate change. In recent years, a variety of methodologies, such as ecohydrological models, the elasticity coefficient method, regression analysis, and hydrological simulation, have been employed to analyze the relationship between runoff sediment changes and the response to climate and human activities in this region [19–21]. Previous studies have provided some insights into the erosion sediment production process and mechanisms in this basin. However, uncertainties still remain regarding the understanding of water sediment changes under the influence of human activities. Currently, there are conflicting viewpoints on the magnitude of changes in runoff and sediment load in the Wuding River Basin, as well as the extent of influence from various factors. Therefore, it is crucial to further clarify and quantitatively evaluate these aspects.

The study focused on the Wuding River Basin, a representative compound erosion area of wind and water in the middle reaches of the Yellow River. Multiple analysis methods were employed, including the M-K test, Pettitt test, Morlet wavelet analysis, double cumulative curve method, and random forest regression analysis. By analyzing long-term time series data of runoff and sediment as well as human activity data in the Wuding River Basin, the cycles and mechanisms of runoff and sediment variations in this basin were clarified. Additionally, a prediction index system for runoff and sediment was established. This system quantitatively identified the main natural and anthropogenic factors influencing changes in runoff and sediment transport in the basin. Furthermore, the study explored the contribution rates of the causes and driving factors of runoff and sediment variations. The aim of this research was to provide a theoretical basis and data support for the comprehensive control of soil erosion in key ecologically vulnerable areas of the Yellow River Basin and the establishment of a scientific runoff and sediment regulation system at the basin level in China.

2. Study Area

The Wuding River is a primary tributary of the Yellow River, located in the southern edge of the Maowusu sandland and the northern part of the Loess Plateau. The main stream has a total length of 491.0 km, and the watershed covers an area of 30,261 km^2 (Figure 1). The Baijiachuan hydrological station, located in the lower reaches of the Wuding River, controls an area of 29,662 km^2, accounting for 98% of the total watershed area.

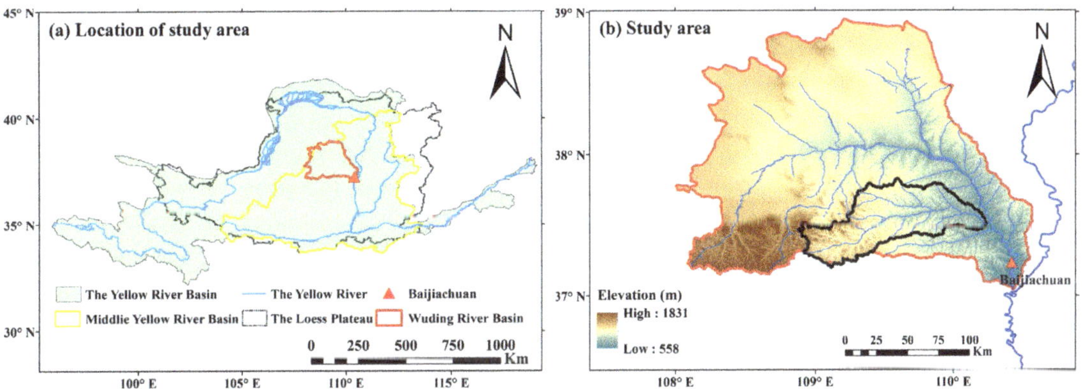

Figure 1. Location and characteristics of the studied watershed.

The Wuding River Basin belongs to the temperate continental arid and semiarid monsoon climate type, characterized by a dry climate, loose soil, and sparse vegetation. According to meteorological data from 1960 to 2020, the average annual precipitation in the Wuding River Basin is 475 mm. The distribution of precipitation within the year is extremely uneven, with over 80% of the rainfall occurring from May to September, mostly in the form of short-duration and high-intensity rainfall. This has resulted in severe soil erosion and intense sediment transport. The area of soil and water loss in the basin is 23,137 km^2, with an average erosion modulus of 6090 t/(km^2·a), making it one of the main sources of coarse sediment in the Yellow River.

The Wuding River Basin has a long history of comprehensive management of small watersheds. Since the initiation of soil erosion control work in 1950, the scope of basin management has been expanding. The 1950s to 1960s marked the initial stage, during which the management scope was limited, and a preliminary completion of 2153 km^2 was achieved. The 1970s to 1980s witnessed the stage of scaled-up management, focusing on channel improvement and the construction of numerous check dams. A total of 5929 check dams were built, accounting for 51.1% of the total number of check dams. In the 1980s, the Wuding River Basin was designated a key area for national soil and water conservation, and a series of comprehensive soil and water conservation measures were carried out with small watersheds as the management units. While intensifying management efforts, advanced practical techniques and achievements were actively promoted and applied. By 2010, the Wuding River Basin had constructed 2322.35 km^2 of farmland, 6234.68 km^2 of soil and water conservation forests, 975.18 km^2 of economic fruit trees, 1775.05 km^2 of grassland, and 360.70 km^2 of afforested areas. The forest and grass coverage rate reached 68.13%. Additionally, 7508 check dams, 19,628 reservoirs and wells, and 11,969 ponds and bunds were constructed. According to the latest statistics, by 2018, the cumulative area of soil erosion control in the Wuding River Basin reached 12,996 km^2, achieving a soil erosion control rate of 50.93%.

3. Data and Methodology

3.1. Data Description

The annual runoff and sediment data from the Baijiachuan hydrological station at the outlet control station of the Wuding River Basin during 1960–2020 were obtained from the "Yellow River Basin Hydrological Yearbook". The daily meteorological data from four meteorological stations (Yulin, Hengshan, Suide, and Jingbian) within the basin during 1960–2020 were obtained from the China Meteorological Data Service Center (http://data.cma.cn (accessed on 15 March 2023)). The digital elevation model (DEM) of the basin with a spatial resolution of 30 m × 30 m was obtained from the Loess Plateau Scientific Data Center of the National Earth System Science Data Sharing Service Platform (http://loess.geodata.cn (accessed on 1 May 2023)). The normalized difference vegetation index (NDVI) used the GIMMS global vegetation index data provided by ECOCAST (https://poles.tpdc.ac.cn/en (accessed on 9 January 2023)) with a spatial resolution of 8 km and a temporal resolution of 15 days. The land use data for 1990–2020 were obtained from the Landsat-derived annual China land cover dataset (CLCD) [22]. The Wuding River Basin covers various counties and cities in Yulin city, Shaanxi Province. The population, GDP, and other socioeconomic statistics were sourced from the "Statistical Yearbook of Shaanxi Province" and the "Statistical Yearbook of Yulin City". The water consumption data for Yulin city were obtained from the "Water Resources Bulletin of Shaanxi Province" (2000–2015). The nighttime light remote sensing data product used was the "Prolonged Artificial Nighttime-light Dataset of China (PANDA)" published by the National Tibetan Plateau Data Center (https://data.tpdc.ac.cn/home (accessed on 2 February 2023)). This dataset covers the time series from 1984 to 2020 with a spatial resolution of 1 × 1 km and a temporal resolution of one year.

3.2. Methodology

This study employed mainstream research methods, such as the Mann–Kendall test, Pettitt test, Morlet wavelet analysis, and double mass curve method, for comparison with previous relevant research results. Additionally, the random forest regression model was used to analyze the primary driving factors influencing the variability of runoff and sediment transport in the Wuding River Basin. Through a comprehensive review of the literature related to runoff and sediment changes in the Wuding River Basin, this study represents the first application of this method in the region.

3.2.1. Mann–Kendall Rank Correlation Trend Test

The Mann–Kendall rank correlation trend test method (hereinafter referred to as the M-K test) is a widely used and effective nonparametric statistical method for analyzing trends in time series data. It is commonly employed in the trend analysis of hydrological, meteorological, and other time series due to its simplicity, robustness, and high level of quantification [23–27]. This method disregards the distribution characteristics of the sample series and is unaffected by a few extreme values, making it suitable for detecting trends in time series [28].

For a time series variable $(X_1, X_2, \ldots, X_n)$, where n represents the length of the time series, the M-K test introduces a statistical parameter denoted as S.

$$S = \sum_{k=1}^{n-1} \sum_{j=k+1}^{n} \operatorname{sgn}(x_j - x_k) \tag{1}$$

where sgn() is the sign function and is shown as follows:

$$\text{sgn}(x_j - x_k) = \begin{cases} 1 & x_j - x_k > 0 \\ 0 & x_j - x_k = 0 \\ -1 & x_j - x_k < 0 \end{cases} \quad (2)$$

When S is a normal distribution, then the variance $Var(S) = n(n-1)(2n+5)/18$. For values of $n > 10$, the statistical measure of the normal distribution is:

$$Z_{MK} = \begin{cases} \frac{S-1}{Var(S)} & S > 0 \\ 0 & S = 0 \\ \frac{S+1}{Var(S)} & S < 0 \end{cases} \quad (3)$$

The trend is examined using the Z_{MK} value. A positive (negative) value of Z_{MK} indicates an upwards (downwards) trend in the series being tested. When the absolute value of the test statistic is greater than 1.28, 1.64, and 2.32, the series passes the significance test at the 90%, 95%, and 99% confidence levels, respectively. The larger the absolute value of the test statistic is, the more significant the trend of the series is.

3.2.2. Theil–Sen Median Estimator Linear Model

The Theil–Sen median estimator linear model, also referred to as Sen's slope estimator, is a robust nonparametric statistical technique utilized for calculating trends. This method is computationally efficient and insensitive to measurement errors and outliers, making it well suited for the trend analysis of lengthy time series data [29,30]. The Theil–Sen model is utilized to determine the slope of the series, denoted as β. The slope β represents the average rate of change and trend in the time series. A positive value of β indicates an upward trend, while a β value of 0 suggests an insignificant trend. Conversely, a negative β value indicates a downward trend. The calculation formula for Sen's slope of a time series $x_t = (x_1, x_2, \ldots, x_n)$ is as follows:

$$\beta = M_f\left(\frac{x_j - x_i}{j - i}\right), \forall j > i \quad (4)$$

where M_f is the median function.

3.2.3. Pettitt Change-Point Test

The Pettitt change-point test is used to determine whether there is a significant change point in a hydrometeorological time series, even when the exact timing of the change is unknown. For a hydrometeorological time series X = ($x_1, \ldots, x_n$), assuming the change point occurs at X_t, the original time series can be divided into two parts: $x_1, x_2, \ldots, x_t$ and $x_{t+1}, x_{t+2}, \ldots, x_n$. The statistic $U_{t,n}$ is defined to assess the possible occurrence of a change point at time t:

$$U_{t,n} = U_{t-1,n} + \sum_{j=1}^{n} \text{sgn}(x_t - x_j) \; t = 2, \ldots, n \quad (5)$$

$U_{i,n} = \sum_{j=1}^{n} \text{sgn}(x_i - x_j)$ and sgn($\cdot$) denote the sign function, which is calculated according to Equation (2).

To determine the probable occurrence time, t, of a mutation point, the statistical measure K_t is defined to locate the most likely mutation point.

$$K_t = \max_{1 \le t \le n} |U_{t,n}| \quad (6)$$

After identifying the mutation point using Equation (6), the significance level P_t is calculated using the following formula:

$$P_t = 2\exp\left(\frac{-6K_t^2}{n^3 + n^2}\right) \tag{7}$$

For a given confidence level α, if $P_t > \alpha$, the null hypothesis is accepted, indicating no significant mutation at time t; if $P_t < \alpha$, the null hypothesis is rejected, indicating a significant mutation at time t. In this paper, a confidence level of $\alpha = 0.5$ was chosen.

3.2.4. Double Mass Curve

The double mass curve method (DMC) is commonly used to examine the consistency and evolution trends of hydrological elements [31]. It plots the cumulative values of one variable over a specific period against the corresponding cumulative values of another variable on a Cartesian coordinate system. The cumulative relationship curve is analyzed to determine the changing trends, timing, and magnitude of the response relationship between the two variables. The double cumulative curve of runoff and sediment transport can be used to study the abrupt changes and trend intensity of sediment transport caused by human activities in rivers [32]. If the slope of the cumulative curve significantly deviates, it indicates that human activities have a significant impact on the changes in runoff and sediment transport. The year corresponding to the deviation point is identified as the year of the abrupt change, and the greater the deviation in slope is, the higher the degree of human interference is [33].

The curve can be represented as two variables, x (t) and y (t). Within a certain observation period length t, there are observed values x_i (t) and y_i (t), where $i = 1, 2, 3, \ldots, n$. The cumulative values of the time series for the variables x (t) and y (t) are calculated as x'_i (t) and $y'_i(t)$, respectively, using the following formulas:

$$x'i(t) = \sum_{i=1}^{n} xi(t), \quad i = 1,2,\ldots,n \tag{8}$$

$$y'i(t) = \sum_{i=1}^{n} yi(t), \quad i = 1,2,\ldots,n \tag{9}$$

The cumulative values of the two variables are plotted on the x-axis and y-axis of a Cartesian coordinate system, resulting in a double cumulative curve. In this study, the cumulative annual runoff is plotted on the x-axis, and the cumulative annual sediment transport is plotted on the y-axis, resulting in a graph of the cumulative curve of runoff and sediment transport.

3.2.5. Copula Function

The theoretical foundation of the copula function lies in the decomposition of an n-dimensional joint distribution function into n marginal distribution functions and a copula function that describes the dependency structure between variables. It can be expressed as follows: let $X = (x_1, x_2, \ldots, x_n)$ be an n-dimensional random variable with marginal distribution functions $F_1, F_2, \ldots, F_n$; then, the n-dimensional copula function C satisfies the following formula:

$$H(x_1, x_2, \ldots, x_n) = C[F_1(x_1), F_2(x_2), \ldots, F_2(x_n)] \tag{10}$$

Function C in the equation represents a joint distribution function that is uniformly distributed in the interval [0, 1]. In hydrological frequency analysis and calculations, commonly used copula functions are the elliptical and Archimedean types (Table 1). Elliptical copula functions have elliptical contour line distributions and are not limited by the dependence structure between variables. Archimedean copula functions, on the other hand, have a simpler form and strong symmetry. In this study, we selected three single-parameter distributions from the most commonly used Archimedean copula functions in hydrological research [34–36], namely, the Gumbel copula, Clayton copula, and Frank copula, to fit the joint distribution function. The fitting performance of these three copula functions was evaluated using the squared Euclidean distance (d_2) and the Akaike information criterion (AIC method) for model selection.

Table 1. Expression of copulas and their parameters.

Classification	Names	Distribution Functions and Parameters
Elliptical	Gaussian	$C(u,v) = \Phi_\theta(\Phi^{-1}(u), \Phi^{-1}(v))$, θ represents a parameter
	Student's t	$C(u,v) = t_{\theta,k}(t_k{}^{-1}(u), t_k{}^{-1}(v))$, θ represents a parameter, k represents the degrees of freedom
Archimedean	Gumbel	$C(u,v) = \exp\left\{-\left[(-\ln u)^\theta + (-\ln v)^\theta\right]^{-\frac{1}{\theta}}\right\}, \theta \geq 1$
	Clayton	$C(u,v) = \left(u^{-\theta} + v^{-\theta} - 1\right)^{-\frac{1}{\theta}}, \theta > 0$
	Frank	$C(u,v) = -\frac{1}{\theta}\ln\left[1 + \frac{(e^{-\theta u}-1)(e^{-\theta v}-1)}{e^{-\theta}-1}\right], \theta \neq 0$

3.2.6. Continuous Wavelet Analysis

Wavelet analysis, a powerful statistical tool, was originally utilized in the field of signal processing and analysis and is currently widely applied in various disciplines, including hydrology and ecosystems [37–39]. Wavelet transformation can be categorized into continuous wavelet transform (CWT) and discrete wavelet transform (DWT). Among these, the continuous wavelet transform is particularly suitable for feature extraction. The equation for the continuous wavelet transform is shown as follows:

The continuous wavelet transforms of a discrete time series $x_n (n = 1, \ldots, N)$ with equal time steps δ_t are defined as the convolution of the wavelet function x_n under scale and translation.

$$W_n^X(s) = \sqrt{\frac{\delta_t}{s}} \sum_{n'=0}^{N-1} x'_n \psi * \left[\frac{(n' - n)\delta_t}{s}\right] \tag{11}$$

In the equation, the symbol * represents the complex conjugate, and N represents the total number of data points in the time series. $(\delta_t/s)^{1/2}$ is a factor used to normalize the wavelet function, ensuring that the wavelet function has unit energy at each wavelet scale s. By transforming the wavelet scale s and localizing it along the time index n, a waveform can be obtained that displays the fluctuation characteristics of the time series at a certain scale and their variation over time, which is known as the wavelet power spectrum.

When conducting a wavelet transformation on a time series, the choice of the mother wavelet is highly important. Typically, when analyzing a time series, it is desirable to obtain smooth and continuous wavelet amplitudes, making nonorthogonal wavelet functions more suitable. Additionally, to obtain information regarding both the amplitude and phase aspects of the time series, complex-valued wavelets should be selected, as complex-valued wavelets have an imaginary component that can effectively express the phase [40]. The Morlet wavelet not only possesses nonorthogonality but also is an exponential complex-valued wavelet modulated by a Gaussian function, with its equation expressed as follows:

$$\psi_0(t) = \pi^{-1/4} e^{i\omega_0 t} e^{-t^2/2} \tag{12}$$

In the equation, t represents time and ω_0 is a dimensionless frequency. When $\omega_0 = 6$, the wavelet scale s is approximately equal to the Fourier period (λ, λ = 1.03 s), allowing for interchangeability between the scale and period terms. This balance in localization between time and frequency is a key feature of the Morlet wavelet. It is evident that the Morlet wavelet contains additional oscillatory information, as the wavelet power can encapsulate both positive and negative peaks within a broad peak. In this study, the Wavelet toolbox in MATLAB software (Version: R2022a, MathWorks, Portola Valley, CA, USA) was utilized to calculate the wavelet coefficients for streamflow and sediment transport. The resulting wavelet transform real part distribution and variance plots were generated using OriginPro software (Version: 2021b, OriginLab, Northampton, MA, USA).

3.2.7. Random Forest Model

The random forest model is a widely used nonparametric regression machine learning algorithm in various fields. It consists of decision trees, which contribute to its high accuracy. To address overfitting, the random forest model utilizes a random sampling method with replacement when selecting feature attributes from the dataset [41]. The determination of the number of decision trees and randomly selected features is based on the model's fitting performance during the benchmark period. Ultimately, the best model is selected based on the coefficient of determination and the Nash efficiency coefficient. The basic form of the random forest model can be expressed as follows:

$$g_i = \frac{1}{n}\sum_{i=1}^{n} h_i(m_i) \tag{13}$$

In the equation, g_i represents the predicted value, n represents the total number of samples, $h_i(m_i)$ represents the training function for each decision tree, and m_i represents the measured data.

This study selects the Bootstrap resampling method for training set sample processing. According to the rules, the sample set is divided into two parts and the decision tree model is built using the binary recursive method. Each decision tree in the model is independent and does not interfere with other trees. Finally, the well-grown decision trees are combined to obtain a classifier, namely, random forest. The classification results of the model data are determined by voting to determine the category of new samples, in order to achieve data prediction. A decision tree is a typical single classifier, which deduces the classification rules of decision trees for the classification of data from a large and disorderly set of samples by the recursive method, and then it analyzes data using these rules. This study uses the Classification and Regression Tree (CART) to implement the node splitting of the decision tree. The CART algorithm measures the data partition standard based on the *Gini* index and uses the feature value with the smallest *Gini* index as the splitting attribute of the node, which can explain the generated rules.

The calculation formula for the *Gini* coefficient is as follows:

$$Gini(T) = 1 - \sum_{i=1}^{k}[p(i|t)]^2 \tag{14}$$

In the equation, T represents the sample; k represents the number of categories in the sample; and $p(i|t)$ represents the probability of category i at node t.

Calculate the index of the partition:

$$Gini(T) = \sum_{i=1}^{m} \frac{n_i}{n} Gini(i) \tag{15}$$

In the equation, m represents the number of child nodes; n_i represents the number of samples at child node i; and n represents the number of samples at the parent node.

During the attribute splitting process, the CART algorithm calculates the parameters, namely, the *Gini* coefficient, based on Formulas (14) and (15). According to the calculation results, the priority attribute for node splitting is selected, which is the attribute with the smallest *Gini* coefficient. Through recursive iteration, the decision tree is continuously updated until a complete decision tree is generated.

Random forest regression analysis enables the ranking of the relative importance of influencing factors, thereby identifying the main controlling factors that significantly affect variations in runoff and sediment transport. This process facilitates the prediction of streamflow and sediment transport. The importance of variables is determined by the increase in the prediction error resulting from permuting the variable data. The percentage reduction in prediction error (IncMSE) serves as an indicator, measuring the extent to which the accuracy of random forest predictions diminishes when a variable's value is changed to a random number. A higher IncMSE value indicates a greater importance of the corresponding influencing factor and a higher contribution to the model [42,43].

4. Results

4.1. Runoff and Sediment Variation Characteristics

The trends of precipitation, annual runoff, and annual sediment transport at the Baijiachuan hydrological station during the period from 1960 to 2020 were examined through the application of the M-K test (Figure 2, Table 2). The outcomes of the statistical analyses revealed that there was relatively limited interannual variability in precipitation. However, both annual runoff and sediment transport exhibited substantial fluctuations on an interannual basis. Over the span of the past six decades, the average multiyear precipitation in the Wuding River Basin amounted to 475 mm, with the highest recorded value reaching 698 mm and the lowest reaching 264 mm. The annual precipitation demonstrated a slight upwards trend, although it did not pass the threshold for statistical significance. Sen's slope analysis indicated that the annual precipitation experienced a gradual increase at a rate of 1.25 mm·a^{-1}. The multiyear average runoff at the Baijiachuan hydrological station stood at 10.55 $\times$ 10^8 m^3, with the maximum value recorded at 20.15 $\times$ 10^8 m^3 and the minimum at 6.086 $\times$ 10^8 m^3. The annual runoff, however, exhibited a significant decreasing trend, with an average reduction of 0.12 $\times$ 10^8 m^3·a^{-1}. The multiyear average sediment transport at the Baijiachuan hydrological station was 0.81 $\times$ 10^8 t, with the maximum recorded value at 3.75 $\times$ 10^8 t and the minimum at 0.0267 $\times$ 10^8 t. The annual sediment transport showed an average decline of 0.02 $\times$ 10^8 t·a^{-1}, and this downwards trend was highly significant at the 0.001 level of significance. Since the 1970s, when soil and water conservation measures in the watershed began to take effect, sediment transport in the basin has gradually diminished as a result of the impact of these measures and reduced water supply.

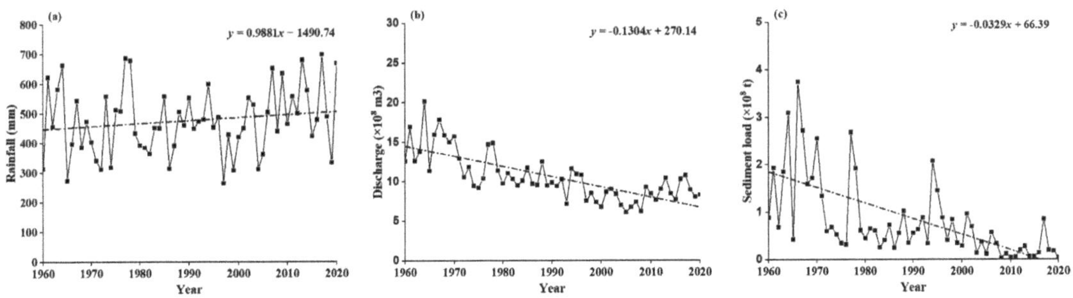

Figure 2. Temporal variations in annual precipitation (**a**), streamflow (**b**), and sediment transport (**c**).

Table 2. M-K test of annual precipitation, streamflow, and sediment transport for 1960–2010.

Characteristic Value of Runoff and Sediment	Z_{MK}	Sen's Slope
Annual rainfall	1.21	1.25
Annual discharge	−6.67 ***	−0.12
Annual sediment load	−6.14 ***	−0.02

Note: *** denotes that the value of Z is significant at the 0.001 level.

Pettitt tests revealed noteworthy change points ($p < 0.001$) for both the yearly runoff and sediment load at the Baijiachuan hydrological station (Table 3). The transition point for annual runoff transpired in 1971, while the transition point for sediment load transpired in 1979. These transition points align with the era of rigorous soil and water conservation measures implemented on the Loess Plateau, thus validating the outcomes of the transition point analysis. By contrasting the interannual and intra-annual fluctuations in runoff and sediment load before and after the transition points in the Wuding River Basin, the following observations were deduced: (I). The average runoff for 1960–1979 and 1980–2020 in the Wuding River Basin was 523.62 m^3 and 488.83 m^3, respectively, denoting a 33.9% decline in runoff from 1980 to 2020 in comparison to 1960–1979. (II). The average sediment loads for 1960–1971 and 1972–2020 in the Wuding River Basin were 1.88×10^8 t and 0.55×10^8 t, respectively, signifying a notable 70.7% decrease in sediment load from 1972 to 2020 in comparison to that of 1960–1971. The reduction in sediment load surpassed the reduction in runoff.

Table 3. Pettitt test of the change point of annual discharge and sediment load.

Statistical Parameter	Annual Discharge	Annual Sediment Load
Change-point year	1979 ***	1971 ***
Prechange year	13.66×10^8 m^3	1.88×10^8 t
Postchange year	9.03×10^8 m^3	0.55×10^8 t
Relative change	−33.9%	−70.7%

Note: *** denotes that the Pettitt test results are significant at the 0.001 level.

By utilizing Morlet continuous wavelet analysis, the temporal variation in runoff and sediment data in the basin was assessed to identify the underlying cycles of runoff and sediment changes. The wavelet coefficients and wavelet variances in annual runoff and sediment transport at the Baijiachuan hydrological station are depicted in Figure 3. The diagram shows two extensive alternations of wet and dry periods in the annual runoff and sediment transport at the Baijiachuan hydrological station. The annual runoff predominantly displays three "wet-dry" cycles: 3 years, 9 years, and 28 years. Among them, the 28-year cycle of annual runoff had relatively stable behavior throughout the entire time domain, and the peak of wavelet variance transpired at this time scale, indicating that the primary period of annual runoff variation at the Baijiachuan hydrological station was 28 years. The annual sediment transport primarily exhibits three cycles: 2 years, 13 years, and 30 years. Among them, the 30-year cycle of annual sediment transport had relatively stable behavior throughout the entire time domain, and the peak of sediment transport variance transpired at the 30-year scale, indicating that 30 years was the dominant period of annual sediment transport variation at the Baijiachuan hydrological station. The temporal characteristics of annual runoff and sediment transport sequences bore similarities; however, there were also disparities, with the sediment transport cycle slightly surpassing the runoff cycle.

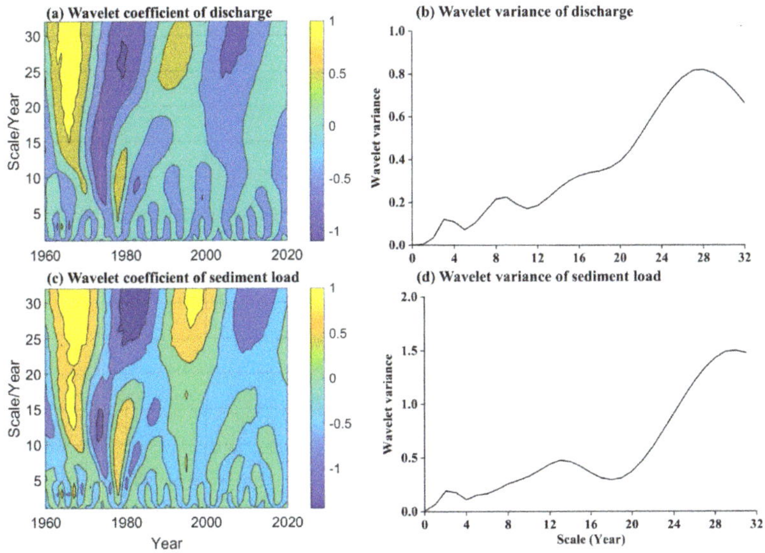

Figure 3. Coefficients and variance in annual runoff and sediment transport at the Baijiachuan Hydrological Station from 1960 to 2020.

4.2. Diagnosing the Variability in the Sediment–Discharge Relationship

Based on the aforementioned analysis findings, the reference period was established from 1960 to 1972 (Period I), followed by 1973 to 2001 (Period II), and culminating with 2002 to 2020 (Period III). The cumulative sediment–discharge curve at the Baijiachuan hydrological station exhibits a distinct convex form (Figure 4), signifying that the fluctuations in runoff and sediment within this basin occurred asynchronously. The extent of sediment attenuation surpassed that of runoff, with a discernible decline in sediment concentration. Furthermore, the cumulative curve manifests inflection points for approximately 1972 and 2000. The findings of relevant studies [20] also indicate a significant decrease in annual runoff in the Wuding River Basin from 1996 to 2007, with a sudden change occurring in 1971. Prior to the abrupt change, the trend in runoff was not significant, but after the change, a significant decrease in runoff was observed. These results are generally consistent with the findings of this study. In the 1970s, large-scale sediment control interventions, including silt dams and terraces, were implemented in the Wuding River basin, resulting in notable reductions in sediment. The cumulative sediment–discharge curve during Period II deviates significantly from Period I, showcasing a more substantial decrease in sediment load in comparison to runoff. After 2000, despite an augmentation in precipitation and subsequent runoff, the implementation of land conversion initiatives to foster forests and grasslands further curtailed sediment yield, leading to a subsequent downwards shift in the cumulative curve.

The period spanning from 1960 to 1972 was utilized as the reference point, and a comparative analysis was conducted on the variations in monthly average runoff and sediment discharge between the intervals of 1973 to 2001 and 2002 to 2020 (Figure 5). In comparison to Period I, both Period II and Period III displayed varying degrees of reduction in the multiyear average monthly runoff, particularly in March and July to September. Furthermore, a discernible trend towards a more equitable distribution of monthly runoff throughout the year was observed, with the proportion of flood season runoff declining from 41.5% in the baseline period to 37.9% in Period II and further declining to 35.9% in Period III. In stark contrast, when the mean monthly sediment discharge between the periods of 1960–1971 and 1972–2020 were contrasted, a substantial decline in sediment discharge was evident for each month; however, the pattern diverged from that of runoff.

Sediment discharge exhibited a greater concentration during the flood season, with the proportion of flood season sediment discharge escalating from 90.4% in Period I to 97.6% in Period II.

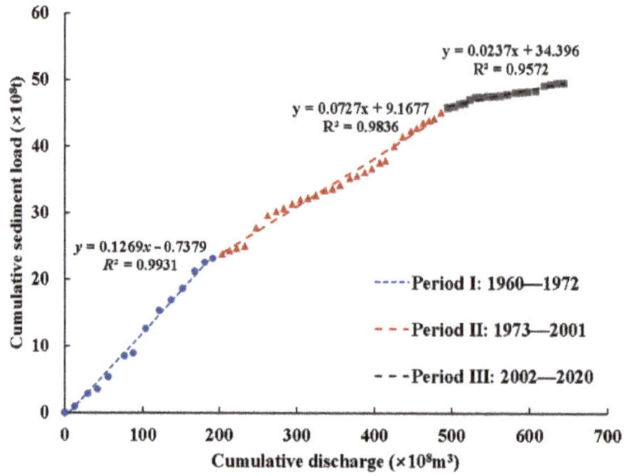

Figure 4. Cumulative curves of annual runoff and sediment discharge at Baijiachuan hydrological station.

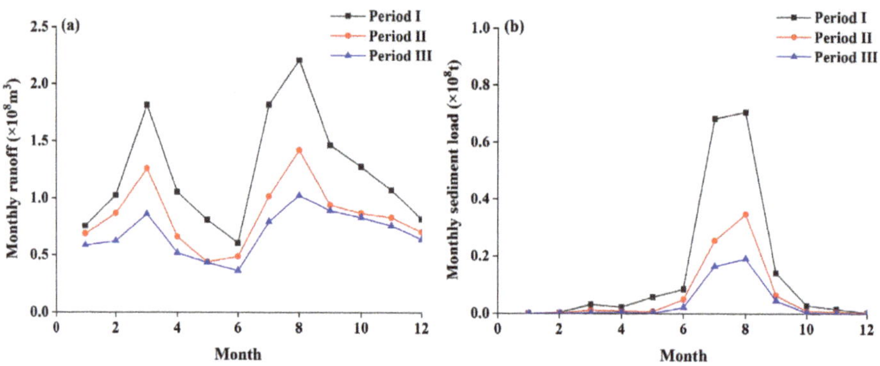

Figure 5. Analysis of the annual distribution characteristics of runoff (**a**) and sediment discharge (**b**) at Baijiachuan hydrological station.

The co-occurrence of runoff and sediment discharge was assessed through the utilization of the widely employed Kolmogorov–Smirnov (K-S) test and Akaike information criterion (AIC) (Table 4). The K-S test yielded p values exceeding 0.05, affirming that the fitting outcomes successfully met the test criteria. Consequently, only the AIC test outcomes are presented in the table to further determine the optimal joint distribution. Table 4 shows that the joint distribution of runoff–sediment discharge at the Baijiachuan hydrological station during Periods I, II, and III adheres to Gaussian, Gumbel, and Clayton distributions, respectively. This finding underscores the dynamic nature of the runoff–sediment relationship, as the joint distribution undergoes alterations. The correlation between runoff and sediment discharge during Period I exhibited a significantly greater magnitude than that of Period II and Period III. Transitioning from Period I to Period III, the correlation coefficient between annual runoff and sediment discharge demonstrated a continuous decrease, which is particularly pronounced in Period III. This observation suggests that the hydrosediment relationship within the watershed becomes increasingly intricate amidst intensified human activities.

Table 4. Goodness-of-fit test of five candidate copulas according to AIC.

Stage	Spearman Coefficient	Gaussian	Student-T	Gumbel	Clayton	Frank
Period I	0.90	−17.53	−15.36	−13.94	−16.21	−17.43
Period II	0.72	−12.91	−11.01	−16.94	−4.91	−13.12
Period III	0.41	−5.81	−3.66	−2.82	−1.23	−8.37

4.3. The Response of Runoff and Sediment to Climate and Human Activities

Table 5 demonstrates a gradual intensification of the impact of human activities on runoff and sediment discharge in the Wuding River Basin from Period I to Period III. This trend coincides with the implementation of extensive soil and water conservation measures in the basin, including the construction of silt dams during the 1970s and the initiation of the Grain for Green program in 1999. These measures have exerted a profound impact on the hydrological and sedimentary dynamics within the basin. As these management practices expanded in scale, they brought about changes in surface morphology, consequently exerting a significant influence on the processes of runoff generation and sediment production. In comparison to Period I, the cumulative reductions in runoff and sediment discharge during Period II in the Wuding River Basin amounted to 4.6×10^8 m^3 and 1.02×10^8 t, respectively. Climate factors accounted for 11.94% and −14.5% of the variations observed in runoff and sediment discharge, respectively, while human activities contributed 88.06% and 114.5%, respectively. Remarkably, during this period, there was a substantial decrease in runoff and sediment discharge despite there being no significant decline in precipitation. Moving into Period III, the total declines in runoff and sediment discharge in the Wuding River Basin reached 6.4×10^8 m^3 and 1.55×10^8 t, respectively. Climate factors contributed −11.9% and −17.7%, while human activities accounted for 111.9% and 117.7% of the variations. Human activities emerged as the primary driver behind the reduction in sediment and water, with a more pronounced impact on sediment discharge, whereas the influence of climate change on sediment discharge within the basin remained relatively weak. The results of relevant studies [21,33] indicate that after 2001, due to the implementation of policies such as land reclamation and afforestation, human activities have intensified. Although there has been an increase in rainfall in the Wuding River Basin since 2007, the runoff and sediment transport continue to decrease. From 2001 to 2020, human activities contributed to over 110% of the reduction in runoff and sediment. Measures like check dams, land reclamation and afforestation, and reservoir construction and irrigation are among the factors causing the decrease in runoff and sediment in the Wuding River Basin, with check dams and reservoirs playing a major role in reducing sediment transport. These findings are consistent with the results of this study.

Table 5. The contributions of climate and human activities to changes in runoff and sediment discharge.

Type	Time	Observation	Decrement	Contribution Rate from Climate Change/%	Contribution Rate from Human Activity/%
Runoff	Period I	14.7×10^8 m^3	-	-	-
	Period II	10.1×10^8 m^3	4.6×10^8 m^3	11.94	88.06
	Period III	8.3×10^8 m^3	6.4×10^8 m^3	−11.9	111.9
Sediment Load	Period I	1.78×10^8 t	-	-	-
	Period II	0.76×10^8 t	1.02×10^8 t	−14.5	114.5
	Period III	0.23×10^8 t	1.55×10^8 t	−17.7	117.7

The impact of human activities on the alterations in runoff and sediment within river basins is intricate and diverse. The influence of human activities on runoff primarily stems from the modification of land use, which in turn affects climate and underlying surface conditions. Conversely, the effect of human activities on sediment transport mainly arises from measures such as the construction of check dams and the restoration of vegetation. Therefore, it is imperative to deeply study the human activities that instigate changes in runoff and sediment to effectively manage soil erosion and achieve the objectives of water conservation, soil preservation, and the mitigation of runoff and sediment transport.

In this study, we meticulously selected 10 indicators that holistically depict the intensity of human activities, encompassing the arable land area, GDP, population, NDVI, forest area, grassland area, water area, nighttime light remote sensing, watershed soil erosion control area, and irrigation water consumption, through an extensive literature review and yearbook consultations. We subsequently investigated the driving impact of human activities on changes in runoff and sediment within the Wuding River Basin. Due to the insufficiency of pre-1990 data pertaining to human activity factors, we constructed a random forest model utilizing runoff and sediment transport as dependent variables for the period of 1990–2015, with human activity indicators serving as independent variables to ensure temporal consistency.

To evaluate the accuracy of the fitted values obtained from the random forest regression model, we computed the R^2 value and mean absolute error (MAE). The analysis and validation of the predicted values derived from the random forest regression model, in comparison with the measured values, are presented in Figure 6. Notably, both the R^2 values of the fitted curves for annual runoff and sediment transport from the random forest regression model and the measured values reached 0.8, indicating a close approximation between the fitting results and the actual values, thus asserting their reliability.

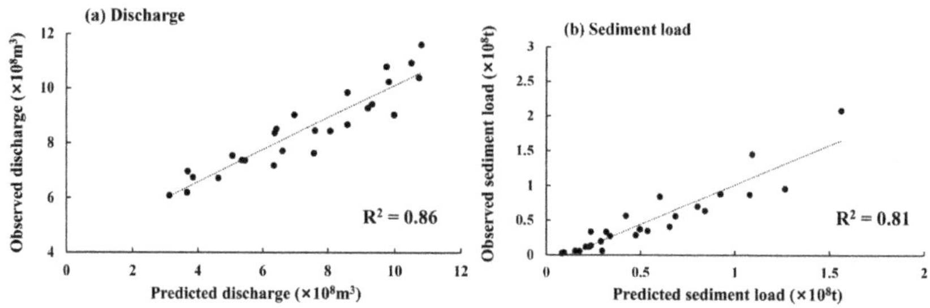

Figure 6. The fitting and validation of the random forest model.

The runoff and sediment transport data were used to fit a random forest regression model. To discern the relative importance of each human activity factor in influencing runoff and sediment transport, the percentage reduction in prediction error (IncMSE) was plotted for the 10 factors, as illustrated in Figure 7. The findings revealed that the NDVI and grassland area had the most profound impact on runoff, with IncMSE values of 14.63% and 14.26%, respectively. Following closely were the forest area (6.90%) and arable land area (5.93%). In terms of sediment transport, the NDVI and forest area emerged as the most significant human activity factors, with IncMSE values of 15.93% and 14.97%, respectively. These were trailed by the soil erosion control area (6.48%) and grassland area (6.07%).

Notably, the NDVI exhibited a significant negative correlation with both runoff and sediment transport, while the arable land area displayed a positive correlation with runoff. The continuous proliferation of forest and grass vegetation intercepted more precipitation, subsequently impeding the flow rate through the interception of dense litter. Moreover, a small portion of the precipitation replenished the groundwater, leading to a consistent decrease in runoff. This perpetual increase in water storage and soil conservation effects attributed to vegetation culminated in a more pronounced reduction in sediment transport.

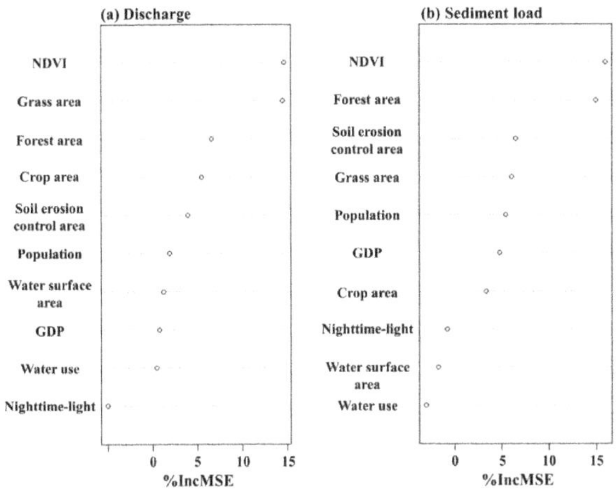

Figure 7. Ranking the importance of factors influencing annual runoff and sediment transport.

5. Discussion

5.1. Variation in Runoff and Sediment in the Wuding River Basin and a Comparative Analysis of Their Influencing Factors

This study unveils noteworthy inflection points in the annual runoff and sediment discharge in the Wuding River Basin over the past six decades, with 1971 marking the change in runoff and 1979 marking the change in sediment discharge. Human activities play a dominant role in diminishing both runoff and sediment, with a greater impact observed on sediment discharge reduction. These findings are consistent with the majority of previous studies [19,21,33]. Previous research has commonly employed attribution analysis and other methodologies to examine natural factors, including climate and precipitation, as well as human factors, such as afforestation and check dam construction, which influence soil erosion. The results demonstrate that human factors play a leading role in the variations in runoff and sediment in the Wuding River Basin. Among these factors, irrigation and afforestation primarily contribute to runoff reduction, while afforestation, grass planting, and check dam construction aid in sediment reduction. The NDVI and cumulative annual check dam control area are significantly negatively correlated with sediment transport. Moreover, the cumulative annual check dam control area has a greater direct impact on sediment transport than the NDVI. It is identified as the primary factor influencing sediment transport variability. Further research indicates that the annual variability of sediment transport in the Wuding River Basin is jointly driven by check dam construction and NDVI changes. The mean jump variability is mainly influenced by precipitation and check dam construction changes. Variance variability is primarily caused by changes in precipitation and NDVI. For annual runoff, trend variability is driven by changes in evaporation, check dam construction, reservoir construction, and agricultural irrigation water consumption. Mean jump variability is mainly influenced by changes in precipitation, reservoir construction, and agricultural irrigation water consumption. Variance variability is primarily caused by changes in precipitation.

This study also identifies the factors with the most significant impact on runoff, such as the NDVI and grassland area, and the most substantial human activity factors that affect sediment discharge, such as the NDVI and forest area. Related studies suggest [44,45] a close negative correlation between annual runoff in the Wuding River Basin and the extent of soil and water conservation measures. The contributions of maximum monthly cumulative precipitation, maximum daily precipitation, annual precipitation, and weighted average of the area of soil and water conservation measures to the variation in annual runoff in the Wuding River Basin are 37.5%, 26.9%, 9.4%, 14.5%, and 11.8%, respectively.

Based on relevant reports on sediment accumulation in the Wuding River Basin, it has been found that there is localized sediment accumulation in some sections due to floods, but there is no large-scale, long-lasting sediment accumulation. This finding suggests a direct relationship between the decrease in sediment production and the reduction in sediment discharge in the Wuding River Basin [46]. In comparison with other research findings, this study reveals that the erosion area and high-intensity erosion area in the Wuding River Basin are still larger than those in the Jing River, Beiluo River, Wei River, and Fen River in the middle reaches of the Yellow River. However, the rate of reduction in the soil erosion area is significantly higher than that in the middle reaches of the Yellow River, especially in the Huangfuchuan River, Jialu River, Jiaohe River, Qiushui River, and Beiluo River, which are key tributaries in the area with high sand and coarse sediment contents. Therefore, the current state of soil erosion and the effectiveness of soil and water conservation in the Wuding River Basin have significant implications for conducting soil and water conservation work in the middle reaches of the Yellow River.

5.2. Limitations of This Study

Due to certain constraints, specifically the limited duration of monitoring basic data for soil and water conservation engineering measures, this study was unable to conduct a thorough investigation into the contribution rates of different types of water conservation measures that promote runoff and sediment reduction. Additionally, it did not quantitatively analyze the impact of extreme rainfall events on changes in runoff and sediment.

The Wuding River Basin lies in a transitional zone, transitioning from arid to semiarid to semihumid regions, which renders the ecological environment extremely delicate. As a result, the pressure exerted by human activities on the ecological environment is escalating. With the ongoing development and progress of the social economy within the basin, the need to strike a balance between the protection of the ecological environment and high-quality economic development has become increasingly prominent. At present, human activities that significantly influence runoff and sediment production in the basin include the construction of check dams, reservoirs, terraced fields, and the conversion of farmland to forests and grasslands. Consequently, future research should focus on investigating the effects and evolutionary trends of different types of soil and water conservation measures in the basin, unveiling profound underlying issues, and offering decision-making support to enhance the effectiveness and sustainability of various soil and water conservation measures in the Wuding River Basin.

The frequent occurrence of extreme rainfall in the Loess Plateau raises important questions that must be addressed in research on the ecological management of soil and water conservation in the Yellow River Basin. It is crucial to determine the resilience of soil and water conservation engineering in the face of extreme rainfall and the potential benefits it can provide during such events. In the middle reaches of the Yellow River, the effectiveness of soil and water conservation measures under extreme rainfall has been a topic of debate. According to some researchers [47], the sediment reduction benefits of soil and water conservation measures in most basins of the Loess Plateau exceed 80%, reaching as high as 96% in some cases, even in the presence of severe rainfall events. However, numerous other researchers argue that the benefits of watershed soil and water conservation under extreme rainfall conditions are limited and may even be negative [48]. Therefore, future studies are urgently needed to address this issue and conduct comprehensive assessments of the sediment reduction benefits of soil and water conservation measures in the Wuding River Basin during extreme rainfall conditions.

6. Conclusions

The Wuding River Basin, located in the middle reaches of the Yellow River, serves as the subject of investigation in this case study. The study utilizes measured data on runoff and sediment from the period spanning 1960 to 2020, aiming to analyze the characteristics of variability and investigate the underlying causes of changes in runoff and sediment. In addition, the study employs quantitative methods to assess the contribution rates of climate change and human activities to these changes and to explore the impact of human activity factors on runoff and sediment. The main findings of this study reveal differences in the periodic characteristics of annual runoff and annual sediment transport in the Wuding River Basin. The variation period of annual sediment transport (30 years) is slightly longer than that of annual runoff (28 years). Human activities play a dominant role in reducing sediment and water in the Wuding River Basin. Grass planting and afforestation contribute significantly to the decrease in runoff, while large-scale siltation dam construction plays a major role in reducing sediment transport. The impact of climate change on sediment transport in the basin is relatively weak. NDVI is significantly negatively correlated with both runoff and sediment transport, while cultivated land area shows a positive correlation with runoff. The continuous increase in forest and grass vegetation results in more precipitation being intercepted by the vegetation canopy and the dense litter intercepting and slowing down the flow rate, with a small amount of precipitation replenishing groundwater, leading to a continuous decrease in runoff. The water storage and soil conservation effect of vegetation continues to increase, resulting in a more significant reduction in sediment. The specific conclusions are as follows:

(1) The annual precipitation in the Wuding River Basin from 1960 to 2020 exhibits a slight upward trend. Due to the implementation of large-scale soil and water conservation measures, both annual runoff and sediment transport show a significant downward trend. Notably, there are distinct change points in both annual runoff and sediment transport, occurring in 1971 and 1979, respectively.

(2) The double cumulative curves reveal a turning point in approximately 1972 and 2000. The decline in sediment transport is much more pronounced than that of runoff, and there is a noticeable decrease in sediment concentration. Comparing the reference period of 1960 to 1972 (Period I), the distribution of monthly runoff tends to be more uniform during the years 1973 to 2001 (Period II) and 2002 to 2020 (Period III), while sediment transport becomes more concentrated during the flood season. Moreover, the joint distribution of the runoff–sediment relationship also undergoes changes before and after these periods.

(3) In comparison to the reference period, climate change contributed 11.94% and −14.5% to runoff and sediment transport during Period II, respectively. Meanwhile, human activities contributed 88.06% and 114.5% to the same parameters. For Period III, climate change's contributions to runoff and sediment transport were −11.9% and −17.7% respectively, while human activities contributed 111.9% and 117.7% to the same parameters. Human activities are predominantly responsible for reducing sediment and water, with a particularly significant impact on sediment transport reduction. In contrast, the impact of climate change on sediment transport in the basin is relatively weak.

(4) The factors that exert the greatest influence on runoff are the NDVI and the area of grassland, with IncMSE values of 14.63% and 14.26%, respectively. Additionally, the area of forest contributes 6.90% and the area of cultivated land contributes 5.93%. In terms of sediment transport, the most significant human activity factors are the NDVI and forest area, with IncMSE values of 15.93% and 14.97%, respectively. Furthermore, the area of soil and water erosion control contributes 6.48% and the grassland area contributes 6.07%.

In future research, there is a need to further collect and filter natural and socio-economic indicators. Additionally, it is necessary to innovate and optimize research methods in order to explore and establish the coupling relationship between "hydro-sediment-ecological-economic". This will provide a better theoretical basis and methodological support for ecological conservation and sustainable social development in the Wuding River Basin.

Supplementary Materials: The following supporting information can be downloaded at: https://www.mdpi.com/article/10.3390/w16010026/s1, Figure S1: A prolonged artificial nighttime-light map of Wuding River Basin from 1995–2015; Table S1: Annual rainfall, discharge and sediment load from 1960–2020; Table S2: Water use, population and GDP of Yulin city from 1991–2015.

Author Contributions: Conceptualization, data curation, investigation, methodology, formal analysis, writing—original draft, writing—review and editing: J.Y.; conceptualization, methodology, supervision: Z.L.; formal analysis, validation, writing—review and editing: B.Z.; validation, project administration, funding acquisition: P.X.; validation, supervision: P.Z.; investigation, validation, writing—review and editing: J.W., W.S., S.M. and Y.Z. All authors have read and agreed to the published version of the manuscript.

Funding: This research was supported by the National Key Research and Development Program of China (Grant No.: 2022YFF1300805), the National Natural Science Foundation of China Major Projects (Grant No.: 42041006), the National Natural Science Foundation of China (Grant No.: U2243210), and the Science and Technology Development Foundation of Yellow River Institute of Hydraulic Research (Grant No.: HKF202312).

Data Availability Statement: The datasets generated during the current study are available from the corresponding author and Supplementary Materials upon reasonable request.

Conflicts of Interest: The authors declare no conflict of interest.

References

1. Sun, P.; Wu, Y.; Yang, Z.; Sivakumar, B.; Qiu, L.; Liu, S.; Cai, Y. Can the Grain-for-Green Program Really Ensure a Low Sediment Load on the Chinese Loess Plateau? *Engineering* **2019**, *5*, 855–864. [CrossRef]
2. Li, J.; Peng, S.; Li, Z. Detecting and Attributing Vegetation Changes on China's Loess Plateau. *Agric. For. Meteorol.* **2017**, *247*, 260–270. [CrossRef]
3. Ma, Y.; Sun, D.; Niu, Z.; Wang, X. Contribution of Climate Change and Human Activities to Runoff and Sediment Discharge Changes Based on Budyko Theory and Water–Sediment Relationships during 1960–2019 in the Taohe River Basin, China. *Atmosphere* **2023**, *14*, 1144. [CrossRef]
4. López-Tarazón, J.A.; Batalla, R.J.; Vericat, D.; Balasch, J.C. Rainfall, Runoff and Sediment Transport Relations in a Mesoscale Mountainous Catchment: The River Isábena (Ebro Basin). *Catena* **2010**, *82*, 23–34. [CrossRef]
5. Aronson, J.; Goodwin, N.; Orlando, L.; Eisenberg, C.; Cross, A.T. A World of Possibilities: Six Restoration Strategies to Support the United Nation's Decade on Ecosystem Restoration. *Restor. Ecol.* **2020**, *28*, 730–736. [CrossRef]
6. Saunders, M.I.; Doropoulos, C.; Bayraktarov, E.; Babcock, R.C.; Gorman, D.; Eger, A.M.; Vozzo, M.L.; Gillies, C.L.; Vanderklift, M.A.; Steven, A.D.L.; et al. Bright Spots in Coastal Marine Ecosystem Restoration. *Curr. Biol.* **2020**, *30*, R1500–R1510. [CrossRef] [PubMed]
7. Bustamante, M.M.C.; Silva, J.S.; Scariot, A.; Sampaio, A.B.; Mascia, D.L.; Garcia, E.; Sano, E.; Fernandes, G.W.; Durigan, G.; Roitman, I.; et al. Ecological Restoration as a Strategy for Mitigating and Adapting to Climate Change: Lessons and Challenges from Brazil. *Mitig. Adapt. Strateg. Glob. Chang.* **2019**, *24*, 1249–1270. [CrossRef]
8. Yin, S.; Gao, G.; Li, Y.; Xu, Y.J.; Turner, R.E.; Ran, L.; Wang, X.; Fu, B. Long-Term Trends of Streamflow, Sediment Load and Nutrient Fluxes from the Mississippi River Basin: Impacts of Climate Change and Human Activities. *J. Hydrol.* **2023**, *616*, 128822. [CrossRef]
9. Zhang, Q.; Xu, C.; Becker, S.; Jiang, T. Sediment and Runoff Changes in the Yangtze River Basin during Past 50 Years. *J. Hydrol.* **2006**, *331*, 511–523. [CrossRef]
10. Dethier, E.N.; Renshaw, C.E.; Magilligan, F.J. Rapid Changes to Global River Suspended Sediment Flux by Humans. *Science* **2022**, *376*, 1447–1452. [CrossRef]
11. Huang, S.; Zheng, X.; Ma, L.; Wang, H.; Huang, Q.; Leng, G.; Meng, E.; Guo, Y. Quantitative Contribution of Climate Change and Human Activities to Vegetation Cover Variations Based on GA-SVM Model. *J. Hydrol.* **2020**, *584*, 124687. [CrossRef]
12. Wang, J.; Wu, T.; Li, Q.; Wang, S. Quantifying the Effect of Environmental Drivers on Water Conservation Variation in the Eastern Loess Plateau, China. *Ecol. Indic.* **2021**, *125*, 107493. [CrossRef]
13. Chen, J.; Yan, F.; Lu, Q. Spatiotemporal Variation of Vegetation on the Qinghai–Tibet Plateau and the Influence of Climatic Factors and Human Activities on Vegetation Trend (2000–2019). *Remote Sens.* **2020**, *12*, 3150. [CrossRef]
14. Liang, W.; Bai, D.; Wang, F.; Fu, B.; Yan, J.; Wang, S.; Yang, Y.; Long, D.; Feng, M. Quantifying the Impacts of Climate Change and Ecological Restoration on Streamflow Changes Based on a Budyko Hydrological Model in China's Loess Plateau: Streamflow Response to Climate Change/Ecological Restoration. *Water Resour. Res.* **2015**, *51*, 6500–6519. [CrossRef]
15. Gu, D.; Guo, J.; Fan, Y.; Zuo, Q.; Yu, L. Simulating and Assessing the Effects of Seasonal Fallow Schemes on the Water-Food-Energy Nexus in a Shallow Groundwater-Fed Plain of the Haihe River Basin of China. *Sci. Total Environ.* **2021**, *267*, 3907–3925. [CrossRef]
16. Chen, Y.; Wang, K.; Lin, Y.; Shi, W.; Song, Y.; He, X. Balancing Green and Grain Trade. *Nat. Geosci.* **2015**, *8*, 739–741. [CrossRef]
17. Wang, S.; Fu, B.; Piao, S.; Lü, Y.; Ciais, P.; Feng, X.; Wang, Y. Reduced Sediment Transport in the Yellow River Due to Anthropogenic Changes. *Nat. Geosci.* **2016**, *9*, 38–41. [CrossRef]

18. Gao, G.; Zhang, J.; Liu, Y.; Ning, Z.; Fu, B.; Sivapalan, M. Spatio-Temporal Patterns of the Effects of Precipitation Variability and Land Use/Cover Changes on Long-Term Changes in Sediment Yield in the Loess Plateau, China. *Hydrol. Earth Syst. Sci.* **2017**, *21*, 4363–4378. [CrossRef]
19. Xu, J.; Wang, H. Influence of Soil and Water Conservation Measures on Reducing in Frequency of Hyperconcentrated Flows in the Wudinghe River Basin. *Environ. Earth Sci.* **2011**, *62*, 1513–1520. [CrossRef]
20. Jiao, Y.; Lei, H.; Yang, D.; Yang, H. Attribution of discharge changes over Wuding River watershed using a distributed eco-hydrological model. *J. Hydroelectr. Eng.* **2017**, *36*, 34–44.
21. Mou, X.; Liu, Q.; Ping, F.; Liang, J. Analysis of changes in annual streamflow and its influence factors in Wuding River basin in the last 50 years (1960–2007). *J. Beijing Norm. Univ. (Nat. Sci.)* **2017**, *53*, 465–471. [CrossRef]
22. Yang, J.; Huang, X. 30 m Annual Land Cover and Its Dynamics in China from 1990 to 2019. Data, Algorithms, and Models. *Earth Syst. Sci. Data* **2021**, *13*, 3907–3925. [CrossRef]
23. Kisi, O.; Ay, M. Comparison of Mann–Kendall and Innovative Trend Method for Water Quality Parameters of the Kizilirmak River, Turkey. *J. Hydrol.* **2014**, *513*, 362–375. [CrossRef]
24. Yücel, A.; Markovic, M.; Atilgan, A.; Rolbiecki, R.; Ertop, H.; Jagosz, B.; Ptach, W.; Łangowski, A.; Jakubowski, T. Investigation of Annual Lake Water Levels and Water Volumes with Şen Innovation and Mann-Kendall Rank Correlation Trend Tests: Example of Lake Eğirdir, Turkey. *Water* **2022**, *14*, 2374. [CrossRef]
25. Zhang, Y.; Cabilio, P.; Nadeem, K. Improved Seasonal Mann–Kendall Tests for Trend Analysis in Water Resources Time Series. In *Advances in Time Series Methods and Applications: The A. Ian McLeod Festschrift*; Springer: Berlin/Heidelberg, Germany, 2016; pp. 215–229.
26. Mustapha, A. Detecting Surface Water Quality Trends Using Mann–Kendall Tests and Sen'S Slope Estimates. *Int. J. Agric. Innov. Res.* **2013**, *1*, 108–114.
27. Agbo, E.P.; Nkajoe, U.; Edet, C.O. Comparison of Mann–Kendall and Şen's Innovative Trend Method for Climatic Parameters over Nigeria's Climatic Zones. *Clim. Dyn.* **2022**, *60*, 3385–3401. [CrossRef]
28. Yao, J.; Li, Z.; Yao, W.; Xiao, P.; Zhang, P.; Xie, M.; Wang, J.; Mei, S. The Compound Response Relationship between Hydro-Sedimentary Variations and Dominant Driving Factors: A Case Study of the Huangfuchuan basin. *Sustainability* **2023**, *15*, 13632. [CrossRef]
29. Theil, H. A Rank-Invariant Method of Linear and Polynomial Regression Analysis. *Indag. Math.* **1950**, *12*, 173.
30. Sen, P.K. Estimates of the Regression Coefficient Based on Kendall's Tau. *J. Am. Stat. Assoc.* **1968**, *63*, 1379–1389. [CrossRef]
31. Mu, X.; Zhang, X.; Gao, P.; Wang, F. Theory of Double Mass Curves and Its Applications in Hydrology and Meteorology. *J. China Hydrol.* **2010**, *30*, 47–51.
32. Zhang, P.; Tian, P.; Mu, X.; Zhao, G.; Zhang, Y. Sediment Load Variations and The Driving Forces In The Typical Drainage Basins of The North-South Transitional Zone of China. *Mt. Res.* **2023**, *41*, 169–179. [CrossRef]
33. Yang, Q.; Gao, H.; Han, Y.; Li, Z.; Lu, K. Variation Characteristics of Baseflow and Its Responses to Precipitation in Wuding River Basin. *J. Soil Water Conserv.* **2023**, *37*, 235–242. [CrossRef]
34. Zhou, Y.; Zhang, D.; Zhou, P.; Wang, Z.; Yang, P.; Jin, J.; Cui, Y.; Ning, S. Copula-Based Bivariate Return Period Analysis and Its Implication to Hydrological Design Event. *J. Am. Water Resour. Assoc.* **2023**, *59*, 571–583. [CrossRef]
35. Mitková, V.B.; Halmová, D.; Pekárová, P.; Miklánek, P. The Copula Application for Analysis of the Flood Threat at the River Confluences in the Danube River Basin in Slovakia. *Water* **2023**, *15*, 984. [CrossRef]
36. Huang, T.; Wu, Z.; Xiao, P.; Sun, Z.; Liu, Y.; Wang, J.; Wang, Z. Possible Future Climate Change Impacts on the Meteorological and Hydrological Drought Characteristics in the Jinghe River Basin, China. *Remote Sens.* **2023**, *15*, 1297. [CrossRef]
37. Sang, Y.-F. A Review on the Applications of Wavelet Transform in Hydrology Time Series Analysis. *Atmos. Res.* **2013**, *122*, 8–15. [CrossRef]
38. Yang, R.; Xing, B. Possible Linkages of Hydrological Variables to Ocean–Atmosphere Signals and Sunspot Activity in the Upstream Yangtze River Basin. *Atmosphere* **2021**, *12*, 1361. [CrossRef]
39. Campozano, L.; Mendoza, D.; Mosquera, G.; Palacio-Baus, K.; Célleri, R.; Crespo, P. Wavelet Analyses of Neural Networks Based River Discharge Decomposition. *Hydrol. Process.* **2020**, *34*, 2302–2312. [CrossRef]
40. Torrence, C.; Compo, G.P. A Practical Guide to Wavelet Analysis. *Bull. Am. Meteorol. Soc.* **1998**, *79*, 61–78. [CrossRef]
41. Gislason, P.O.; Benediktsson, J.A.; Sveinsson, J.R. Random Forests for Land Cover Classification. *Pattern Recognit. Lett.* **2006**, *27*, 294–300. [CrossRef]
42. Liu, X.; Zhang, Y.; Zhang, L.; Fang, X.; Deng, W.; Liu, Y. Aggregate-associated Soil Organic Carbon Fractions in Subtropical Soil Undergoing Vegetative Restoration. *Land Degrad. Dev.* **2023**, *34*, 4296–4306. [CrossRef]
43. Liu, X.; Zhang, Y.; Xiao, T.; Li, P.; Zhang, L.; Liu, Y.; Deng, W. Runoff Velocity Controls Soil Nitrogen Leaching in Subtropical Restored Forest in Southern China. *For. Ecol. Manag.* **2023**, *548*, 121412. [CrossRef]
44. Xu, J. Variation in Annual Runoff of the Wudinghe River as Influenced by Climate Change and Human Activity. *Quatern. Int.* **2011**, *244*, 230–237. [CrossRef]
45. Xu, J. The Wind-Water Two-Phase Erosion and Sediment-Producing Processes in the Middle Yellow River Basin, China. *Sci. China Ser. D-Earth Sci.* **2000**, *43*, 176–186. [CrossRef]
46. Xu, J. Implication of Relationships among Suspended Sediment Size, Water Discharge and Suspended Sediment Concentration: The Yellow River Basin, China. *Catena* **2002**, *49*, 289–307. [CrossRef]

47. Liu, X.; Li, H.; Li, X.; Gao, Y. Analysis on the cause of sharp decrease of runoff and sediment from Kuye River in the Loess Plateau. *J. Hydraul. Eng.* **2022**, *53*, 296–305. [CrossRef]
48. Wang, Y.; Liu, W.; Li, H.; Zhang, X. The Flow-sediment Relationship and Its Response to Watershed Management in Yanwachuan Watershed, Loess Plateau Gully Region, China. *J. Nat. Resour.* **2015**, *30*, 1403–1413.

water

MDPI

Article

Response of Sap Flow Trends of Conifer and Broad-Leaved Trees to Rainfall Types in Sub-Humid Climate Region of China

Yongxiang Cao [1,2], Yushi Wang [3], Naichang Zhang [1], Chendong Ning [1], Yu Bai [1] and Jianbo Jia [4,*]

1 Power China Northwest Engineering Corporation Limited, Xi'an 710065, China; caoyongx@nwh.cn (Y.C.); zhangnc111123@163.com (N.Z.); ningcd123124@163.com (C.N.); jotham880303@126.com (Y.B.)

2 Shaanxi Union Research Center of University and Enterprise for River and Lake Ecosystems Protection and Restoration, Xi'an 710065, China

3 Beijing Forestry University, Beijing 100083, China; wangys123124@163.com

4 Central South University of Forestry & Technology, Changsha 410004, China

* Correspondence: jiajianbo@csuft.edu.cn

Abstract: Sap flow is one of the most important physiological water transport processes of trees, and the characteristics of sap flow are greatly affected by the spatial and temporal distribution of water in the SPAC (soil–plant–atmosphere continuum). However, different precipitation characteristics have great influence on the water environment of forest trees, which causes considerable differences in sap flow. Therefore, researching the response of sap flow to precipitation type is the key to accurately determining plant transpiration in semi-arid areas. We used K-means clustering analysis to divide the rainfall during the study period into three rainfall types (the highest rainfall amount and intensity (types I), medium rainfall amount and intensity, with a long duration (types II); and the lowest rainfall amount and intensity (types III)) based on the rainfall amount and intensity in order to compare the differences in the response of sap flow trends and influencing factors of *Pinus tabulaeformis* and *Robinia pseudoacacia* under different rainfall types. The results showed that, under the daily scale average sap flow of *P. tabulaeformis* and *R. pseudoacacia*, rainfall type II decreased significantly relatively to rainfall types I and III ($p < 0.05$). In rainfall type II, The sap flow characteristics of *R. pseudoacacia* were positively correlated with solar radiation ($p < 0.05$), while those of *P. tabulaeformis* were positively correlated with temperature, solar radiation, and VPD ($p < 0.01$). The sap flow of *P. tabulaeformis* and *R. pseudoacacia* were significantly positively correlated with temperature, solar radiation, VPD, and soil moisture content ($p < 0.01$) and negatively correlated with relative humidity ($p < 0.05$) in rainfall type III. The hourly sap flow of *P. tabulaeformis* and *R. pseudoacacia* on rainfall days was higher than before the rainfall. Rainfall type I promoted the daily sap flow of both species, and the proportion of the sap flow in daytime was also higher. On rainy days, the sap flow rates of rainfall type I and III showed a "midday depression". In type I rainfall events, the sap flow "midday depression" after rainfall occurred an hour ahead compared to the sap flow "midday depression" before rainfall. In type II rainfall events, the daytime sap flow rates of *P. tabulaeformis* and *R. pseudoacacia* were obviously inhibited, but the nighttime sap flow rate increased. In type III rainfall events, the sap flow before rainfall presented a unimodal curve versus time. The daily average sap flow of *R. pseudoacacia* was more susceptible to rainfall type II, while *P. tabulaeformis* was more susceptible to rainfall types I and III. The sap flow rate of *R. pseudoacacia* decreased on rainy days. The results show that the effects of different rainfall types on the sap flow trends of *P. tabulaeformis* and *R. pseudoacacia* were different. They revealed the responses of their sap flow trends to meteorological factors under different rainfall types, which provided basic data and theoretical support for further predicting the sap flow trends on rainy days, clarifying the effects of rainfall amount, rainfall duration, and rainfall intensity on sap flow trends and accurately estimating the transpiration water consumption of typical tree species in the sub-humid climate regions of China.

Keywords: sap flow; rainfall types; K-means clustering; sub-humid climate

1. Introduction

Water plays an important role in vegetation distribution and plant life activities [1]. Precipitation, as a crucial element of climate conditions, is straightforwardly linked to the growth of plants, biodiversity, and the stability of forest ecosystems, fundamentally impacting almost all hydrological processes within the forest ecosystem [2–4]. Under the action of transpiration pull, plants absorb moisture in the soil through roots, transport it to various parts through ducts, and output it into the atmosphere through the transpiration of the leaves [5,6]. Sap flow is not only one of the most important physiological processes of trees, but it also reflects the plant's water transport capacity and sensitivity to environmental changes [7,8]. At the same time, sap flow is a key link in the SPAC (soil–plant–atmosphere continuum), which transports the soil water absorbed in roots and determines the amount of tree transpiration [9–11]. This can be used to analyze the characteristics of water consumption and the water transport mechanism of trees [12,13]. Climate change affects global hydrological processes and changes the spatial and temporal patterns of continental precipitation [14–16]. The most important characteristics are the decrease in precipitation frequency, the uneven spatial distribution of precipitation, the increase in extreme precipitation, and the increase in seasonal drought in sub-humid climate regions [8,17]. Precipitation is an important supplement to soil water, and changing temporal and spatial precipitation patterns also affect plant water use and canopy transpiration [18,19]. The extent to which vegetation is affected by rainfall patterns is related to different strategies through which tree species deal with drought [20,21]. Therefore, it is of great importance to research the correlation between precipitation patterns and soil water use of trees in seasonal arid areas under climate change [22,23].

The sap flow characteristics of different tree species are greatly different under the influence of physiological structures, water conditions, and meteorological factors [24,25]. In the long term, sap flow is mainly affected by root-related factors, while in the short term, it is more sensitive to meteorological factors [26–28]. Therefore, the time-delay effect of environmental factors can be used to simulate sap flow characteristics [29]. Previous studies have shown that the characteristics of sap flow show different variation patterns under the influence of weather [30–32]. On sunny days, the sap flow tends to be high in the daytime and low at night, and the variation trend is a single-peak curve or double-peak curve [8]. On rainy days, on the other hand, the curves fluctuate greatly or show gentle trends [33]. Therefore, the sap flow is one of the most important physiological water transport process of trees, and the characteristics of sap flow are greatly affected by the water environment [34,35]. However, different precipitation characteristics have a great influence on the water environment of forest trees, which leads to significant differences in sap flow. The response of sap flow to rainfall types is the key to accurately calculating plant transpiration [36,37].

In order to study the transpiration characteristics of trees, most studies on the relationship between sap flow and environmental factors were mainly conducted on typical sunny days [34,38]. Other studies have found that other rainfall parameters, such as rainfall duration and rainfall intensity, also affect plant transpiration in dryland ecosystems [35]. However, there have been few studies on the response of the sap flow characteristics of trees to rainfall types. The main objectives of this study were: (i) to explore the differences in the response of sap flow characteristics to different rainfall types and (ii) analyzing the difference in sap flow between conifers and broad-leaf trees under different rainfall types, as well as its influencing factors. Our purpose was to provide data supporting the study of vegetation transpiration and hydrological process. Our study aimed to investigate the effects of different rainfall events on sap flow for the purpose of providing water use efficiency, which is a supplement to the research on vegetation transpiration and hydrological processes.

2. Materials and Methods

2.1. Study Site Description

This study was performed in Jiu Feng National Forestry Park, Haidian District, about 30 km northwest of Beijing, China (39°540 N, 116°280 E) (Figure 1). It is a semi-humid region situated 140 m above the mean sea level, with a 10–25° sloping terrain. The mean daily temperatures range between 23 and 28 °C from May to mid-October, and between −5 and +25 °C throughout the winter, with a mean annual temperature of 12 °C. The mean annual precipitation is 630 mm, and the maximum rainfall occurs between June and September. In the study area, *P. tabulaeformis* and *R. pseudoacacia* are the dominant conifer and broad-leaved tree species. Most of these species were planted in the 1950s and 1960s. Lolium perenne, the main herbaceous vegetation (mean 70% ground cover), is widely distributed in the study area. The main shrubs were *Grewia biloba*, *Vitex negundo*, and *Broussonetia papyrifera*, whose densities were 3–5 plants/m^2. The soil in the study area was cinnamon soil, with an average slope of approximately 10% in the northeastern direction.

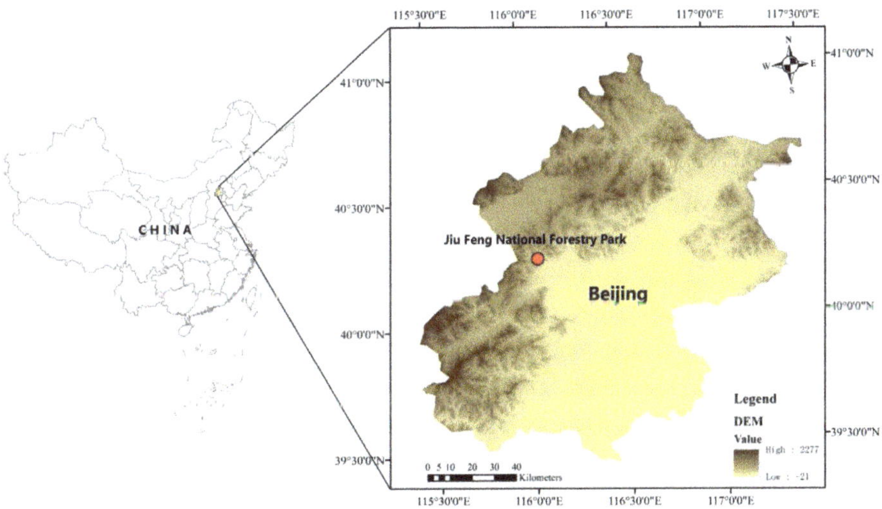

Figure 1. The location of Jiu Feng National Forestry Park in Beijing, China.

Three fixed observation stands (20 m × 20 m) of *P. tabulaeformis* and three fixed observation stands (20 m × 20 m) of *R. pseudoacacia* were selected and numbered. According to the average DBH and height, three standard sample trees with good growth and straight trunks without disease or insect pests were selected in each observation stand. The conditions of the selected plots were similar, e.g., in terms of altitude, slope, and aspect, which was advantageous for comparing the difference in sap flow between the two species under the same environmental conditions. The basic information on the plots and trees is shown in Table 1.

Table 1. Information of observation stands and average values of sample trees.

Species Stand	Density/(Tree/hm^2)	Average Age/a	Indices	Standard Sample Trees No.1	No.2	No.3
P. tabuliformis stand	1200	25	Height/m	8.5	5.3	6.2
			diameter/cm	18.47	16.57	13.06
R. pseudoacacia stand	900	43	Height/m	9.6	9.1	9.4
			diameter/cm	23.89	17.83	14.65

2.2. Sap Flow Measurements

A thermal dispersive sap flow measurement system (FLGS-TDP XM1000, Dynamax, Houston, USA) was used for continuous monitoring (the monitoring period lasted from April 2020 to September 2021). As for the calibration of Granier sensors, we have already considered radial calibration and tested the accuracy of the flow when installing the Granier sensors. Two kinds of probes (20 mm and 30 mm) were selected according to the xylem catheter in the radial position to ensure the accurate position of the probe. The collection frequency was set to 15 min to record once, and the sap flow rate was calculated according to the empirical formula proposed by Granier [38].

$$J_s = 0.0119 \times \left(\frac{d_{T_{max}} - d_T}{d_T} \right)^{1.231} \quad (1)$$

where J_s is the sap flow rate ($g \cdot cm^{-2} \cdot s^{-1}$), d_{Tmax} is the maximum temperature difference between the upper and lower probes within a day (°C), and d_T is the instantaneous temperature difference at the time of measurement (°C).

The daily sap flow ($g \cdot d^{-1}$) of the whole tree was calculated according to the average sap flow rate and the sapwood area (cm^2) of the sample tree, and the formula was as follows:

$$SF = J_s \times A_s \times 3600 \times 24 \quad (2)$$

Our research focuses on the influence of different rainfall events on sap flow, that is, the difference analysis of sap flow before rainfall and after rainfall. According to some rainfall events research literature [39,40], we defined the interval between two rainfall events as more than 24 h; the sap flow before rainfall (or pre-rainfall) refers to the sap flow within 24 h before a rainfall event begins, and the sap flow after rainfall (or rainfall) refers to the sap flow within 24 h before a rainfall event stops. The sapwood area was calculated according to sample trees with a similar DBH and height in the fixed observation stand, and wood cores were taken with growth cones to measure DBH and sapwood thickness to calculate the sapwood area. For basic information on the sample trees and sapwood area calculation, we refer the reader to our previously published literature [39].

2.3. Soil Water and Meteorological Factor Measurements

According to the previous studies on the distribution range of rhizosphere and the depth of root water absorption of two species of trees [40], we measured the average soil water content (SWC) of 0–40 cm soil layers using soil water sensors (ECH2O-TE System, Decagon Devices, Pullman, Washington, USA). At the same time, meteorological data, which included the characteristic values of precipitation (rainfall R_a, rainfall intensity R_i, rainfall duration R_d), net radiation (R_n), air temperature (T_a), relative humidity of the air (RH, %) and wind speed (W_S), were recorded synchronously by the HOBO automatic small weather station (U30, Onset, MA, USA). The data were recorded at 15 min intervals. The vapor pressure deficit (VPD) was calculated from the air temperature and relative humidity, and the formula is as follows:

$$VPD = 0.61078 \times e^{\frac{17.27 \times T_a}{T_a + 240.97}} \times (1 - RH) \quad (3)$$

2.4. Cluster Analysis of Rainfall Events

Compared with traditional rainfall classification, the clustering classification based on rainfall characteristic values can more accurately depict the rainfall process. The measurement period lasted 316 days (from April 21 to 28 September 2020 and from 4 April to 5 September 2021). Therefore, 82 rainfall events (growing seasons for 2020 and 2021) were characterized by three precipitation characteristic values, which included the rainfall amount (R_a(mm)), rainfall duration (R_d(h)), and rainfall intensity (Ri($mm \cdot h^{-1}$)), were evaluated by the K-means clustering analysis method (Table 2). The classification met the criteria for one-way analysis of variance (ANOVA) at the level of significance ($p < 0.05$), and the three clusters kept the overall intra-group variance to a minimum. Type I had the highest rainfall amount and intensity and medium rainfall duration; type II had a medium rainfall amount and the lowest intensity, but the longest rainfall duration; and type III had the lowest rainfall amount and medium rainfall intensity, but the shortest rainfall duration.

Table 2. Statistical characteristics of different rainfall types.

Rainfall Types	Frequency	Rainfall Characteristic	Mean ± Standard Deviation	Variation Coefficient (%)
Type I	3 (3.67%)	Rainfall amount/mm	77.60 ± 74.10	95.43
		Rainfall duration/h	7.83 ± 6.77	86.37
		Rainfall intensity/mm h^{-1}	13.27 ± 6.84	51.56
Type II	28 (34.15%)	Rainfall amount/mm	15.64 ± 14.55	93.03
		Rainfall duration/h	13.40 ± 5.19	38.71
		Rainfall intensity/mm h^{-1}	1.27 ± 1.37	107.35
Type III	51 (62.20%)	Rainfall amount/mm	2.37 ± 3.60	152.40
		Rainfall duration/h	2.00 ± 2.09	104.25
		Rainfall intensity/mm h^{-1}	1.48 ± 1.88	126.60

2.5. Data Analysis

Origin Pro 18, R and Minitab 19 were used to calculate the mean, standard deviation, coefficient of variation, and other statistical data of various indicators under different rainfall types. ANOVA was used to evaluate the statistical differences of daily meteorological factors under different rainfall types and the statistical differences of sap flow before and during rainfall between different rainfall types. Pearson correlation analysis was used to test the correlation between sap flow and influencing factors.

3. Results

3.1. Precipitation Characteristics and Meteorological Factors

Comparing the amount of rainfall and the duration and intensity of the 82 rainfall events in the two growing seasons, as well as the classification results of the K-means clustering analysis, it can be seen that rainfall events of type I occurred in July, August, and September of the first growing season (Figure 2). The amount and intensity were higher than the other types, but the occurrence frequency was low (3 events, 3.67%). Types II and III appeared alternately in two growing seasons, and the frequency of type III (51, 62.2%) was higher than that of type II (28, 34.15%). Type II was characterized by the lowest rainfall intensity and longer rainfall duration (over 8 h), while type III had the shortest rainfall duration (under 5 h) along with the lowest rainfall amount (less than 6 mm).

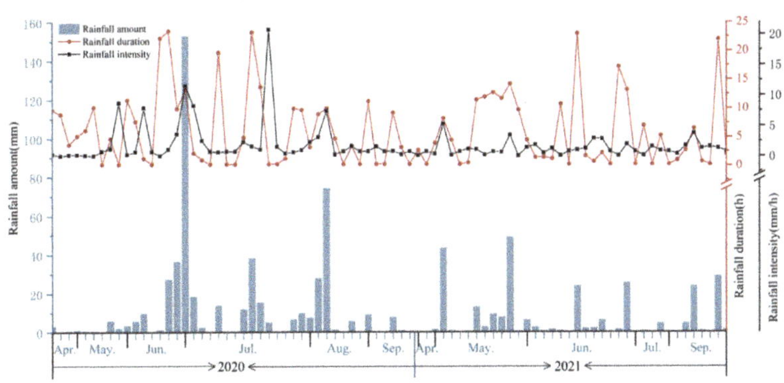

Figure 2. Precipitation characteristics of 82 rainfall events during two growing seasons.

The range of variation for T_a (5.81–31.54 °C), Rn (0–494.36 $W \cdot m^{-2}$), RH (7.13–100%), and VPD (0–4.30 kpa) during the two growing seasons showed obvious seasonal variations (Figure 3). The lowest values of monthly mean T_a (13.28 °C), RH (38.06%), and VPD (0.99 kpa) appeared in April, and the highest values appeared in July and August (26.67 °C, 63.80%, 2.14 kpa). The lowest value of Rn appeared in September (94.03 $W \cdot m^{-2}$) and the highest value occurred in May (145.40 $W \cdot m^{-2}$).

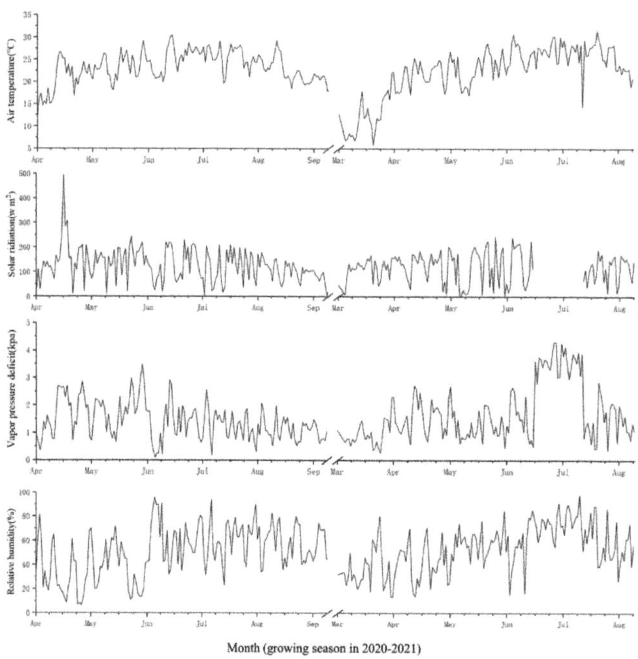

Figure 3. Meteorological trends during two growing seasons.

The daily mean meteorological factors of the three rainfall types are shown in Table 3 (Rn in July of the second growing season was missing due to equipment failure). The variation range of daily mean T_a, Rn, RH, and VPD on rainy days were 7.16–29.26 °C, 4.1–240.62 $W \cdot m^{-2}$, 15.30–94.22%, and 0.18–2.65 kpa, respectively. The daily Rn of type I was significantly higher than that of type III ($p < 0.05$), but there were no significant differences in the T_a, Rn, or VPD of the three rainfall types.

Table 3. Characteristics of daily mean meteorological factors within different rainfall types during the growing seasons.

Rainfall Types	T_a (°C)	Rn ($w \cdot m^{-2}$)	RH (%)	VPD (kpa)
Type I	23.65 ± 3.76 a	94.30 ± 106.80 a	49.53 ± 15.91 a	1.51 ± 0.20 a
Type II	20.05 ± 4.49 a	45.71 ± 41.31 ab	50.96 ± 19.49 a	1.19 ± 0.60 a
Type III	22.20 ± 4.84 a	95.31 ± 70.99 b	56.65 ± 17.08 a	1.17 ± 0.54 a

Note(s): Different letters indicate significant differences in daily mean meteorological factors among the different rainfall types.

3.2. Daily Sap Flow Characteristics within Three Rainfall Types

The characteristics of sap flow under three rainfall types are shown in Figure 4. Type I only appeared in the growing season in 2020. Types II and III appeared to be evenly distributed in the two growing seasons, and the daily mean sap flow of *P. tabulaeformis* and *R. pseudoacacia* showed no obvious changes under the three rainfall types. The highest daily sap flow values of *P. tabulaeformis* and *R. pseudoacacia* occurred in type I (25,986 g·d^{-1} and 42,931 g·d^{-1}, respectively), and the lowest values occurred in the type II (6905 g·d^{-1} and 14,656 g·d^{-1}, respectively). There were significant differences in the average daily sap flow of *P.tabulaeformis* before and after type II ($p < 0.05$), but no significant difference in the other two rainfall types.

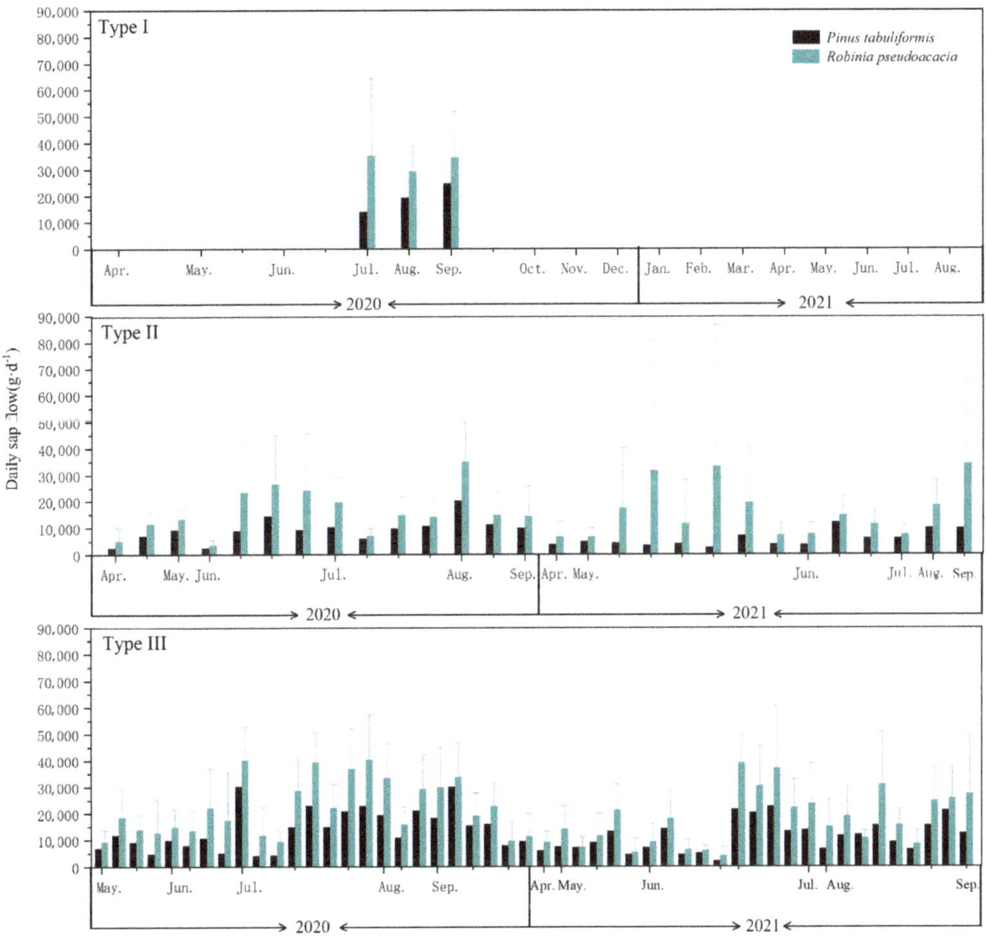

Figure 4. Daily sap flow characteristics of *P. tabuliformis* and *R. pseudoacacia* under three rainfall types in two growing seasons (the value was measured 24 h after rain).

Both before and after rainfall, the average daily sap flow of *P. tabulaeformis* in type II was significantly lower than that in types I and III ($p < 0.05$). Before rainfall, the average daily sap flow of *R. pseudoacacia* showed no significant differences among the three rainfall types. After rainfall, the average daily sap flow of *R. pseudoacacia* in type I was significantly higher than that in type II ($p < 0.05$), and that in type II was significantly higher than that in type III ($p < 0.05$). The daily sap flow of *R. pseudoacacia* was higher than that of *P. tabulaeformis* (Figure 5). Under type I, the daily values of *P. tabulaeformis* and *R. pseudoacacia* after rainfall increased by 12.6% and 9.3% relatively to before rainfall, and by 85.2% and 98.7% under type II. But under type III, the daily values decreased by 29.1% and 21.0%, respectively.

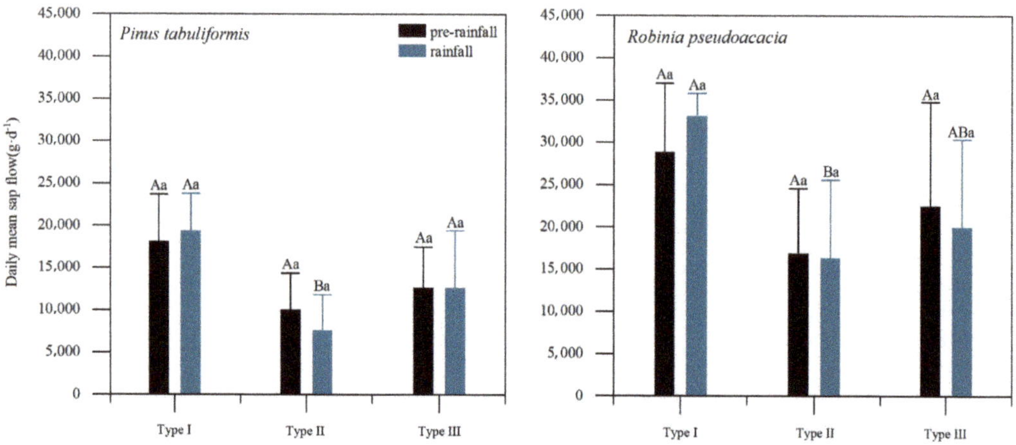

Figure 5. Comparison of daily pre-rainfall and after rainfall sap flow characteristics of *P. tabuliformis* and *R. pseudoacacia* under three rainfall types (different letters indicate significant differences in daily mean sap flow between the different rainfall types).

The correlation between the sap flows of *P. tabulaeformis* and *R. pseudoacacia* for types II and III, as well as the influencing factors, are analyzed in Table 4. The sap flow of *P. tabulaeformis* was positively correlated with temperature, soil water content, and solar radiation, and VPD ($p < 0.01$) for types II and III, and was negatively correlated with relative humidity for type III ($p < 0.01$). The correlation between the sap flow of *R. pseudoacacia* and the influencing factors was not consistent between the two rainfall types. Under type II, the daily sap flow of *R. pseudoacacia* was only positively correlated with solar radiation ($p < 0.05$), while under type III, the daily sap flow of *R. pseudoacacia* was positively correlated with temperature, solar radiation, VPD, and soil moisture content ($p < 0.01$). It was negatively correlated with relative humidity ($p < 0.05$).

Table 4. Correlation coefficients between daily sap flow and influencing factors of *P. tabuliformis* and *R. pseudoacacia* with rainfall types II and III.

Species	Rainfall Types	T	SR	RH	VPD	SWC
P.tabuliformis	Type II	0.60 **	0.73 **	−0.25	0.52 **	0.31
	Type III	0.37 **	0.68 **	−0.41 **	0.63 **	0.52 **
R.pseudoacacia	Type II	0.37	0.46 *	0.17	0.09	0.15
	Type III	0.50 **	0.72 **	−0.34 *	0.66 **	0.62 **

Note: * Correlation is significant at the 0.05 level (2-tailed); ** Correlation is significant at the 0.05 level (2-tailed).

The linear relationship between the daily sap flows of *P. tabulaeformis* and *R. pseudoacacia* and the influencing factors varied for different rainfall types, as shown in Figure 6. The daily sap flows of *P. tabulaeformis* and *R. pseudoacacia* increased with the increase in solar radiation under the two rainfall types. Under type II, there was no obvious correlation between the daily sap flow of *R. pseudoacacia* and temperature or VPD, but under type III, with the increase in temperature, solar radiation, VPD, and soil water content, the increase in the daily sap flow of *R. pseudoacacia* was higher than that of *P. tabulaeformis*, and with the increase in relative humidity, the decrease in the daily sap flow of *R. pseudoacacia* was also higher than that of *P. tabulaeformis*.

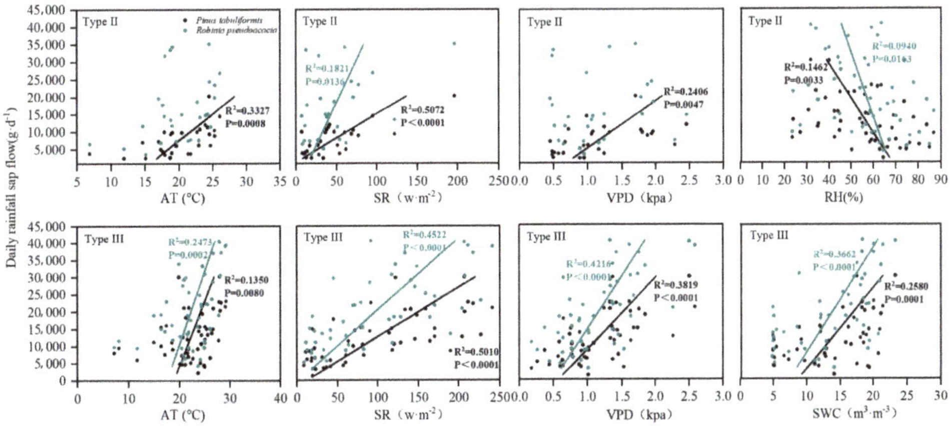

Figure 6. Correlation between daily sap flow and influencing factors of *P. tabuliformis* and *R. pseudoacacia* with rainfall types II and III.

3.3. Hourly Sap Flow Characteristics within Three Rainfall Types

The hourly variation in sap flow for the three rainfall types is shown in Figure 7. The sap flow rate was higher after rainfall than before rainfall for the three rainfall types, and the difference in sap flow was the greatest for type II. The hourly sap flow of the two species showed significant diurnal variation, and the characteristics of sap flow before the rainfall showed a single peak curve only when type III decreased. The daytime sap flow of *P. tabulaeformis* and *R. pseudoacacia* accounted for 78.78–92.53% and 84.44–90.85% (from 5:00 to 19:00), respectively.

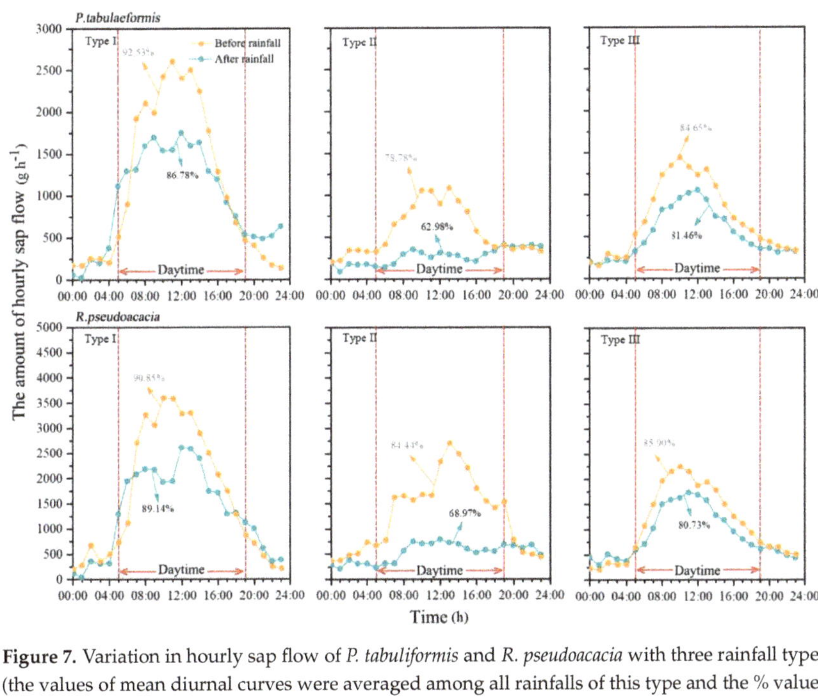

Figure 7. Variation in hourly sap flow of *P. tabuliformis* and *R. pseudoacacia* with three rainfall types (the values of mean diurnal curves were averaged among all rainfalls of this type and the % values were the daytime sap flow of two special trees (from 5:00 to 19:00) accounted for whole day (24 h)).

In the case of type I, the before rainfall sap flow of the two species was higher than that on the day of the rainfall for a short time at sunrise, and the "siesta" phenomenon of a transient decrease in sap flow occurred from 10:00 to 11:00. In the case of type III, the "siesta" phenomenon occurred at 12:00.

Three rainy days with similar sap flow characteristics were screened to illustrate the hourly sap flow characteristics of the two species (Figure 8). Compared with type III, the sap flow of *R. pseudoacacia* decreased, while *P. tabulaeformis* was relatively stable. The rapid recovery of sap flow after type I rainfall indicates that the recovery of sap flow has little relationship with the interval time after rainfall. In type II, the recovery of sap flow after rainfall was slow and the recovery range was low, and the sap flows of the two species were low, and *R. pseudoacacia* decreased more significantly than *P. tabulaeformis*. The decrease in the sap flow of *R. pseudoacacia* was more obvious than that of *P. tabulaeformis*. After rainfall, the nighttime sap flows of both species increased. In type III, the recovery of sap flow after rainfall was fast and the sap flow of *R. pseudoacacia* also increased after rainfall, and that of *P. tabulaeformis* was relatively stable. Therefore, the sap flow of *R. pseudoacacia* was more susceptible to rainfall than *P. tabulaeformis*, and the two species were different under the same rainfall types, with different rainfall distribution periods.

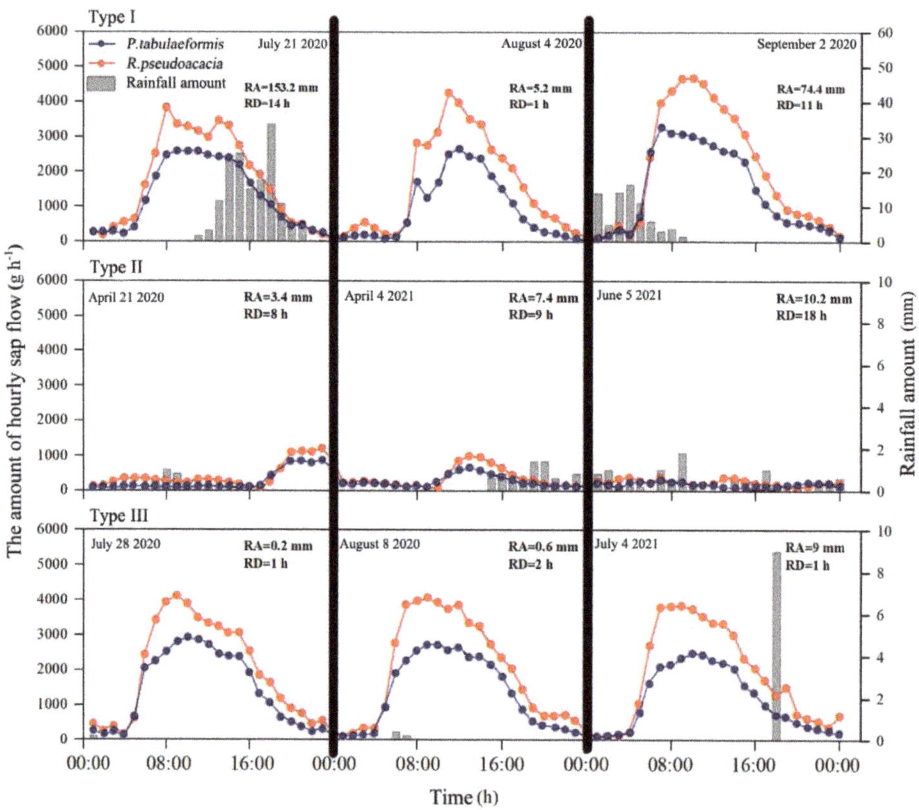

Figure 8. Variations in hourly sap flow of *P. tabuliformis* and *R. pseudoacacia* with three rainfall types (the three rainy days for type I were 21 July 2020, 4 August 2020 and 2 September 2020; the three rainy days for type II were 21 April 2020, 4 April 2021 and 5 June 5 2021; the three rainy days for type III were 28 July 2020 and 8 August 2020, and 4 July 2021.The time of all sap flow values was the corresponding time 24 h after rainy day).

4. Discussion

4.1. Response of Sap Flow Characteristics to Other Factors under Three Rainfall Types at the Daily Scale

The average daily solar radiation of type I rainfall was significantly higher than that of type III rainfall, while other meteorological factors (average daily temperature, relative humidity, and VPD) showed no significant difference among the three rainfall types. The highest daily sap flow values of the two species occurred with type I rainfall. At the daily scale, solar radiation and VPD were the main influencing factors of sap flow characteristics. The reason for this is that solar radiation changed the hydrothermal state of the soil, vegetation, and atmosphere, and then affected the distribution characteristics of the soil water content, soil water potential, and VPD. These factors have important effects on sap flow. Some studies have shown that a 74.37% variation in sap flow for broad-leaved trees in the Beijing mountain area of China was caused by solar radiation and temperature [35,38,41,42]. The daily sap flow rate of *R. pseudoacacia* was higher than that of *P. tabulaeformis*, which may be related to the difference in DBH between the two tree species and the difference in the physiological structures of conifer species. In this study, the tree height and DBH of *R. pseudoacacia* were higher than those of *P. tabulaeformis*. Previous studies have found that the xylem sapwood area affects the sap flow rate of the trunk, and the sap flow flux and sap flow rate of the large-diameter class were significantly higher than those of the small-diameter class. Broad-leaved tree species have more efficient water transport structures and larger leaves, and their sap flow characteristics are more responsive to meteorological factors [43].

The daily sap flows of *P. tabulaeformis* and *R. pseudoacacia* decreased significantly under type II rainfall, which may be because type II rainfall was characterized by a longer rainfall duration (more than 8 h). The long-term rainfall was usually accompanied by higher relative humidity, lower temperature, and solar radiation, which attenuated the transpiration pull and extended the drying time of the leaves. The increasing number of stomata and prolonged closure time reduced the transpiration rate of the trees [44]. In addition, the sap flow on the day after rainfall increased compared with the day before rainfall for the three rainfall types. Under type II, the average daily sap flows of *P. tabulaeformis* and *R. pseudoacacia* after rainfall increased by 85.2% and 98.7% compared with those before rainfall, a greater difference than that of types I and III rainfall. This result was contrary to other sap flow studies in sub-humid climate regions, where it has been found that the sap flow of trees before rainfall is higher than that after rainfall. The reason may be that the study samples were located in a relatively arid soil rock mountain area [45]. The lower soil water and higher transpiration force caused the trees to close their stomata to prevent water loss before rainfall, and the sap flow rate decreased. Rainfall replenished the soil water, and the increased water accelerated the transpiration rate of the trees. Changes in the soil water content led to greater nighttime sap flow on rainy days than on sunny days [42].

Type I rainfall is a special rainfall type in the study area, mainly composed of extremely heavy rainfall with a low frequency (three times). Types II and III rainfall were more common in the study area. There were differences in the response of the sap flow characteristics of *P. tabulaeformis* and *R. pseudoacacia* to meteorological factors under the two rainfall types. Under type II rainfall, the sap flow characteristics of *P. tabulaeformis* were positively correlated with temperature, solar radiation, and VPD, while those of *R. pseudoacacia* were only positively correlated with solar radiation. Under type III rainfall, the sap flow characteristics of the two species were also significantly positively correlated with soil moisture content and negatively correlated with relative humidity. This indicates that, compared with typical sunny days, on rainfall days, the influence of various meteorological factors on the sap flow characteristics of trees is restricted by the rainfall types. For type II rainfall, the rainfall duration was longer, and lower rainfall intensity usually leads to higher relative humidity all day. The sap flow characteristics of the trees were more affected by solar radiation. Under type III rainfall, the rainfall duration was shortest and intensity was the highest. The tree sap flow characteristics were more susceptible to the influence of relative humidity and soil water status. Previous studies have shown that the root distribution of *P. tabulaeformis* and *R. pseudoacacia* was shallow, and in dry soil and rock mountains, high-intensity rainfall exerts a significant impact on the sap flow characteristics of shallow-rooted tree species by supplementing the soil water in shallow layers.

4.2. Response of Sap Flow Characteristics to Other Factors under Three Rainfall Types at the Hourly Scale

Under three rainfall types, the hourly sap flow amounts of *P. tabulaeformis* and *R. pseudoacacia* after rainfall were higher than before rainfall. Type I rainfall significantly promoted the sap flow of these two species. Its daytime sap flow rate was also higher, which may have been due to the high intensity of the rainfall supplementing shallow soil water, while the short-term changes in sap flow were more susceptible to environmental factors. VPD tended towards 0 during rainfall, resulting in the rate of sap flow tending towards 0. As for the flow under type I rainfall, the sap flow also increased from 0 earlier, which may be related to the higher solar radiation possessed by type I rainfall, or because the higher solar radiation possessed by the type I rainfall start time of the sap flow lagged behind the solar radiation. The time lag between the trunk flow and canopy transpiration was less, almost negligible.

Both types I and III rainfall had the phenomena of a midday depression, mainly because of the stronger solar radiation of these two types. The root water absorption from the soil was less than the transpiration consumption. The strongest solar radiation plants closed their leaf stomata to prevent water loss, which was consistent with the change in solar radiation. As for type I at 11:00 and 10:00, the flow fluctuation was greater, the midday depression on rainfall days occurred at 12:00, and the rainfall flow was a single peak curve with no "midday depression" phenomenon. The characteristics of type I rainfall flow before and after showed the biggest difference, and the daytime flow before rainfall was significantly lower than that with other rainfall types. The flow fluctuation on rainfall days was larger, which may be because the class rainfall lasted the longest, and the change in environmental factors was relatively complex. Plants adapt to environmental changes through their water capacity adjustment. Research has shown that unsteady flow characteristics were mainly the result of their own regulation. Contrasting the three types of rainfall represents the change in flow characteristics. For two tree species, type II rainfall had decreased. The sap flow rate in the rainfall period was the result of transpiration tension and soil water supplementation. The rainfall occurred in the daytime, although two species of sap flow at night have significantly improved, and night hydrating may occur.

4.3. The Difference in Sap Flow Characteristics between Conifer and Broad-Leaved Species under Three Rainfall Types

The effects of different types of rainfall species on the rate of increase in sap flow were different, with different influences on the soil moisture status and tree transpiration rate. Rainfall significantly promoted an increase in sap flow rate, promoting the sap flow effect of *R. pseudoacacia* over *P. tabulaeformis*. The type I rainfall promotion effect was small, but still greater than *R. pseudoacacia*, which may suggest that different species of trees' sap flow characteristics respond differently to the three rainfall types.

The study shows that the flow characteristics of broad-leaf species are more susceptible to environmental factors, but this study found that, under different rainfall types, the sap flow was higher than *P. tabulaeformis*. However, *P. tabulaeformis* was more sensitive to meteorological factors, which may be related to the responses of the two species to different rainfall types, and the transpiration rate was more susceptible to environmental factors other than solar radiation.

The daytime sap flow decreased after type I rainfall, and the suppression of the *R. pseudoacacia* sap flow rate was more obvious, indicating that the *R. pseudoacacia* flow rate on rainfall days was more likely to receive inhibition. The sap flow of *R. pseudoacacia* rebounded significantly after rainfall, while the sap flow of *P. tabulaeformis* was relatively stable under the three types of rainfall. This may be related to the physiological structures of *R. pseudoacacia* and *P. tabulaeformis*. The leaf structure can affect the plant's interception of precipitation, thus affecting the time consumed by leaf drying and stomatal opening and closing [46].

5. Conclusions

The daily sap flow of *R.pseudoacacia* and *P. tabulaeformis* was significantly reduced under the type II rainfall ($p < 0.05$). The sap flow characteristics of *R. pseudoacacia* were positively correlated with solar radiation ($p < 0.05$), while those of *P. tabulaeformis* were significantly positively correlated with air temperature, solar radiation, and VPD ($p < 0.01$); the sap flow characteristics of *R. pseudoacacia* and *P. tabulaeformis* were significantly positively correlated with air temperature, solar radiation, VPD, and soil water content ($p < 0.01$), and were negatively correlated with the relative humidity ($p < 0.05$).

The sap flows of both *R. pseudoacacia* and *P. tabulaeformis* after rainfall were higher than before rainfall at the daily scale. Type I rainfall improved the sap flow of the two tree species, and the proportion of the daytime sap flow was higher, whilst the sap flow of *R. pseudoacacia* and *P. tabulaeformis* showed a "midday depression" phenomenon after rainfall types I and III. For type I, the "midday depression" phenomenon after rainfall occurred an hour earlier than that before rainfall type III. The flow of type III showed a unimodal curve, while the type II rainfall significantly inhibited the daytime flow of *R. pseudoacacia* and *P. tabulaeformis*. The flow rate rebounded at night after rainfall.

The daily sap flow of *R. pseudoacacia* was more easily affected by the type of rainfall, while *P. tabulaeformis* was more susceptible to types I and III. *P. tabulaeformis* was more sensitive to changes in meteorological factors, and the sap flow rate of *R. pseudoacacia* was more easily inhibited on rainy days. *P. tabulaeformis* was relatively stable.

The main factors affecting the sap flow characteristics differed under different time scales and rainfall types. When studying tree trunk sap flow characteristics in arid environments, the rainfall types should be considered in addition to conventional meteorological factors to clarify the influence of various rainfall parameters on the sap flowing from the tree trunk. The results further reveal the hydrological processes and some physiological hydrological driving mechanisms of forest ecosystems in northern China, and provide a technical reference for the accurate estimation of forest ecosystem water carrying capacity and sustainable forest management under seasonal drought.

Author Contributions: Y.C. conducted data curation, methodology and writing—original draft. Y.W. conducted data curation, formal analysis, software and writing—original draft. N.Z., C.N. and Y.B. conceived investigation and project administration. J.J. conducted methodology, writing—original draft and writing—review and editing. All authors have read and agreed to the published version of the manuscript.

Funding: This research was funded by Natural and Science Basic Research Program of Shaanxi Province (2022JQ-295), Hunan Province Natural Science Foundation (2023JJ31003), Changsha Natural Science Foundation Project (KQ2208413), and Hunan Province key research and development project (2023SK2055). The authors gratefully acknowledge all financial support for this study.

Data Availability Statement: All relevant data are within the manuscript.

Conflicts of Interest: Yongxiang Cao, Naichang Zhang, Chendong Ning, and Yu Bai were employed by the company of Power China Northwest Engineering Corporation Limited. The remaining authors declare that the research was conducted in the absence of any commercial or financial relationships that could be construed as potential conflicts of interest.

References

1. Nadezhdina, N. Sap flow index as an indicator of plant water status. *Tree Physiol.* **1999**, *19*, 885. [CrossRef] [PubMed]
2. Kray, J.A.; Cooper, D.J.; Sanderson, J.S. Groundwater use by native plants in response to changes in precipitation in an intermountain basin. *J. Arid. Environ.* **2012**, *83*, 25–34. [CrossRef]

3. Liu, H.; Yin, Y.; Wang, Q.; He, S. Climatic effects on plant species distribution within the forest–steppe ecotone in northern China. *Appl. Veg. Sci.* **2015**, *18*, 43–49. [CrossRef]
4. Tao, Z.; Wang, H.; Liu, Y.; Xu, Y.; Dai, J. Phenological response of different vegetation types to temperature and precipitation variations in northern China during 1982–2012. *Int. J. Remote Sens.* **2017**, *38*, 3236–3252. [CrossRef]
5. Macias, F.M. Satellite images show China going green. *Nature* **2018**, *553*, 411–413. [CrossRef] [PubMed]
6. Zhang, M.F.; Wei, X.H. Deforestation, forestation and water supply. *Science* **2021**, *371*, 990–991. [CrossRef] [PubMed]
7. Kathy, S.; De, P.; Raoul, L.; Vanrolleghem, P.A. A mathematical model linking tree sap flow dynamics to daily stem diameter fluctuations and radial stem growth. *Tree Physiol.* **2006**, *26*, 257.
8. Jia, G.; Liu, Z.; Chen, L.; Yu, X. Distinguish water utilization strategies of trees growing on earth-rocky mountainous area with transpiration and water isotopes. *Ecol. Evol.* **2017**, *7*, 10640–10651. [CrossRef]
9. Jian, S.; Zhang, X.; Wu, Z. Water use pattern of Pinus tabulaeformis in the semiarid region of Loess Plateau, China. *For. Syst.* **2016**, *25*, 077–084. [CrossRef]
10. Li, M.; Xia, J.; Chen, Z.; Meng, D.; Xu, C. Variation analysis of precipitation during past 286 years in Beijing area, china, using non-parametric test and wavelet analysis. *Hydrol. Process.* **2013**, *27*, 2934–2943. [CrossRef]
11. Liu, J.; Bloomberg, M.; Li, G.; Liu, Y. Effects of copper root pruning and radicle pruning on first-season field growth and nutrient status of Chinese cork oak seedlings. *New For.* **2016**, *47*, 715–729. [CrossRef]
12. Graham, E.A.; Mulkey, S.S.; Kitajima, K.; Phillips, N.G.; Wright, S.J. Cloud cover limits net CO_2 uptake and growth of a rainfallforest tree during tropical rainfally seasons. *Proc. Natl. Acad. Sci. USA* **2018**, *100*, 2.
13. Zhao, C.; Si, J.; Feng, Q.; Yu, T.; Li, W. Stem Sap Flow Research: Progress and Prospect. *J. Northwest For. Univ.* **2015**, *30*, 98–105.
14. Seneviratne, S.I.; Lüthi, D.; Litschi, M.; Schär, C. Land-atmosphere coupling and climate change in Europe. *Nature* **2006**, *443*, 7108. [CrossRef] [PubMed]
15. Shi, P.; Sun, X.; Wang, M.; Li, N.; Wang, J.; Jin, Y.; Yin, W. Climate change regionalization in China (1961–2010). *Sci. China Earth Sci.* **2014**, *44*, 2294–2306. [CrossRef]
16. Nie, Y.P.; Chen, H.S.; Wang, K.L.; Tan, W.; Deng, P.Y.; Yang, J. Seasonal water use patterns of woody species growing on the continuous dolostone outcropsand nearby thin soils in subtropical China. *Plant Soil* **2011**, *341*, 399–412. [CrossRef]
17. Ren, G.; Ren, Y.; Zhan, Y.; Sun, X.; Liu, Y.; Chen, Y.; Wang, T. Spatial and temporal patterns of precipitation variability over mainland China: II: Recent trends. *Adv. Water Sci.* **2015**, *26*, 451–465.
18. Chen, S.; Fu, Y.H.; Geng, X.; Hao, Z.; Tang, J.; Zhang, X.; Xu, Z.; Hao, F. Influences of shifted vegetation phenology on runoff across a hydroclimatic gradient. *Front. Plant Sci.* **2022**, *12*, 802664. [CrossRef]
19. D'Odorico, P.; Caylor, K.; Okin, G.S.; Scanlon, T.M. On soil moisture-vegetation feedbacks and their possible effects on the dynamics of dryland ecosystems. *J. Geophys. Res. Biogeosci.* **2015**, *112*, G04010. [CrossRef]
20. Liu, X.; Zhang, Y.; Xiao, T. Runoff velocity controls soil nitrogen leaching in subtropical restored forest in Southern China. *For. Ecol. Manag.* **2023**, *548*, 121412. [CrossRef]
21. Liu, X.; Zhang, Y.; Zhang, L. Aggregate-associated soil organic carbon fractions in subtropical soil undergoing vegetative restoration. *Land. Degrad. Dev.* **2023**, *34*, 4296–4306. [CrossRef]
22. Zhang, X.L.; Lv, P.C.; Xu, C.; Huang, X.R.; Rademacher, T. Dryness decreases average growth rate and increases drought sensitivity of Mongolia oak trees in north China. *Agric. For. Meteorol.* **2021**, *308*, 108611. [CrossRef]
23. Ziv, A.; Jean-Christophe, D.; Ram, O.; Way, D.A.; Menachem, M. Growth and physiological responses of isohydric and anisohydric poplars to drought. *J. Exp. Bot.* **2015**, *14*, 4373–4381.
24. Tyree, M.T. A dynamic model for water flow in a single tree: Evidence that models must account for hydraulic architecture. *Tree Physiol.* **1988**, *4*, 195–217. [CrossRef]
25. Schaeffer, S.M.; Williams, D.G.; Goodrich, D.C. Transpiration of cottonwood/willow forest estimated from sap flux. *Agric. For. Meteorol.* **2000**, *105*, 257–270. [CrossRef]
26. Bosch, D.D.; Marshall, L.K.; Teskey, R. Forest transpiration from sap flux density measurements in a Southeastern Coastal Plain riparian buffer system. *Agric. For. Meteorol.* **2014**, *187*, 72–82. [CrossRef]
27. Angadi, S.V.; Entz, M.H. Root system and water use patterns of different height sunflower cultivars. *Agron. J.* **2002**, *94*, 136–145. [CrossRef]
28. Wu, H.W.; Li, X.Y.; Jiang, Z.Y.; Chen, H.Y.; Zhang, C.C.; Xiao, X. Contrasting water use pattern of introduced and native plants in an alpine desert ecosystem, northeast Qinghai–tibet plateau, China. *Sci. Total Environ.* **2016**, *542*, 182–191. [CrossRef]
29. Wang, D.; Gao, G.; Li, J.; Yuan, C.; Lü, Y.; Fu, B. Sap flow dynamics of xerophytic shrubs differ significantly among rainfall types in the Loess Plateau of China. *J. Hydrol.* **2020**, *585*, 124815. [CrossRef]
30. Johnson, D.M.; McCulloh, K.A.; Woodruff, D.R.; Meinzer, F.C. Hydraulic safety margins and embolism reversal in stems and leaves: Why are conifers and angiosperms so different? *Plant Sci.* **2012**, *195*, 48–53. [CrossRef]
31. Jia, G.; Yu, X.; Deng, W. Seasonal water use patterns of semi-arid plants in China. *For. Chron.* **2013**, *89*, 169–177. [CrossRef]
32. Li, B.F.; Chen, Y.N.; Shi, X.; Chen, Z.S.; Li, W.H. Temperature and precipitation changes in different environments in the arid region of northwest China. *Theor. Appl. Climatol.* **2013**, *112*, 589–596. [CrossRef]
33. Huang, L.; Zhang, Z. Effect of rainfall pulses on plant growth and transpiration of two xerophytic shrubs in a revegetated desert area: Tengger Desert, China. *Catena* **2012**, *137*, 269–276. [CrossRef]

34. Chen, D.; Wang, Y.; Liu, S.; Wei, X.; Wang, X. Response of relative sap flow to meteorological factors under different soil moisture conditions in rainfallfed jujube (*Ziziphus jujuba* Mill.) plantations in semiarid Northwest China. *Agric. Water Manag.* **2014**, *136*, 23–33. [CrossRef]
35. Huxman, T.E.; Snyder, K.A.; Tissue, D.; Leffler, A.J.; Schwinning, S. Precipitation pulses and carbon fluxes in semiarid and arid ecosystems. *Oecologia* **2004**, *141*, 254–268. [CrossRef] [PubMed]
36. David, T.S.; Pinto, C.A.; Nadezhdina, N.; Kurz-Besson, C.; Henriques, M.O.; Quilhó, T.; Cermak, J.; Chaves, M.M.; Pereira, J.S. Root functioning, treewater use and hydraulic redistribution in Quercus suber trees: A modelingapproach based on root sap flow. *For. Ecol. Manag.* **2013**, *307*, 136–146. [CrossRef]
37. Liu, Z.; Yu, X.; Jia, G. Contrasting water sources of evergreen and deciduous tree species in rocky mountain area of Beijing, China. *Catena* **2017**, *150*, 108–115. [CrossRef]
38. Granier, A.; Huc, R.; Barigah, S.T. Transpiration of natural rainfall forest and its dependence on climatic factors. *Agric. For. Meteorol.* **1996**, *78*, 19–29. [CrossRef]
39. Jia, J.; Yu, X.; Li, Y. Response of forestland soil water content to heavy rain pattern in Beijing Mountain, northern China. *J. For. Res.* **2016**, *27*, 541–550. [CrossRef]
40. Zhang, Y.; Yang, M.; Pan, X.; Song, L.; Yu, L. Influence of Rainfall Division Method on Capture Ratio of Rainfall. *China Water Wastewater* **2019**, *35*, 122–127.
41. Liu, Z.; Ye, L.; Jiang, J.; Liu, R.; Xu, Y.; Jia, G. Increased uptake of deep soil water promotes drought resistance in mixed forests. *Plant Cell Environ.* **2023**, *46*, 3218–3228. [CrossRef] [PubMed]
42. Szutu, D.J.; Papuga, S.A. Year-round transpiration dynamics linked with deep soil moisture in a warm desert shrubland. *Water Resour. Res.* **2019**, *55*, 7. [CrossRef]
43. Kume, T.; Kuraji, K.; Yoshifuji, N.; Morooka, T.; Sawano, S.; Chong, L. Estimation of canopy drying time after rainfall using sap flow measurements in an emergent tree in a lowland mixed-dipterocarp forest in Sarawak, Malaysia. *Hydrol. Process.* **2006**, *20*, 565–578. [CrossRef]
44. Yue, G.; Zhao, H.; Zhang, T.; Zhao, X.; Niu, L.; Sam, D. Evaluation of water use of Caragana microphylla with the stem heat-balance method in Horqin Sandy Land, Inner Mongolia, China. *Agric. For. Meteorol.* **2008**, *148*, 11. [CrossRef]
45. Wen, X.; Zhang, S.; Sun, X. Water vapor and precipitation isotope ratios in Beijing. *J. Geophys. Res.* **2010**, *115*, 133–134. [CrossRef]
46. Yuan, C.; Gao, G.; Fu, B. Comparisons of stem flow and its bio-/abiotic influential factors between two xerophytic shrub species. *Hydrol. Earth Syst. Sci.* **2017**, *21*, 3. [CrossRef]

water

MDPI

Article

Operational Risk Assessment of Check Dams in Ningxia Considering the Impact of Extreme Precipitation in the Future

Yujie Yang, Shengdong Cheng *, Zongping Ren, Zhanbin Li and Lu Jia

State Key Laboratory of Ecological Water Conservancy in Northwest Arid Areas, Xi'an University of Technology, Xi'an 710048, China; 1220412028@stu.xaut.edu.cn (Y.Y.); renzongping@163.com (Z.R.); zhanbinli@162.com (Z.L.); xiamuxingyi94@163.com (L.J.)

* Correspondence: chengsd@xaut.edu.cn

Abstract: To analyze the operation risk of check dams under extreme precipitation conditions, taking Ningxia area as an example, this paper carried out a risk assessment of check dams under extreme precipitation conditions in Ningxia through data collection, hydrological statistics, numerical simulation, and other methods. The conclusions are the following: (1) By the end of 2020, about 40% of the silt reservoir capacity of check dams in various water and soil conservation zones in Ningxia has been accumulated. During 1966–2020, the extreme precipitation and frequency of extreme precipitation in Ningxia increased while the intensity of extreme precipitation decreased. The extreme precipitation in Ningxia increased year by year and lasted longer. (2) Under two future scenarios of RCP4.5 (the full name of RCP is Representative Concentration Pathway) and RCP8.5, the extreme precipitation threshold in Ningxia is gradually decreasing from south to north. Extreme precipitation in the future will bring high risk to the operation of check dams in Ningxia. The results of this paper can provide a scientific basis for the operation and management of check dams in Ningxia.

Keywords: extreme precipitation; Ningxia check dams; operational risk assessment; climate model

1. Introduction

As an effective soil and water conservation facility, check dams can effectively intercept and store floods, cement sand, and silt for farmland in the Loess Plateau region with severe soil erosion [1–4]. However, due to the fact that most of the existing medium-sized and above check dams were built in the 1960s and 1970s, their supporting flood discharge facilities are not complete, and the flood control standards are not high, which cannot meet the flood control requirements of current engineering construction. For example, in July 2017, the Wuding River Basin in Shaanxi Province experienced a once-in-a-century rainfall with a maximum daily rainfall of 218.4 mm, resulting in the destruction of 337 silt dams in Suide County by floods [5]. The various situations of precarious check dams pose a threat to the safe operation of the project to varying degrees. If the removal or reinforcement measures are not timely, not only will the main building of the project be damaged, but it will also endanger the life and property safety of downstream residents [6].

In recent years, with the increase in temperature and precipitation in the northwest region of China [7,8], the heat resources in the Ningxia Hui Autonomous Region located in the eastern part of the region have been ever more abundant. At the same time, the amount of evapotranspiration has also increased, resulting in an increase in extreme precipitation events in Ningxia, and the precipitation pattern has undergone subtle changes [9–11]. The non-stationary changes caused by extreme precipitation may cause a certain degree of damage to the infrastructure of water conservancy construction projects that have been in use for a long time, such as earth and rock dams and check dams, and even pose a risk of failure [12]. Therefore, predicting whether impending climate change will affect its reliability and other safety issues is urgent.

The future situation of extreme precipitation is no longer optimistic, but there are few scholars evaluating its impact on dam operation risk. Mallakpour et al. in the United States [13] used the calculation method of historical and predicted flood return periods to analyze the potential changes in the hydrological failure risk of major dams in California under the background of climate warming, and estimated the probability of the hydrological failure of dams. As for Chinese dams, some scholars [12] have calculated the degree to which extreme precipitation intensity exceeds the dam design intensity through non-stationary GP distribution, in order to analyze the potential impact level (R) of non-stationary changes in extreme precipitation on the operational risk of Chinese dams.

Since most of these studies have not mentioned the impact of future extreme precipitation on the operational risk of check dams, this article refers to the above research methods and ideas; that is, taking Ningxia check dams as the research object, using the CMIP5 model to simulate future extreme precipitation data, using the non-stationary GP distribution to analyze the future changes of extreme precipitation in Ningxia under different climate scenarios, and evaluating the operational risks of Ningxia check dams under extreme precipitation conditions and providing valuable suggestions for the management and maintenance of check dams in the Ningxia region in the future. I hope to be able to provide assistance in the operation and management of check dams in Ningxia, to ensure their safety and functionality, and to cope with future climate change.

2. Materials and Methods

2.1. Study Area

The Ningxia Hui Autonomous Region (hereinafter referred to as Ningxia) is located in the middle and upper reaches of the Yellow River in northwest China, covering a total area of 66,400 km^2 with a distance of 456 km from north to south and about 250 km from east to west. Situated at the junction of the Loess Plateau and the Inner Mongolian Plateau, the average altitude of the whole territory is generally more than 1000 m, and the overall terrain is high in the south and low in the north. The average annual precipitation in the entire region ranges from 150 mm to 600 mm, with an average annual precipitation of around 300 mm. The geographical location and elevation of Ningxia are presented in Figure 1.

The soil erosion in Ningxia roughly includes two types: hydraulic erosion and wind erosion. As of the end of 2020, the total area of soil erosion in the entire region was 15,687.4 km^2, accounting for 23.6% of the total land area of the region, and a decrease of 201.1 km^2 compared to 2019 and a decrease of 1.3%. Among them, the hydraulic erosion area is 10,680.9 km^2, accounting for 68.1% of the soil erosion area, and the wind erosion area is 5006.6 km^2, accounting for 31.9% of the soil erosion area. According to the functional zoning of soil and water conservation, Ningxia is divided into 7 functional zones for soil and water conservation, namely Zone I—Key Prevention Area for Helan Mountain and Water Erosion, Zone II—Potential Key Prevention Area for Wind Erosion in Yinchuan Plain, Zone III—Key Control Area for Wind and Water Erosion in Arid Grassland of Hilly Plateau, Zone IV—Key Prevention Area for Liupan Mountain and Water Erosion, Zone V—Key Control Area for Water Erosion in Loess Hilly Gully and Residual Plateau, Zone VI—Key Control Area for Water Erosion in Loess Hilly and Gully, and Zone VII—Key Management Area for Water Erosion Intersection in Huangtu Hilly Gully and Phoenix (Figure 2). In order to develop more scientific measures for the operation and management of silt dams in various regions, the precipitation and silt dam situation will be analyzed in the context of soil and water conservation zoning in Ningxia.

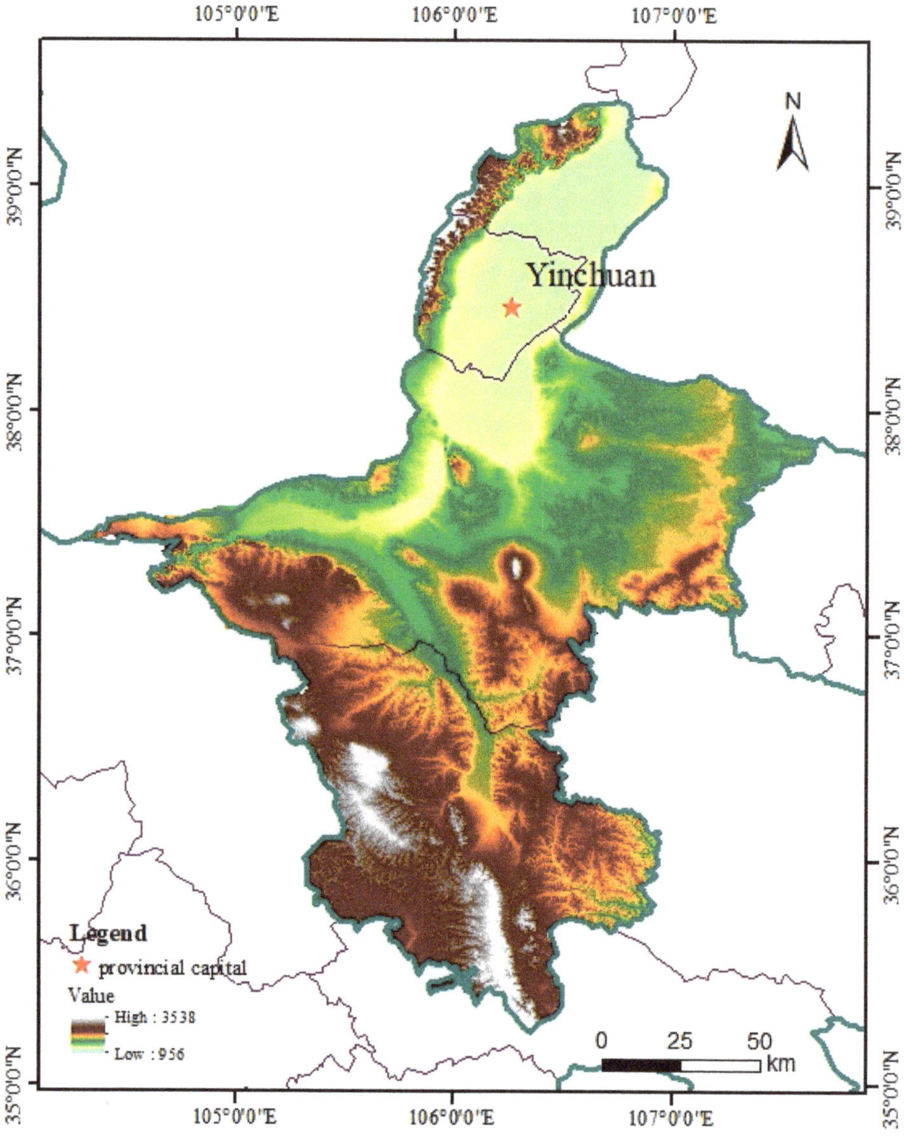

Figure 1. Geographical location and elevation of Ningxia.

2.2. Data Sources

The diverse data information of check dams in the Ningxia region mainly comes from the latest survey and statistics of check dams in Ningxia (2020), organized and compiled by the Soil and Water Conservation Monitoring Station of Ningxia Hui Autonomous Region.

The daily precipitation data of 20 stations in Ningxia from 1966 to 2020 are sourced from the China Meteorological Science Data Sharing Service Network (http://cdc.cma.gov.cn/) National meteorological station data provided. For future precipitation, accessed on 5 June 2020 to obtain the CMIP5 model data from https://esgf-node.llnl.gov/projects/esgf-llnl/, and the YSU-RSM climate model with good applicability to the Ningxia region is selected [14]. The content covers historical simulations from 1980 to 2005 and daily precipitation data under RCP4.5 (the full name of RCP is Representative Concentration Pathway, i.e., by 2100, greenhouse gas concentration corresponds to a radiation forcing of 4.5 W/m^2) and RCP8.5 (i.e., by 2100, greenhouse gas concentration corresponds to a radiation forcing of 8.5 W/m^2) scenarios from 2006 to 2050.

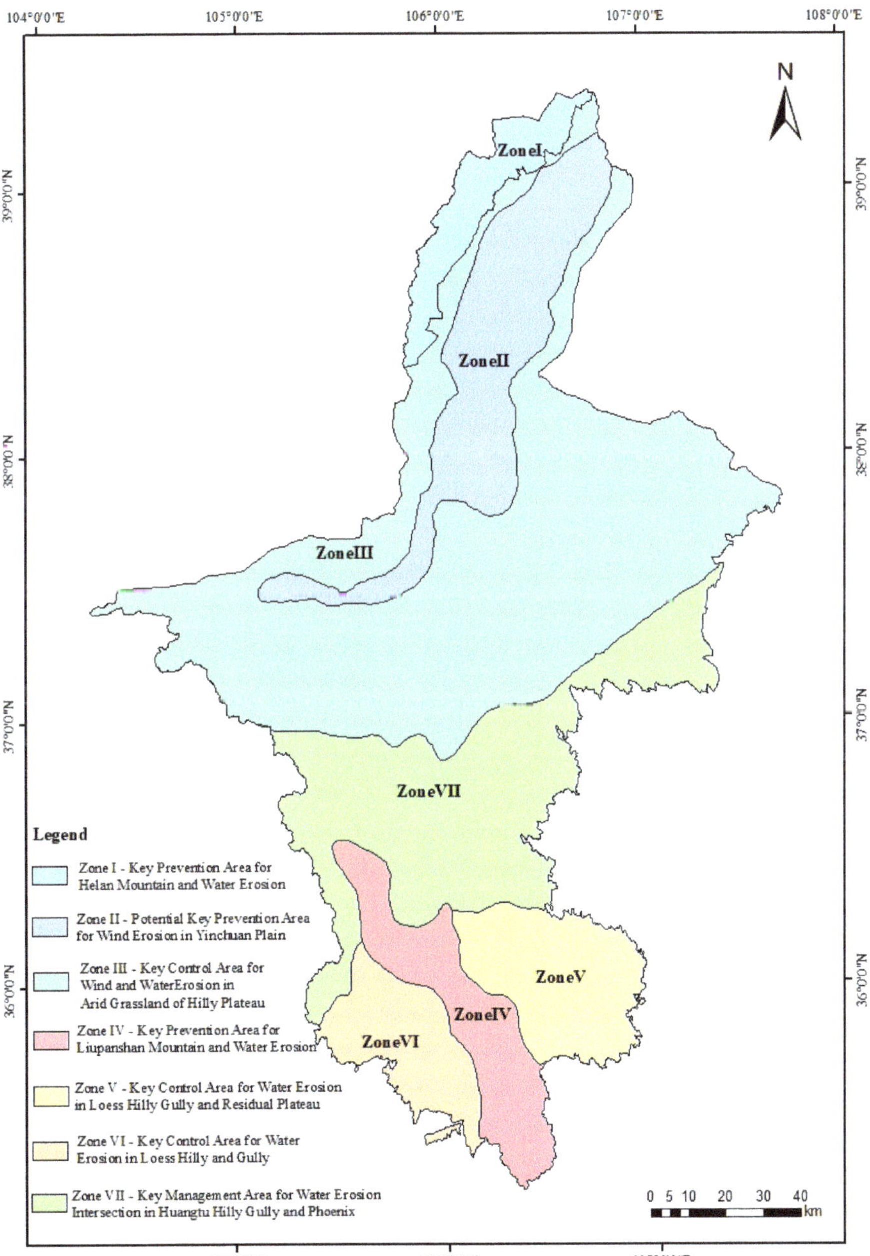

Figure 2. Functional zoning map of soil and water conservation in Ningxia (including county and district names).

2.3. Analysis Methods

2.3.1. Analysis Methods for Extreme Precipitation Events

1. Determination of extreme precipitation events and extreme precipitation indicators

This article defines extreme precipitation using 95% percentile values [15,16]. Consider adopting the following three indicators to describe the basic characteristics of extreme precipitation, and the definitions of each indicator are shown in Table 1.

Table 1. Extreme precipitation indicators and their definitions.

Serial Number	Symbol	Indicator Name	Define	Unit
1	P	Extreme precipitation	The total annual precipitation of extreme precipitation events	mm
2	F	Extreme precipitation frequency	The number of days that extreme precipitation events occur each year	d
3	I	Extreme precipitation intensity	The ratio of extreme precipitation to extreme precipitation days	mm/d

2. Trend analysis of changes

Use the linear regression method to analyze the trend of extreme precipitation events in the Ningxia region. The linear regression equation is shown in Equation (1) [17]:

$$\hat{y} = \hat{a} + \hat{b}t \tag{1}$$

In the formula, $\hat{y}$ is the extreme precipitation indicators; $\hat{a}$ is the regression intercept; and the regression coefficient $\hat{b}$ represents the variable $\hat{y}$ trend inclination. When $\hat{b}$ is positive, it indicates that as time t increases, $\hat{y}$ shows an increasing trend; when $\hat{b}$ is negative, it indicates that as time t increases, $\hat{y}$ shows a decreasing trend; and t is the time range.

3. Deviation correction of pattern data

This article adopts the Delta bias correction method to correct the bias of pattern data [18,19]. The formula is shown in Equation (2):

$$P_{station,f,daily} = P_{GCM,f,daily} \cdot \frac{\overline{P_{station,h,monthly}}}{\overline{P_{GCM,h,monthly}}} \tag{2}$$

In the formula, $P_{station,f,daily}$ are the daily precipitation data of GCM under future scenarios after down scaling; $P_{GCM,f,daily}$ are the original daily precipitation data of GCM under future scenarios; $\overline{P_{station,h,monthly}}$ is the annual monthly average precipitation of meteorological stations in historical periods; and $\overline{P_{GCM,h,monthly}}$ is the original monthly average precipitation of GCM during the historical period.

2.3.2. Indicators and Calculation Methods for Evaluating Operational Risks

1. Evaluation indicators

This article uses the potential impact level R to conduct an operational risk assessment on check dams in the Ningxia region. The definition of this evaluation index is shown in Equation (3) [12]:

$$R(t) = \frac{N(t) - S}{S} \times 100\% \tag{3}$$

In the formula, $R(t)$ is the potential impact level, %; $N(t)$ is the precipitation intensity at time t calculated from the local non-stationary GP distribution, mm; and S is the design strength of the dam body calculated from the locally stable GP distribution, mm.

From Equation (3), we can see that compared to the design intensity, the greater the extreme precipitation intensity in the future, the greater the impact on the check dam.

Therefore, the potential impact level $R(t)$ can more intuitively represent the possible influence of extreme precipitation changes on the operation process of check dams.

The critical value of the potential impact level $R(t)$ can be determined based on the anti-skid safety factor of the check dam [12]. Considering that the storage capacity of check dams will show significant siltation over time, the design standards for various types of check dams should be adjusted in future scenarios. The diagnostic values of the potential impact levels $R(t)$ for different types of check dams are listed in Table 2.

Table 2. List of critical values for potential impact levels *R* (*t*) of different types of check dams.

Type of Check Dam	Original Design Standard (Return Period T_{R0}/a)	The Anti-Skid Safety Factor for Normal Use	Adjusted Design Standards (Return Period T_R/a)	*R*(*t*) Critical Value (%)
Key dam	20–30	1.25	10–20	20
Medium sized dam	20–10	1.20	10–20	20
Small dam	10–20	1.20	5–10	15

2. Method of calculation

- Stable GP distribution

For the calculation of the design strength *S* of the dam body, it can be seen from Equation (3) that the locally stable GP distribution is used to analyze the design strength of different types of check dams near numerous stations in Ningxia. Simulate future extreme precipitation from 2016 to 2050 using two scenarios in the YSU-RSM climate model for each site. Based on the adjusted design standards for different types of check dams, calculate the corresponding design values at the design frequency, which will serve as the design strength of the check dams. Due to the use of the ultra-quantitative method [18] for extreme precipitation data, the relationship between the design frequency and the recurrence period of the ultra-quantitative method is shown in Equation (4):

$$T_R = \frac{1}{\bar{k} \cdot p} \tag{4}$$

In the formula, T_R is the return period of the ultra-quantitative method, a; $\bar{k}$ means the average number of selected data per year for the ultra-quantitative method; and p is the design frequency, %.

- Non-stationary GP distribution

According to Equation (3), the upcoming extreme precipitation intensity $N(t)$ of the different types of check dams near each station can be calculated using local non-stationary GP distribution. Similar to the method of calculating the design strength S of the dam body, it is assumed that the model simulates the extreme precipitation from 2016 to 2050, which follows a fixed parameter GP distribution every five years. The design values of different types of check dams at corresponding design frequencies are analyzed every five years, and this is used as the extreme precipitation intensity of the check dams during these five years.

3. Results and Analysis

3.1. Spatial Characteristics of the Distribution of Check Dams in the Ningxia Region

3.1.1. Spatial Distribution Characteristics of Check Dams

As of the end of 2020, there were a total of 1119 check dams built in Ningxia, of which 324 were backbone dams with a storage capacity of over 500,000 square meters, accounting for 29.0% of the total number of check dams. There are 369 medium-sized check dams with a storage capacity between 100,000 and 500,000 square meters, accounting for 33.0% of the total number of check dams. There are 426 small check dams with a storage capacity between 10,000 and 100,000 square meters, accounting for 38.1% of the total number of check dams. The distribution of check dams in Ningxia is illustrated in Figure 3. It can be seen that check dams are mainly distributed in Zone III (Yanchi County, Shapotou District, and Lingwu City), Zone V (Yuanzhou District and Pengyang County), Zone VI (Xiji County and Longde County), and Zone VII (Tongxin County and Haiyuan County).

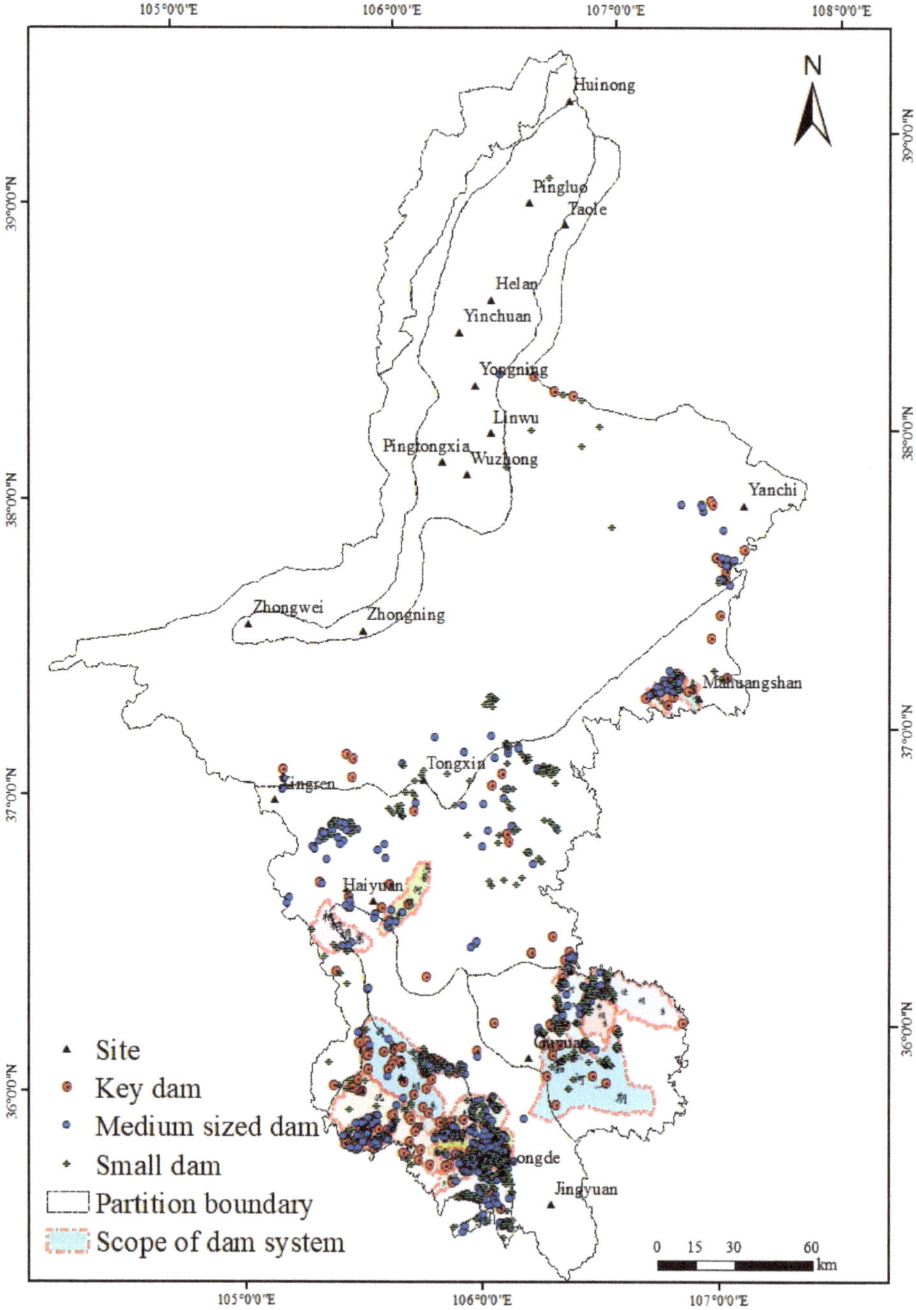

Figure 3. Distribution map of check dams in Ningxia.

The distribution of backbone dams, medium-sized check dams, and small check dams in various cities and counties (districts) is shown in Figure 4, which shows the number of check dams constructed according to the classification of soil and water conservation zones. Overall, the maximum number of check dams in Zone VI is 448, followed by 319 in Zone V and 187 in Zone VII, while the minimum number of check dams in Zone III is 156. It can be inferred that the key area for the construction of check dams in Ningxia is the loess hill and gully areas.

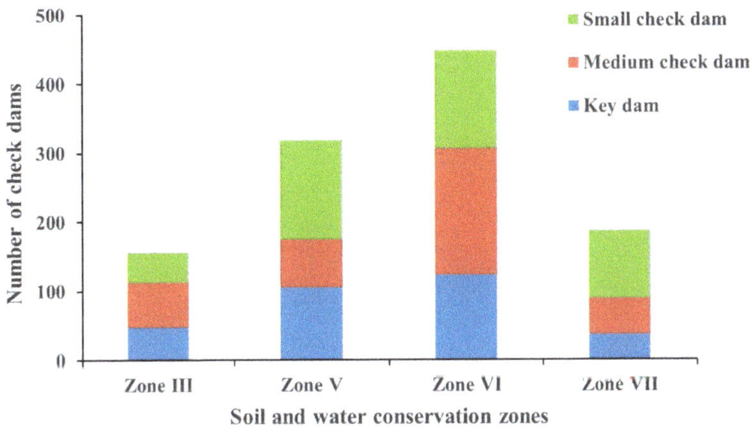

Figure 4. Statistics of the number of check dams constructed in different soil and water conservation zones in Ningxia.

3.1.2. Current Situation of Sedimentation in Check Dams

Analyzing the siltation situation of the silt dam is beneficial for understanding the operational status of the silt dam, providing technical support and a decision-making basis for the planning and design, risk prevention and control, operation management, and benefit evaluation of the silt dam. The total control area of backbone dams and medium-sized check dams in Ningxia is 4561.0 km^2, with a designed total storage capacity of 412.245 million m^3, including a flood control storage capacity of 243.078 million m^3, siltation storage capacity of 169.167 million m^3, siltation storage capacity of 63.264 million m^3, and remaining siltation storage capacity of 105.903 million m^3. In the current situation of siltation in countless districts of Ningxia (Figure 5), the proportion of the control area in Zone III is relatively large. However, for aggregate storage capacity, flood control storage capacity, siltation storage capacity, and already silted storage capacity, the proportion of Zones III and VI is relatively large. Overall, the silted storage capacity of the check dams in each zone accounts for about 40% of the silted storage capacity, while the remaining silted storage capacity accounts for about 60%.

3.2. Analysis of Extreme Precipitation Trends in the Ningxia Region

3.2.1. Analysis of Historical Precipitation Data

1. Threshold distribution

The distribution map of extreme precipitation thresholds during the historical period of Ningxia (Figure 6) shows that the areas with higher thresholds are mainly located in the southeast of Zone IV and Zone V, while the smaller areas are in the west of Zone II and Zone III. The general distribution trend is slightly decreasing from southeast to northwest. In Zone IV and V, the extreme precipitation thresholds for Jingyuan and Guyuan are 23.2 mm and 19.9 mm, respectively, while in Qingtongxia in Zone II, Zhongwei in Zone III, and Xingren in Zone VII, they are all below 15.29 mm. This spatial distribution is basically consistent with the distribution of the average annual precipitation in the climate (Figure 7).

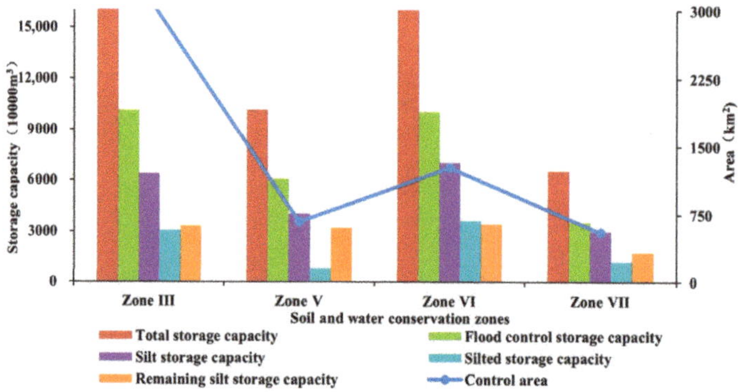

Figure 5. Statistical chart of the current siltation status indicators of siltation dams in various districts of Ningxia.

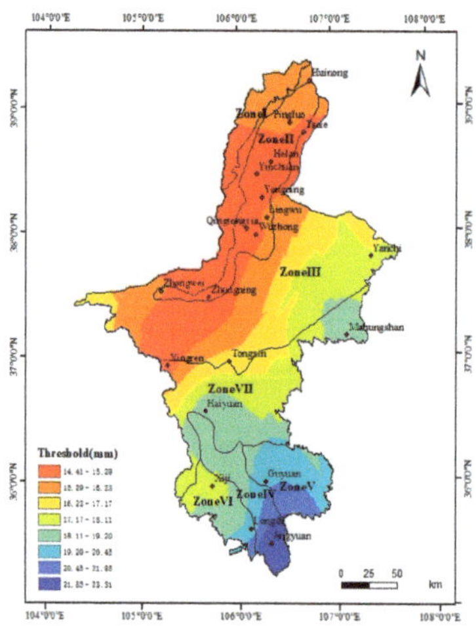

Figure 6. Extreme precipitation threshold distribution map.

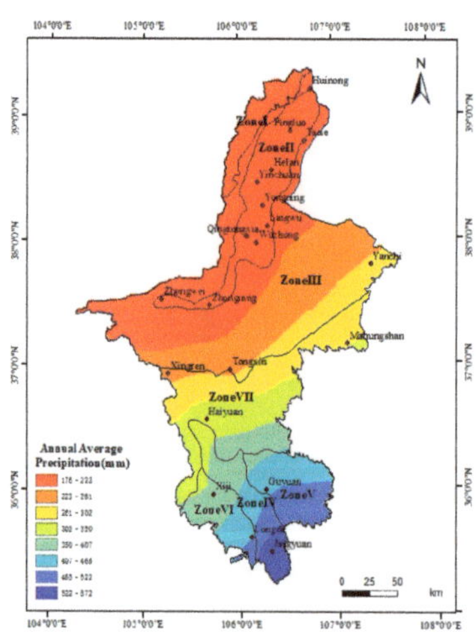

Figure 7. Distribution map of annual average precipitation in Ningxia from 1966 to 2020.

2. The spatial distribution of extreme precipitation index trends

It can be observed in Figure 8a that the extreme precipitation in most regions of Ningxia from 1966 to 2020 was mainly increasing, with a change range of −0.32 to 0.85 mm/a. Among them, Zhongning in Area II, Jingyuan in Area IV, Xiji in Area VI, Ma Mount Huangshan in Area VII, and other regions showed a slightly decreasing trend, with a relatively large decrease range of 0.32 mm/a in Zhongning, and an increasing trend in other regions. Pingluo in Area II was the largest, reaching 0.85 mm/a. According to the approximate distribution in the graph, except for the VI zone where the extreme precipitation from 1966 to 2020 showed a decreasing trend, all other zones had stations with an increasing tendency.

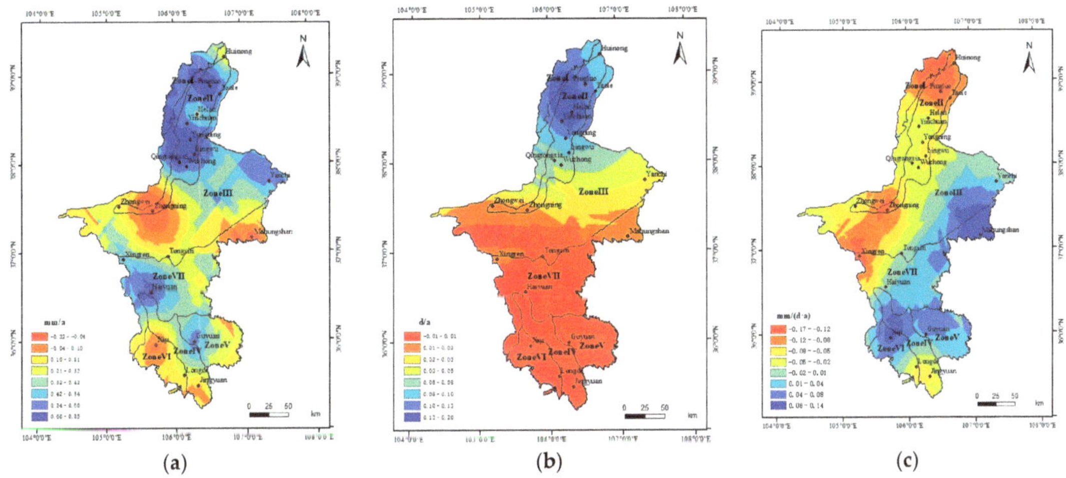

Figure 8. Distribution map of extreme precipitation index tendency in Ningxia from 1966 to 2020: (**a**) extreme precipitation; (**b**) extreme precipitation frequency; (**c**) extreme precipitation intensity.

The distribution of the extreme precipitation frequency tendency rate (Figure 8b) is approximately the same as the distribution of the extreme precipitation tendency rate. Except for Zhongning in Zone II, Xiji in Zone VI, and Ma Mount Huangshan in Zone VII, the distribution of the extreme precipitation frequency tendency rate shows an increasing trend in most other regions, with a trend rate of −0.01~0.20 d/a. Helan in Zone II has the most obvious increasing trend, with an increase rate of 0.20 d/a.

From Figure 8c, we can get the tendency rate of extreme precipitation intensity in Ningxia. Except for Zhongning and Helan in Area II, and Ma Mount Huangshan in Area VII, the trend is increasing, and most other areas are decreasing, in which Pingluo in Area II decreases by 0.17 mm/(d·a), indicating that the extreme precipitation and the frequency of extreme precipitation events in most water and soil conservation areas in Ningxia are increasing from 1966 to 2020, but the extreme precipitation intensity is decreasing.

3.2.2. Prediction of Future Change Trends

1. Result analysis under RCP4.5 scenario
    - Threshold distribution

As shown in Figure 9, the extreme precipitation threshold of each partition station exceeds 19.46 mm, and the distribution range is consistent with the historical extreme precipitation threshold, which gradually increases from the northwest of Ningxia to the southeast; the central guard in Zone II is the smallest, 19.46 mm, but larger than its historical analysis of 14.56 mm, and the largest Jingyuan in Zone IV is 38.42 mm, which is also greater than its historical analysis of 23.2 mm. This indicates that under the RCP4.5 scenario, the threshold for extreme precipitation in diverse regions of Ningxia is increasing.

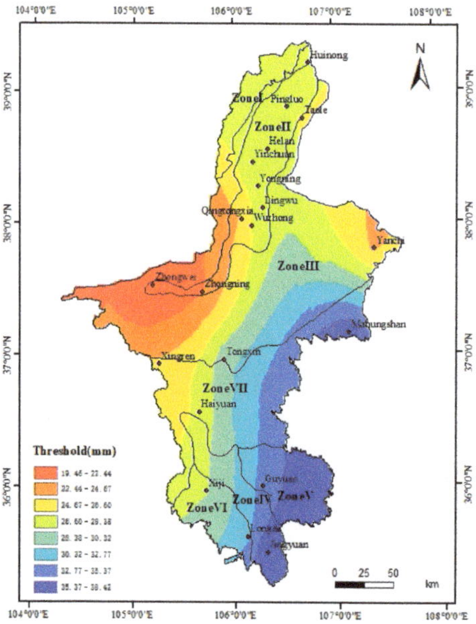

Figure 9. Distribution of extreme precipitation threshold under RCP4.5 scenario.

- The spatial distribution of extreme precipitation index trends

From 2006 to 2050, the utmost precipitation in most areas of Ningxia showed an increasing trend (see Figure 10a), with a range of −1.18 to 5.55 mm/a. Among them, the stations in Zone II showed a decreasing trend, with Wuzhong showing a relatively larger decrease of 1.18 mm/a, while the other zones showed an increasing trend. Jingyuan in Zone IV had the largest increase of 5.55 mm/a. From the rough distribution in the figure, the tendency rate of extreme precipitation under the RCP4.5 scenario also shows a gradually increasing distribution from north to south.

The extreme precipitation frequency trend of most stations in the region is positive (Figure 10b), ranging from −0.020 to 0.117 d/a. Among them, Helan, Yongning, Lingwu, and Wuzhong in Zone II have negative values, while the trend rates of additional stations are greater than 0. The inclination rate in the central and northern regions (from Zone I to Zone III) is relatively low, and except for the larger values at stations such as Yinchuan, Zhongwei, and Zhongning in Zone II, the rest are generally less than 0.036 d/a. The frequency tendency of extreme precipitation in the southern region (from Zone IV to Zone VII) is generally greater than 0.036 d/a. In the RCP4.5 scenario, there is a slight increase in the frequency of extreme precipitation at each station, and it gradually increases from north to south.

The trend rate of extreme precipitation intensity at each partition site ranges from −0.22 to 0.74 mm/(d·a) (Figure 10c). Among them, the trend rates of Wuzhong and Zhongning in Zone II, as well as Yanchi and Tongxin in Zone III, are relatively small, ranging from −0.22 to −0.10 mm/(d·a), showing a decreasing trend. However, Xingren in Zone VII is relatively enormous, at 0.74 mm/(d·a). Therefore, it can be concluded that the trend rates of extreme precipitation intensity in each partition of Ningxia are relatively muted, In the RCP4.5 scenario, there is a gradually increasing distribution trend from east to west.

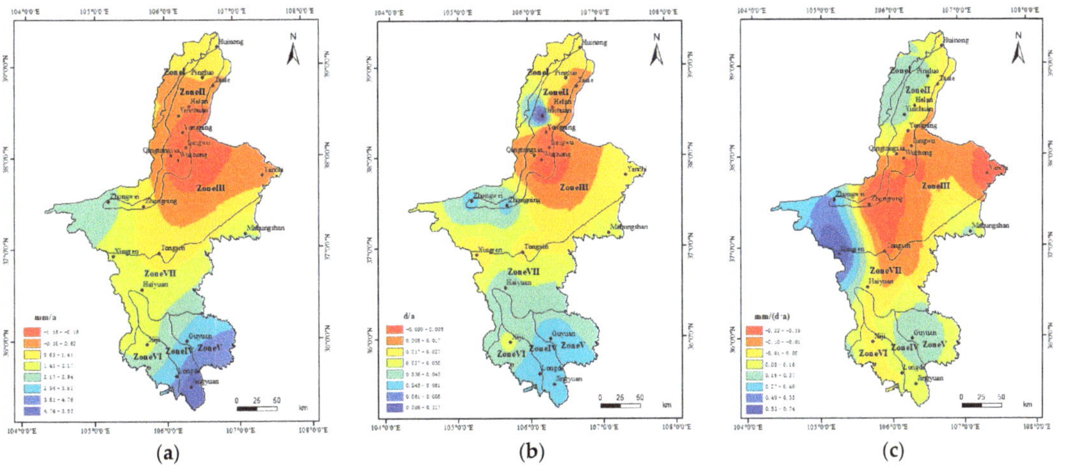

Figure 10. Distribution map of extreme precipitation index tendency in Ningxia from 2006 to 2050 under RCP4.5 scenario: (**a**) extreme precipitation; (**b**) extreme precipitation frequency; (**c**) extreme precipitation intensity.

2. Result analysis under RCP8.5 scenario
    - Threshold distribution

As shown in Figure 11, it can also be seen that the threshold values of each partition exceed the maximum historical threshold value of 19.46 mm, and the distribution situation is basically the same as the historical threshold, which increases gradually from northwest to southeast. Therefore, the extreme precipitation thresholds of various regions in Ningxia gradually increase under the RCP8.5 scenario.

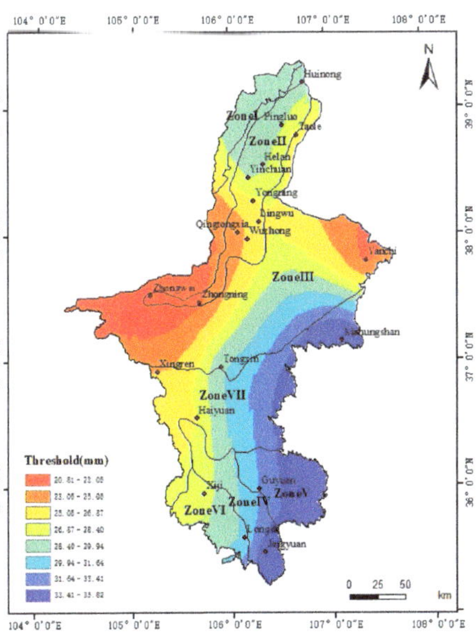

Figure 11. Distribution of extreme precipitation threshold under RCP8.5 scenario.

- The spatial distribution of extreme precipitation index trends

The tendency rate of extreme precipitation at each station in each partition is assured (Figure 12a), ranging from 0.31 to 4.85 mm/a. Among them, Zhongwei and Zhongning in Zone II, Huinong in Zone III, and Xingren in Zone VII are smaller, ranging from 0.31 to 1.00 mm/a, while Jingyuan in Zone IV and Haiyuan in Zone VII are larger, ranging from 4.00 to 4.85 mm/a.

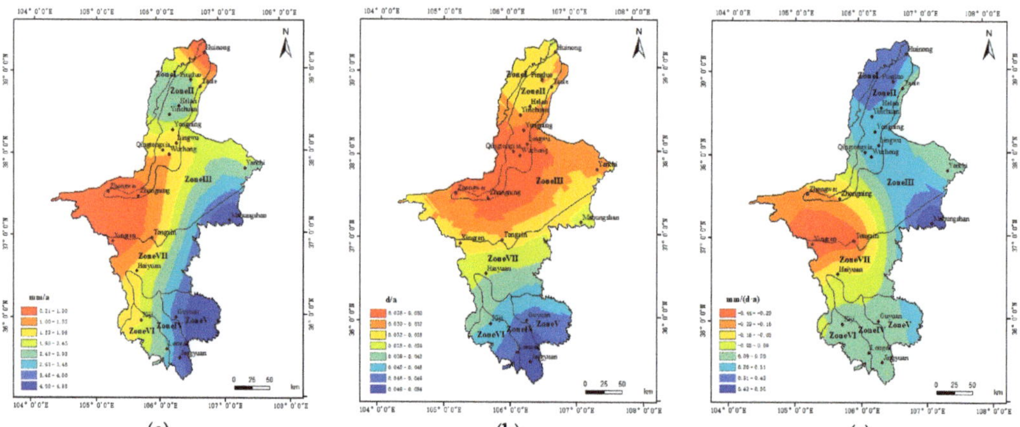

(a) (b) (c)

Figure 12. Distribution map of extreme precipitation index tendency in Ningxia from 2006 to 2050 under RCP8.5 scenario: (**a**) extreme precipitation; (**b**) extreme precipitation frequency; (**c**) extreme precipitation intensity.

The tendency rate of extreme precipitation frequency is also positive (Figure 12b), with stations south of Yongning in Zone II having a smaller tendency rate, ranging from 0.026 to 0.030 d/a, while Jingyuan in Zone IV and Longde in Zone VI have a larger tendency rate, ranging from 0.049 to 0.054 d/a. Under the RCP8.5 scenario, the frequency of extreme precipitation at various sub stations in Ningxia also shows an increasing trend.

The variation range of the extreme precipitation intensity tendency rate at each station is −0.44~0.55 mm/(d·a) (Figure 12c), in which Tongxin in Area III and Xingren in Area VII are smaller, between −0.44 and ~−0.29 mm/(d·a), while Ma Mount Huangshan in Area VII is the largest, at 0.55 mm/(d·a). Therefore, the extreme precipitation intensity changes in discrete zones under the RCP8.5 scenario are not entirely the same, showing a gradually increasing trend from west to east.

3.3. Risk Assessment of Check Dam Operation under Extreme Precipitation

Based on the evaluation index of the probable impact level $R(t)$ in Equation (4), analyze the operational risks of different types of siltation dams in different regions of Ningxia from 2016 to 2050 under different scenarios of RCP4.5 and RCP8.5. The latitudinal distribution map of the check dam is shown in Figure 3.

3.3.1. Evaluation Results under RCP4.5 Scenario

As shown in Figure 13, due to the lack of the construction of check dams in the northern part of Ningxia (the northern part of Zone I, II, and III), the $R(t)$ value in this area remains around 0, while it far exceeds 0 from 2026 to 2030. It is not excluded that there may be deviations in the selected evaluation methods or model simulation data. It is worth noting that the $R(t)$ near the inner guard in Zone III has been consistently above 50%, far exceeding 20%. From 2016 to 2030, the $R(t)$ in the southern part of Ningxia (from Zone IV to Zone VII) moderately increased, reaching a maximum of 35% by 2030 and gradually decreasing thereafter, but also exceeding 20%. In addition, considering that the extreme precipitation intensity in the central region of Ningxia is slightly higher than that in the southern region, from 2016 to 2050, the $R(t)$ value in the central region of Ningxia is generally higher than that in the southern region. In short, in the RCP4.5 scenario, there are certain operational risks for different types of check dams in various zones south of Zhongwei in Ningxia.

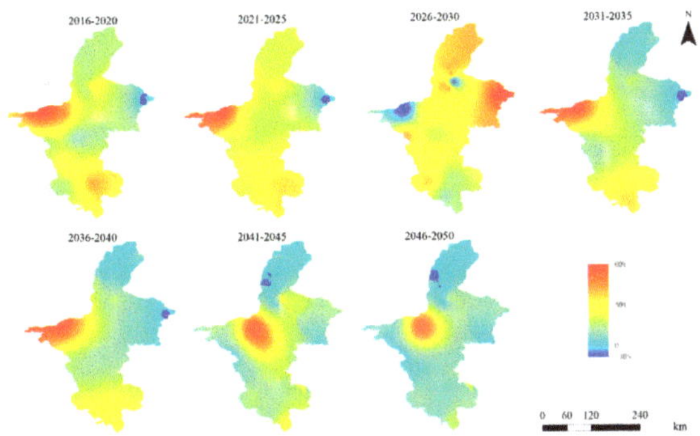

Figure 13. Changes in $R(t)$ value of potential impact level of check dams in Ningxia region under RCP4.5 scenario.

3.3.2. Evaluation Results under RCP8.5 Scenario

From Figure 14, it can be seen that the northern part of Ningxia (the northern part of Zone I, II, and III) shows a value of around 0, while the risk of the principal region varies greatly at different ages. This may also be due to the weaker applicability of the evaluation method in the fundamental region compared to the northern and southern regions. From 2021 to 2050, the $R(t)$ of most areas in the southern region (from Zone IV to Zone VI) has been almost consistently above 50%. Since 2031, the $R(t)$ of the southern and seventh regions of Zone III has also been between 10 and 20%, indicating that under the RCP8.5 scenario, the operation of different types of check dams in southern Ningxia, especially from Zone IV to Zone VI, will be affected. Overall, the operational risk of silt dams in the Ningxia region is higher under the RCP8.5 scenario.

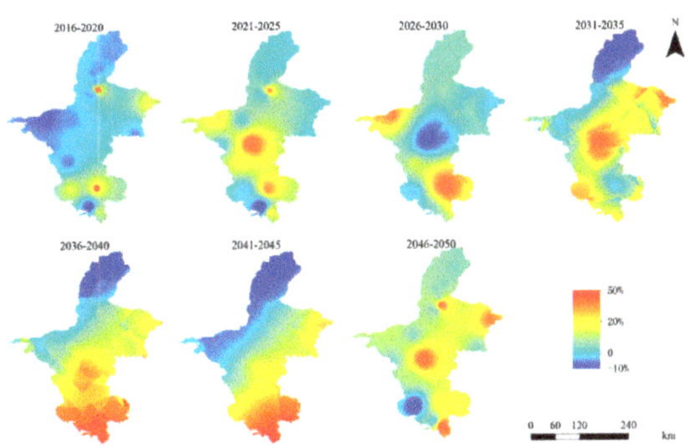

Figure 14. Changes in $R(t)$ value of potential impact level of check dams in Ningxia region under RCP8.5 scenario.

4. Discussion

4.1. Analysis of Historical Extreme Precipitation Trends

The extreme precipitation threshold in Ningxia from 1966 to 2020, determined using the percentile method, mainly shows a marginally decreasing distribution from southeast to northwest. Zhang, M.J. et al. [20] also reached similar conclusions that the annual precipitation in Ningxia from 1951 to 2008 increased from north to south. Scholars such as Chen, X.G. et al. [21] have shown that during the 45 years from 1961 to 2005, the number of days with precipitation above 25.0 mm in the summer and autumn seasons in Ningxia increased, and the frequency distribution of precipitation showed a clear trend towards a higher level of precipitation, which is consistent with the analysis results in the article.

In addition, the distribution of historical extreme precipitation thresholds in Ningxia is very close to the distribution of average annual climate precipitation, which is consistent with the study by Li, F. et al. [22]. In the view of Yang, Y. et al. [23], climate and terrain may be important factors affecting the north–south differences in extreme precipitation events, or they may be influenced by particularly significant human activities.

From the research of Fan, K. [24] and Xin, Z.B. et al. [25], it can be concluded that Ningxia belongs to an arid and semi-arid region. Once a long period of heavy rainfall occurs, it means that the surface runoff generated by natural precipitation increases or flood disasters intensify, which easily leads to urban waterlogging and soil erosion in the region. Therefore, analyzing the trend of extreme precipitation changes in the region and implementing corresponding protection and management measures can significantly reduce the losses caused by natural catastrophes.

4.2. Operational Risk Assessment of Check Dams

Ma, L. et al. [26,27] found that under the condition of a rainstorm, the blocking dam has an efficient sediment-retaining effect, but the dam with a small reservoir capacity very easily has a large siltation depth, which poses a threat to the dam body, indicating that extreme precipitation will cause dam break danger to the warping dam. Zu, Q. et al. [28] also pointed out that different types of check dams have altered susceptibility to burst debris flows caused by precipitation of different durations and frequencies. Overall, the longer the precipitation duration, the lower the precipitation frequency, and the higher the reservoir capacity, the higher the susceptibility level of debris flows formed by check dams, and the higher the risk of dam failure. Therefore, when extreme precipitation occurs, various types of check dams are given the opportunity to collapse. These studies [29] further confirm the rationality of the operative risk assessment results of check dams under extreme precipitation conditions in this paper.

At the same time, Gao, Y. et al. [30] and others pointed out that the extreme value of rainstorm is one of the important bases for the design flood of large-scale water conservancy projects. China's Code for Design Flood Calculation of Water Resources and Hydropower Projects (SL44-2006) emphasizes that attention should be paid to the collection of extremely heavy rainstorm data in the region and adjacent areas, and the design results should be compared with the rainstorm records to check whether the design data are safe and reliable. For certain areas lacking flow data, the method of calculating design floods based on rainfall can also provide a basis for design. Therefore, the study of the rainstorm extreme value is of great significance for the planning, design, and safe operation of water conservancy projects.

5. Conclusions

(1) As of the end of 2020, there were a total of 1119 check dams in Ningxia, including 324 backbone dams, 369 medium-sized check dams, and 426 small check dams. These completed check dams are mainly located in the soil and water conservation functional zones (Zone III, V, VI, and VII) of the Loess Hilly Gully. The silted storage capacity of each zone's check dams accounts for about 40% of the silted storage capacity, and the siltation situation is relatively serious.

(2) During the period of 1966 to 2020, the threshold for extreme precipitation in the entire Ningxia region gradually decreased from southeast to northwest, with an increase in extreme precipitation of about 0.27 mm/a, an increase in extreme precipitation frequency of about 0.10 d/a, and a decrease in extreme precipitation intensity of about −0.02 mm/(d·a). The extreme precipitation in the entire region increased year by year and its duration was prolonged.

(3) Under the two future scenarios of RCP4.5 and RCP8.5, the distribution of extreme precipitation thresholds in Ningxia is basically the same as in bygone periods. In the RCP4.5 scenario, the extreme precipitation increases approximately by 2.19 mm/a, the frequency of extreme precipitation increases roughly by 0.05 d/a, and the intensity of extreme precipitation increases approximately by 0.26 mm/(d·a). In the RCP8.5 scenario, the increase in extreme precipitation is slightly greater than in RCP4.5, while the increase in extreme precipitation frequency and intensity is slightly smaller than in RCP4.5.

(4) In the RCP4.5 scenario, the check dams south of Zhongwei in Zone III and Zone VII will be significantly affected ($R(t)$ much greater than 20%); in the RCP8.5 scenario, there is a certain degree of risk ($R(t)$ of over 20%) in the operation of check dams from Zone IV to Zone VI. Therefore, in different future scenarios, the potential impact level $R(t)$ of the siltation dams in various regions of Ningxia is much higher than the critical value of 20% (15%), and extreme precipitation will bring high risks to the siltation dams in the entire region.

Author Contributions: Conceptualization, Y.Y. and S.C.; methodology, Z.R.; software, Y.Y.; validation, Y.Y., S.C. and Z.L.; formal analysis, L.J.; investigation, S.C.; resources, Z.R.; data curation, Y.Y.; writing—original draft preparation, Y.Y.; writing—review and editing, S.C.; visualization, Y.Y.; supervision, Z.R.; project administration, S.C.; funding acquisition, Z.L. All authors have read and agreed to the published version of the manuscript.

Funding: The study was financially supported by the National Key Research and Development Program of China (Grant No. 2022YFF1300803) and the National Natural Science Foundations of China (No.52179043).

Data Availability Statement: Data will be made available upon request.

Conflicts of Interest: The authors declare that they have no known competing financial interests or personal relationships that could have appeared to influence the work reported in this paper.

References

1. Xu, X.Z.; Zhang, H.W.; Zhang, O.Y. Development of check-dam systems in gullies on the Loess Plateau, China. *Environ. Sci. Policy* **2004**, *7*, 79–86. [CrossRef]
2. Chen, Z.Y.; Li, Z.B.; Wang, Z.Y. Some thoughts on the strategic positioning of check dam construction in the Loess Plateau. *Soil. Water. Conserv. China* **2020**, *9*, 32–38. [CrossRef]
3. Ministry of Water Resources of the People's Republic of China. *Zhongguo Heliu Nisha Gongbao*; China Water & Power Press: Beijing, China, 2020.
4. Gao, H.D.; Jian, L.L.; Li, Z.B.; Xu, G.C.; Zhao, B.H. Mechanism underlying impact of check damson runoff based graph theory in the hilly-gully loess region. *Sci. Soil Water Conserv.* **2015**, *13*, 1–8. [CrossRef]
5. Yu, T.; Li, Z.B.; Chen, Y.T.; Yuan, S.L.; Wang, W. Analysis of structural characteristics of typical check dam system in the third subregion of loess hilly region. *Res. Soil Water Conserv.* **2019**, *26*, 26–30. [CrossRef]
6. Wei, X.; Li, Z.B.; Wu, J.H.; Li, B.B.; Du, Z. A Discussion on some deological problems in research of water damage hazards of check dam. *Res. Soil Water Conserv.* **2007**, *14*, 235–237.
7. Zhang, Q.; Yang, J.H.; Wang, W.; Ma, P.L.; Lu, G.Y.; Liu, X.Y.; Yu, H.P.; Fang, F. Climatic Warming and Humidification in the Arid Region of Northwest China: Multi-Scale Characteristics and Impacts on Ecological Vegetation. *J. Meteorol. Res.* **2021**, *35*, 113–127. [CrossRef]
8. Zhang, Q.; Yang, J.H.; Duan, X.Y.; Ma, P.L.; Lu, G.Y.; Zhu, B.; Liu, X.Y.; Yue, P.; Wang, Y.H.; Liu, W.P. The eastward expansion of the climate humidification trend in northwest China and the synergistic influences on the circulation mechanism. *Clim. Dyn.* **2022**, *59*, 2481–2497. [CrossRef]
9. Ma, M.Y.; Wang, Q.M.; Liu, R.; Zhao, Y.; Zhang, D.Q. Effects of climate change and human activities on vegetation coverage change in northern China considering extreme climate and time-lag and -accumulation effects. *Sci. Total Environ.* **2023**, *860*, 160527–160540. [CrossRef]
10. Yang, S.P.; Zhao, G.P.; Mu, J.H.; Su, Z.S.; Ma, L.W.; Chen, X.J. Extreme Climatic Events and Its Effect in Ningxia. *J. Desert Res.* **2008**, *28*, 1169–1173.
11. Song, M.W.; Jiang, X.H.; Lei, Y.X.; Zhao, Y.R.; Cai, W.J. Spatial and temporal variation characteristics of extreme hydrometeorological events in the Yellow River Basin and their effects on vegetation. *Nat. Hazards* **2023**, *116*, 1863–1878. [CrossRef]
12. Huang, H.F.; Cui, H.J.; Ge, Q.S. Will a nonstationary change in extreme precipitation affect dam security in China. *J. Hydrol.* **2021**, *603*, 126859–126870. [CrossRef]

13. Iman, M.; Amir, A.K.; Mojtaba, S. Climate-Induced Changes in the Risk of Hydrological Failure of Major Dams in California. *Geophys. Res. Lett.* **2019**, *46*, 2130–2139. [CrossRef]
14. Du, Y.; Wang, D.G.; Zhu, J.X. Research on the Evolution of Warming and Humidification in Northwest China Based on CMIP5. *J. Water Resour. Water Eng.* **2021**, *32*, 61–69.
15. Zhang, B.; Liu, X.F.; Zheng, G.F.; Wang, F.; Wang, S.Y. Variation of the days of extreme precipitation in Ningxia in summer and its causes. *J. Atmos. Sci.* **2018**, *41*, 176–185. [CrossRef]
16. Kharin, V.V.; Zwiers, F.W.; Zhang, X.; Wehner, M. Changes in temperature and precipitation extremes in the CMIP5 ensemble. *Clim. Chang.* **2013**, *119*, 345–357. [CrossRef]
17. Qin, P.H.; Xie, Z.H.; Zou, J.; Liu, S.; Chen, S. Future Precipitation Extremes in China under Climate Change and Their Physical Quantification Based on a Regional Climate Model and CMIP5 Model Simulations. *Adv. Atmos. Sci.* **2021**, *38*, 460–479. [CrossRef]
18. Hay, L.E.; Wilby, R.L.; Leavesley, G.H. A Comparison of Delta Change and Downscaed GCM Scenarios for Three Mountainous Basins in the United States. *J. Am. Water. Resour. Assoc.* **2010**, *36*, 387–397. [CrossRef]
19. Gao, F.Q.; Zhao, X.Q. Delta method in large deviations and moderate deviations for estimators. *Ann. Stat.* **2011**, *39*, 1211–1240. [CrossRef]
20. Zhang, M.J.; Wang, B.L.; Wei, J.L.; Wang, S.J.; Ma, Q.; Li, X.F. Extreme event changes of air temperature in Ningxia in recent 50 years. *J. Nat. Disasters* **2012**, *21*, 152–160. [CrossRef]
21. Chen, X.G.; Conway, D.; Chen, X.J.; Zheng, G.F. Analysis of the Trend of Extreme Precipitation Events in Ningxia from 1961 to 2005. *Adv. Clim. Chang. Res.* **2008**, *4*, 156–160.
22. Li, F.; Zhang, M.J.; Li, X.F.; Wang, S.J.; Wang, Q.; Liu, W.L. Research on spatiotemporal evolution of extreme precipitation in Ningxia during 1962–2011. *J. Nat. Disasters* **2013**, *22*, 171–180. [CrossRef]
23. Yu, Y.; Hua, T.; Chen, L.; Zhang, Z.; Pereira, P. Divergent changes in vegetation greenness, productivity, and rainfall use efficiency are characteristic of ecological restoration towards high-quality development in the Yellow River Basin, China. *Engineering* **2023**. [CrossRef]
24. Fan, K.; Shi, H.L.; Chen, Y.X. Analysis on Characteristics of Precipitation Variation in Zhongwei City in Recent 50 Years. *Agric. Sci. Technol.* **2011**, *12*, 1917–1920. [CrossRef]
25. Xin, Z.B.; Xie, Z.R.; Wang, W. Character of Precipitation in Ningxia and Its Response to ENSO. *Sci. Geol. Sin.* **2005**, *25*, 49–55.
26. Ma, L.; Zhang, Q.F.; Ni, Y.X.; Li, M.; Lu, X.Z. Study on Siltation Depth in Warping Dam Under Different Silting Conditions in Chabagou Watershed. *Res. Soil Water Conserv.* **2022**, *29*, 18–22, 30. [CrossRef]
27. Baggio, T.; D'Agostino, V. Simulating the effect of check dam collapse in a debris-flow channel. *Sci. Total Environ.* **2022**, *816*, 151660–151671. [CrossRef] [PubMed]
28. Zu, Q.; Chen, Z.Y.; Yu, S.; Huang, X.P.; Yu, K.X.; Wang, Z.J.; Wang, L. Risk Analysis of Check Dam System Under Extreme Rainfall in Small Watershed. *J. Soil Water Conserv.* **2022**, *36*, 30–37. [CrossRef]
29. Liu, X.J.; Zhang, Y.; Xiao, T.Q.; Li, P.; Liu, Y.Q.; Deng, W.P. Runoff velocity controls soil nitrogen leaching in subtropical restored forest in Southern China. *For. Ecol. Manag.* **2023**, *548*, 121412–121420. [CrossRef]
30. Gao, Y.; Feng, Q.; Liu, W.; Lu, A.G.; Wang, Y.; Yang, J.; Cheng, A.G.; Wang, Y.M.; Su, Y.B.; Liu, L.; et al. Changes of daily climate extremes in Loess Plateau during 1960–2013. *Quat. Int.* **2015**, *371*, 5–21. [CrossRef]

water

MDPI

Article

Rainfall Runoff and Nitrogen Loss Characteristics on the Miyun Reservoir Slope

Na Wang [1,2,†], Lei Wang [1,2,†], Liang Jin [2,†], Jiajun Wu [1], Min Pang [1,2], Dan Wei [2], Yan Li [2], Junqiang Wang [3], Ting Xu [3], Zhixin Yang [1,*] and Jianzhi Xie [1,*]

[1] College of Resources and Environmental Sciences, Agricultural University of Hebei, Baoding 071000, China; wn15512202703@163.com (N.W.); yzswanglei@163.com (L.W.); w1550791604@163.com (J.W.); 13716069539@163.com (M.P.)

[2] Institute of Plant Nutrition, Resources and Environment, Beijing Academy of Agricultural and Forestry Sciences, Beijing 100097, China; jinliang19762003@aliyun.com (L.J.); wd2087@163.com (D.W.); li.yan622@163.com (Y.L.)

[3] Qiqihar Branch of Heilongjiang Academy of Agricultural Sciences, Qiqihar 161000, China; august-wjq@163.com (J.W.); xuting830630@126.com (T.X.)

* Correspondence: yangzhixin@126.com (Z.Y.); xjianzhi@126.com (J.X.)

† These authors contributed equally to this work.

Abstract: Rainfall intensity and slope gradient are the main drivers of slope surface runoff and nitrogen loss. To explore the distribution of rainfall runoff and nitrogen loss on the Miyun Reservoir slopes, we used artificial indoor simulated rainfall experiments to determine the distribution characteristics and nitrogen migration paths of surface and subsurface runoff under different rainfall intensities and slope gradients. The initial runoff generation time of subsurface runoff lagged that of surface runoff, and the lag time under different rainfall intensity and slope conditions ranges from 3.97 to 12.62 min. Surface runoff rate increased with increasing rainfall intensity and slope gradient; compared with a rainfall intensity of 40 mm/h, at a slope of 15°, average surface runoff rate at 60 and 80 mm/h increased by 2.38 and 3.60 times, respectively. Meanwhile, the subsurface runoff rate trended upwards with increasing rainfall intensity, in the order 5 > 15 > 10°. It initially increased and then decreased with increasing slope gradient, in the order 5 > 10 > 15°. Total nitrogen (TN) loss concentration of surface runoff shows a decrease followed by a stabilization trend; the concentration of TN loss decreases with decreasing rainfall intensity, and the stabilization time becomes earlier and is most obvious in 5° slope conditions. TN loss concentration in subsurface runoff decreased with increasing rainfall intensity, i.e., 40 > 60 > 80 mm/h. The surface runoff rainfall coefficient was mainly affected by rainfall intensity, a correlation between αs and slope gradients S was not obvious, and the fitting effect was poor. The subsurface runoff rainfall coefficient was mainly affected by slope gradient, the R^2 of all rainfall intensities was <0.60, and the fitting effect was poor. The main runoff loss pathway from the Miyun Reservoir slopes was surface runoff, which was more than 62.57%. At the same time, nitrogen loss was subsurface runoff, more than 51.14%. The proportion of surface runoff to total runoff increases with the increase of rainfall intensity and slope, with a minimum of 62.57%, and the proportion of nitrogen loss from subsurface runoff also decreases with increasing rainfall intensity but does not change with slope gradient. The order of different runoff modulus types was mixed runoff (surface and subsurface runoff occur simultaneously) > surface runoff > subsurface runoff. The surface and mixed runoff modulus increased significantly with increasing rain intensity under different rain intensities and slope gradients. Overall, rainfall intensity significantly affected slope surface runoff, and slope gradient significantly affected nitrogen loss.

Keywords: rain intensity; slope gradient; runoff; total nitrogen

1. Introduction

Soil erosion occurs ubiquitously around the world and is a shared environmental problem, although it is more severe in parts of China, the United States, Russia, Australia, India, and Europe [1]. China is deeply affected by soil erosion, with about 5.29 gt of soil lost or displaced annually. Human activities have accelerated soil erosion, resulting in the erosion rate being much greater than natural soil formation. This has led to problems such as cultivated field quality degradation, reduced crop yield, and ecological environment quality decline, which seriously threatens food security and ecosystem stability [2,3]. The water quality safety of Miyun Reservoir plays an important role in Beijing. Strengthening the protection of its water resources is vital for a safe urban and rural water supply, smooth city operation, peaceful citizens, peaceful lives and work, and the sustainable development of the economy and society. The Miyun Reservoir area terrain is undulating, and soils are thin and shallow [4]. Furthermore, the soils have weak fertilizer and water sorption capacity, which makes them particularly prone to soil erosion and nutrient loss, as well as eutrophication of the surrounding water bodies, which poses significant ecological risks. Soil nitrogen is the most important indicator of soil nutrient levels [5] and also the main factor causing eutrophication/pollution of water bodies [6]. However, there have been few studies on soil nitrogen loss from slopes in the Miyun Reservoir area, especially the temporal and spatial effects of surface and subsurface runoff on soil nitrogen loss processes and mechanisms. The pathways and processes of soil nitrogen loss under different rainfall intensities and slope gradients have not been adequately addressed.

Rainfall intensity is the main influencing factor of surface erosion, and generally, the surface erosion rate is directly proportional to the intensity of rainfall [7]. Deng et al. (2018) studied erosive weathered granite slopes and found that the proportion of total runoff that was subsurface was greater than slope surface runoff [8]. Simultaneously, with increased rainfall intensity, slope surface runoff also increased, and the interaction between raindrop impact and runoff aggravated the erosion process. Studies have noted that rainfall intensity has the greatest impact on slope surface runoff in the Miyun Reservoir area, accounting for 21.84% of the total. Initial runoff generation time is also advanced with increased rain intensity, and the two are clearly negatively correlated [9]. However, most researchers believe that the greater the slope gradient, the greater the runoff [10]. Tan et al. found that when the slope gradient increases, runoff velocity in the fine gullies generated by the slope accelerates, runoff erosion intensity increases, and runoff and sediment yield increases [11]. Others have suggested that the slope rainfall-bearing capacity decreases with increased slope gradient, so factors that promote increased runoff are canceled [12]. Additionally, it has been documented that slope surface runoff and sediment yield increase initially, then decrease, and then increase with increased slope gradient [13].

The migration of nutrients such as nitrogen and phosphorus carried by runoff sediment during soil erosion is the key cause of water eutrophication, which is an important ecological threat in China [14]. Soil nutrients from the slope field mainly enter the downstream water body through slope surface runoff, subsurface runoff, and erosion sediment. Deng et al. and Liu et al. found that the greater the rainfall intensity, the more nitrogen is lost in runoff [15,16]. Chen et al. showed that under the same rainfall intensity conditions, runoff and nitrogen loss are directly proportional to rainfall time [17]. In simulation experiments, Wu et al. found that slope surface and subsurface runoff TN loss increased with increased rainfall intensity or slope gradient. Subsurface runoff is the main source of TN loss from slopes, and the loss ratio can reach 91.26 to 99.61%. The proportion of TN loss in slope surface runoff to total TN loss on slopes increases with increasing rain intensity [18]. Qian et al.'s nitrogen loss study showed that in the early stage of runoff, nitrogen loss was significant and then gradually decreased and tended to be stable. Dissolved nitrogen was the main source of runoff at the beginning, and the proportion of insoluble nitrogen increased with rainfall extension [19]. Presently, research on nutrient transport through surface runoff is relatively common, but research on nutrient transport through subsurface runoff is relatively rare.

Therefore, this study selected the slope of Miyun Reservoir as the research object; using artificial rainfall simulation, we explored runoff and nitrogen loss characteristics under different rainfall intensities and slope gradients and analyzed the contribution of surface and subsurface runoff to soil nitrogen loss, clarified the impact mechanism of rainfall intensity and slope on runoff nitrogen loss. To provide some reference for the prevention and reduction of agricultural non-point source pollution and water eutrophication on the slope of Miyun Reservoir.

2. Materials and Methods

2.1. Test Equipment and Materials

The simulated rainfall equipment adopts the QYJY-502 model produced by China Xi'an Qingyuan Measurement and Control Technology Co., Ltd. (Qingyuan, China), with a rainfall height of 5 m and a rainfall adjustment accuracy of 7 mm/h. The simulated rainfall instrument has an effective rainfall area of 11 m × 8 m, a rainfall intensity variation range of 15–200 mm/h, and a rainfall uniformity coefficient > 0.80. The simulated rainfall equipment has been recalibrated before the experiment. The experimental soil box is a fixed variable slope steel box with a slope adjustment range of 0–30°. The length, width, and height of the soil box are 2.0 m × 0.5 m × 0.6 m in sequence. The middle and bottom of the steel trough are equipped with drainage holes for collecting subsurface runoff. The top of the steel trough has a triangular outlet for collecting surface runoff (Figure 1).

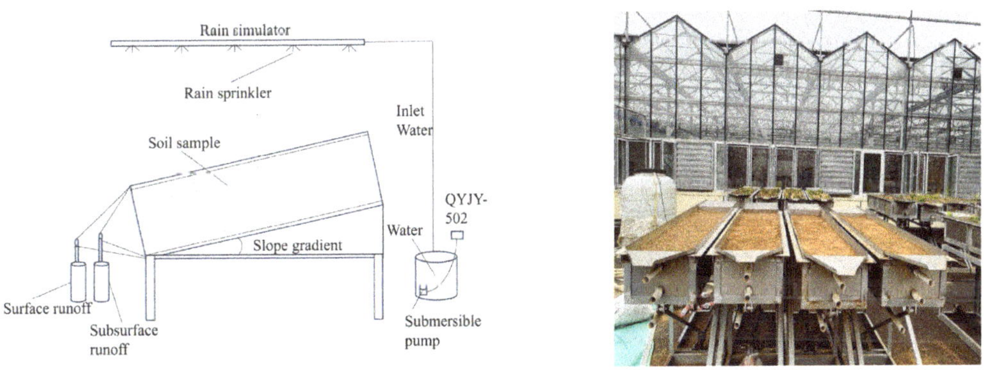

Figure 1. Schematic diagram and physical diagram of experimental structure.

2.2. Test Soil

Test soil was collected from Taishitun Town (117°6′42.08″ E, 40°32′22.02″ N), Miyun District, Beijing (Figure 2). The climate type belonged to the warm temperate semi-humid monsoon climate. The average annual rainfall of multiple years was about 624 mm; June to August is the rainy season. The soil in trial areas mainly belongs to cinnamon soil. The soil type was sandy loam soil. The basic physiochemical properties of soil are shown in Table 1 [20]. To ensure the soil did not change, the sample and bag soil were placed at a depth of 50 cm with 5 cm intervals. Based on field measurements of soil bulk density at different depths, control the volume and quality of each layer of filled soil sample to ensure that the compactness and bulk density of the soil sample are close to the natural soil. Once it was close to natural soil bulk density, the artificial simulated rainfall test was carried out.

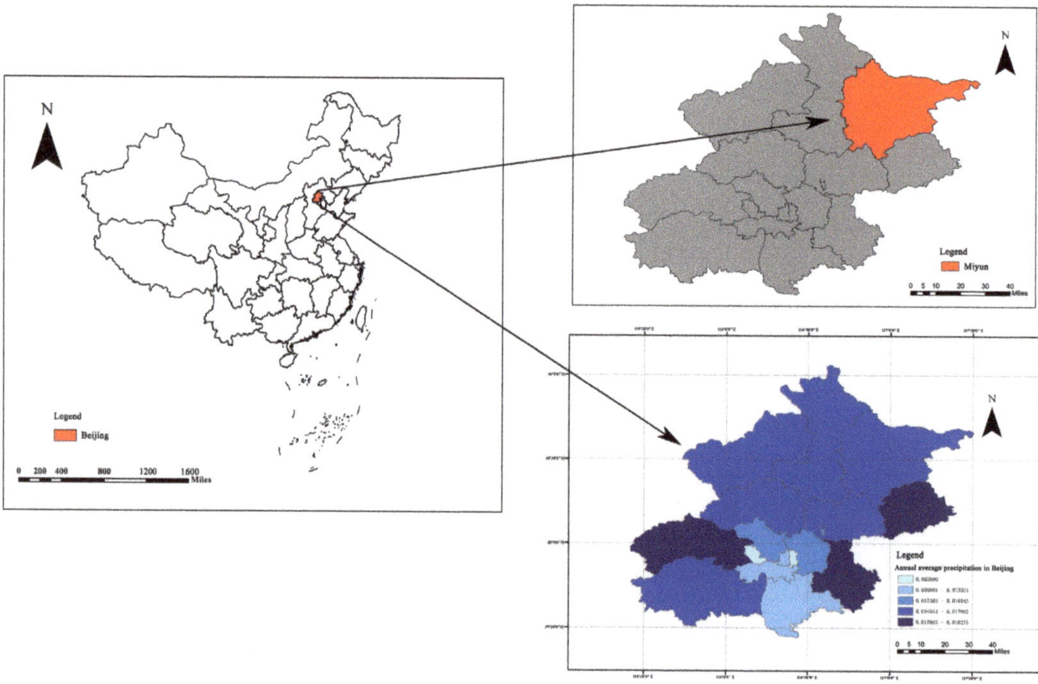

Figure 2. Location of the study area.

Table 1. The soil's basic physiochemical properties.

pH	Soil Organic Matter (g/kg)	TN (g/kg)	Available Phosphorous (mg/kg)	Available Potassium (mg/kg)	Sand Content (%)	Silt Content (%)	Clay Content (%)
6.33	9.97	0.448	4.55	45.9	75.12	16.5	8.38

2.3. Experimental Design

The test was conducted from June to October 2022 at the Beijing Academy of Agriculture and Forestry Sciences simulation test base, Haidian District, Beijing, China. Based on field investigation results, the slope around Miyun Reservoir is mostly between 3° and 20°; experimental slopes (S) are set at 5°, 10°, and 15°. By analyzing the year-round rainfall data in Miyun District and based on the local rainfall intensity characteristics, the experimental rainfall intensity (I) is set to 40, 60, and 80 mm/h. Design combinations (paired with 5°, 10°, and 15° slopes and rainfall intensities of 40, 60, and 80) for a total of 9 trials. In all treatments, the soil was bare. To ensure test accuracy, each treatment was repeated 3 times. Before the experiment begins, rainfall intensity calibration should be carried out, and the experiment should start when the rainfall intensity and uniformity are greater than 90%. Accurately record the initial time of surface and subsurface runoff. Each rainfall time is 60 min, with a sampling interval of 5 min. Slope surface and subsurface runoff samples collected from each rainfall were brought back to the laboratory and allowed to stand at room temperature (25 °C) for 4 to 5 h for precipitation. After precipitation, the amount of sediment was measured, and the nutrient content of the supernatant was immediately determined. Finally, total nitrogen (TN) in runoff was detected by potassium persulfate oxidation–ultraviolet spectrophotometry, and all tests were completed within 48 h. Determination of soil particle size by sieving and settling method. The soil density is determined using the ring knife method, and the soil moisture content is determined using the drying method. The soil organ organic matter is determined using the dichromate volumetry–outside heating method; available phosphorous is determined using the sodium hydroxide melting–Mo-Sb colorimetry, and available potassium is determined by sodium hydroxide melting–flame photometry [21].

2.4. Data Acquisition and Analysis

Data is analyzed using Excel 16.0 for statistics and Origin 2022 for plotting. The least significant difference (LSD) method in the Statistical Package for the Social Sciences 19 (SPSS) was used to test for significant differences in TN loss from surface and subsurface runoff at $p < 0.05$. The following formula was used to calculate the contribution rate of each factor (PF) [22]:

$$P_F = \frac{SS_F - (DOF_F \times V_{Er})}{SS_T} \times 100$$

where P_F is the factor contribution rate (%); SS_F is type III sum of squares; DOF_F is the factor degrees of freedom; V_{Er} is the sum of squares of the error; and SS_T is the sum of squares of the total dispersion. SS_F, SS_T, DOF_F, and V_{Er} values were obtained directly from the analysis of variance using SPSS 19.0 software.

3. Results and Analysis

3.1. Influence of Rainfall Intensity and Slope on Surface Runoff

Due to the influence of topographic conditions, land use patterns, and other factors, initial runoff generation time will change, so understanding initial runoff generation time during rainfall is vital for calculating soil erosion rate [23]. For the same slope gradients, initial surface runoff generation time advanced with increased rain intensity, in the order 40 > 60 > 80 mm/h (Table 2). At a slope gradient of 5°, for example, initial surface runoff generation time declined by 3.4 min and 0.87 min with increased rainfall intensity. Other slopes also showed similar changes, but with the increase of slope gradient, showed a gradual decrease. Subsurface runoff exhibited the same trend as surface runoff. With increasing rainfall intensity under the same slope gradient, more rainfall infiltrates into the soil groove per unit of time, and the soil quickly reaches saturation, which is conducive to slope surface and subsurface runoff generation. Under the same rainfall intensity, the initial surface runoff generation time advances with increasing slope gradient, in the order 5 > 10 > 15°, and the initial surface runoff generation time decreases gradually with increasing rainfall intensity. The initial subsurface runoff generation time was consistent with that of surface runoff. However, the decrease in slope gradient was not obvious. The initial subsurface runoff generation time lagged behind surface runoff, which is because soil water content begins to produce subsurface runoff when it reaches saturation, and the lag time decreases with increasing rainfall intensity and slope gradient.

Table 2. Initial surface and subsurface generation times of different treatments (rainfall intensities and slope gradients) (mins).

Types of Runoff	Rainfall Intensity (mm/h)	Slope Gradient (°)		
		5	10	15
Surface Runoff	40	6.23	4.76	3
	60	3.83	1.75	0.97
	80	1.96	1.12	0.24
Subsurface Runoff	40	18.85	14.42	10.79
	60	13.33	9.14	7.55
	80	10.92	8.65	4.21
Surface Runoff - Subsurface Runoff	40	12.62	9.66	7.79
	60	9.5	7.39	6.58
	80	8.96	7.53	3.97

The surface runoff rate increased with increasing rain intensity (Figure 3). Compared with a rainfall intensity of 40 mm/h at a slope of 15°, the average surface runoff rate at 60 and 80 mm/h increased by 2.38 and 3.60 times, respectively. Under the same rainfall intensity, the surface runoff rate increased with increasing slope gradient. At 60 mm/h, for example, the average surface runoff of 5, 10, and 15° slopes was 2.90, 3.2, and 3.6 L, respectively. Subsurface runoff rate showed an upward trend with rainfall (Figure 2). Under the same rainfall intensity, the subsurface runoff rate was 5 > 15 > 10°; as the slope increases, it tends to rise first and then decrease, which indicates a critical slope between 5 and 10°. Under the same slope gradients, the subsurface runoff rate decreased with increasing rainfall intensity, in the order 5 > 10 > 15°.

There is a significant difference in surface runoff under different rainfall intensities ($p < 0.05$), and there is no significant difference in subsurface runoff at 5° and 10° under 60 mm/h and 80 mm/h. There are also significant differences in surface runoff between slopes under the same rainfall intensity ($p < 0.05$), while there was no significant difference in subsurface runoff at 40 mm/h and 60 mm/h at 5° and 10° slopes ($p < 0.05$) (Table 3).

Table 3. Significant analysis of runoff under different slope gradient and rainfall intensity.

Slope Gradient S/(°)	Surface Runoff (L)			Subsurface Runoff (L)		
	40 mm/h	60 mm/h	80 mm/h	40 mm/h	60 mm/h	80 mm/h
5	22.43 ± 0.32 Cc	52.3 ± 0.82 Bc	87.2 ± 0.82 Ac	7.17 ± 0.7 Bb	8.27 ± 0.61 Ab	11.37 ± 0.7 Ac
10	25.93 ± 0.61 Cb	58.4 ± 0.85 Bb	93.37 ± 0.85 Ab	6.3 ± 0.66 Bab	7.3 ± 0.8 Bab	9.63 ± 0.61 Ab
15	27.77 ± 0.35 Ca	65.1 ± 0.7 Ba	98.3 ± 0.79 Aa	5.77 ± 0.55 Ba	6.67 ± 0.75 Aa	7.4 ± 0.75 Aa

Note: value = mean ± standard error (SE). Different capital letters in the same slope gradient denote significant differences between different rainfall intensities of the same slope gradient ($p < 0.05$). Different lowercase letters in the same column denote significant differences in different slope gradients of the same rainfall intensity ($p < 0.05$).

3.2. Effects of Rainfall Intensity and Slope Gradient on Nitrogen Loss on Slopes

Surface runoff N loss concentration initially decreased and then tended to stabilize with time under different treatments, indicating that the initial rainfall period is critical for nitrogen loss from surface runoff (Figure 4). This is because surface runoff initially removes some particulate nitrogen that is easily washed away [24]. Thus, surface runoff TN concentration will be relatively high at the beginning and show a sharp downward trend. Then, runoff scours the nitrogen on the soil surface and takes it away, so surface runoff TN eventually tends to be stable under different rainfall intensities. The lower the rain intensity, the earlier the TN loss concentration tends to stabilize, and this was most obvious in 5° slope conditions because higher surface runoff means faster runoff rates, resulting in shorter contact time between runoff and soil and, thus, lower surface runoff TN concentration. Under the same slope gradients, increased rainfall intensity significantly increased surface runoff TN loss concentration, but under the same rainfall intensity, surface runoff TN loss concentration increases showed a variation pattern of 10° > 15° > 5°.

Under the same rainfall intensity, subsurface runoff TN loss concentration initially decreased and then stabilized after runoff occurrence on both 5 and 10° slope gradients, and all of them tended to be stable after 60 min of rainfall. When the slope gradient was 15°, subsurface runoff TN loss concentration under different rainfall intensities initially rose and then decreased in the early stage, and then rose slightly and stabilized in the mid to late stage. This is maybe because the greater the slope gradient, the more water in the soil can readily move, and the more contact with the soil particles, the easier it is to dissolve nitrogen fixed on the soil particle surfaces and be lost with water runoff so that subsurface runoff TN concentration is higher [24]. Under the same slope gradients, TN loss concentration in the subsurface runoff decreased with increased rainfall intensity, i.e., 40 > 60 > 80 mm/h. Under 40 and 60 mm/h, with extended rainfall, subsurface runoff gradually decreased, then increased, and finally stabilized. However, under 80 mm/h,

TN concentration loss decreased with extended rainfall time. This is because subsurface runoff at the beginning was low; moreover, free TN concentration in the soil was higher, and the greater the rain intensity, the more subsurface runoff, causing a dilution effect and a decrease in subsurface runoff TN concentration with increasing rain intensity.

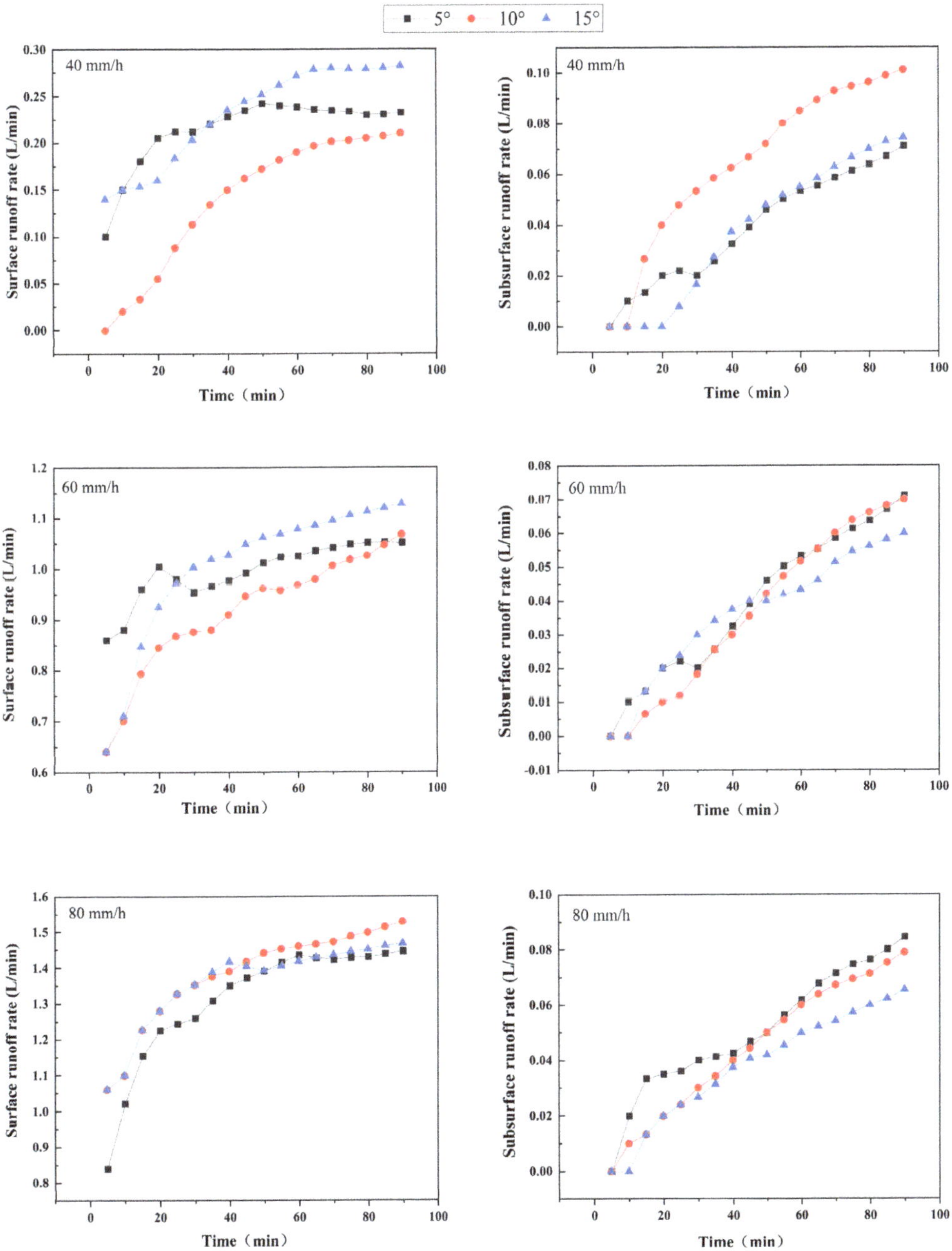

Figure 3. Runoff rate under different slope gradient and rainfall intensity.

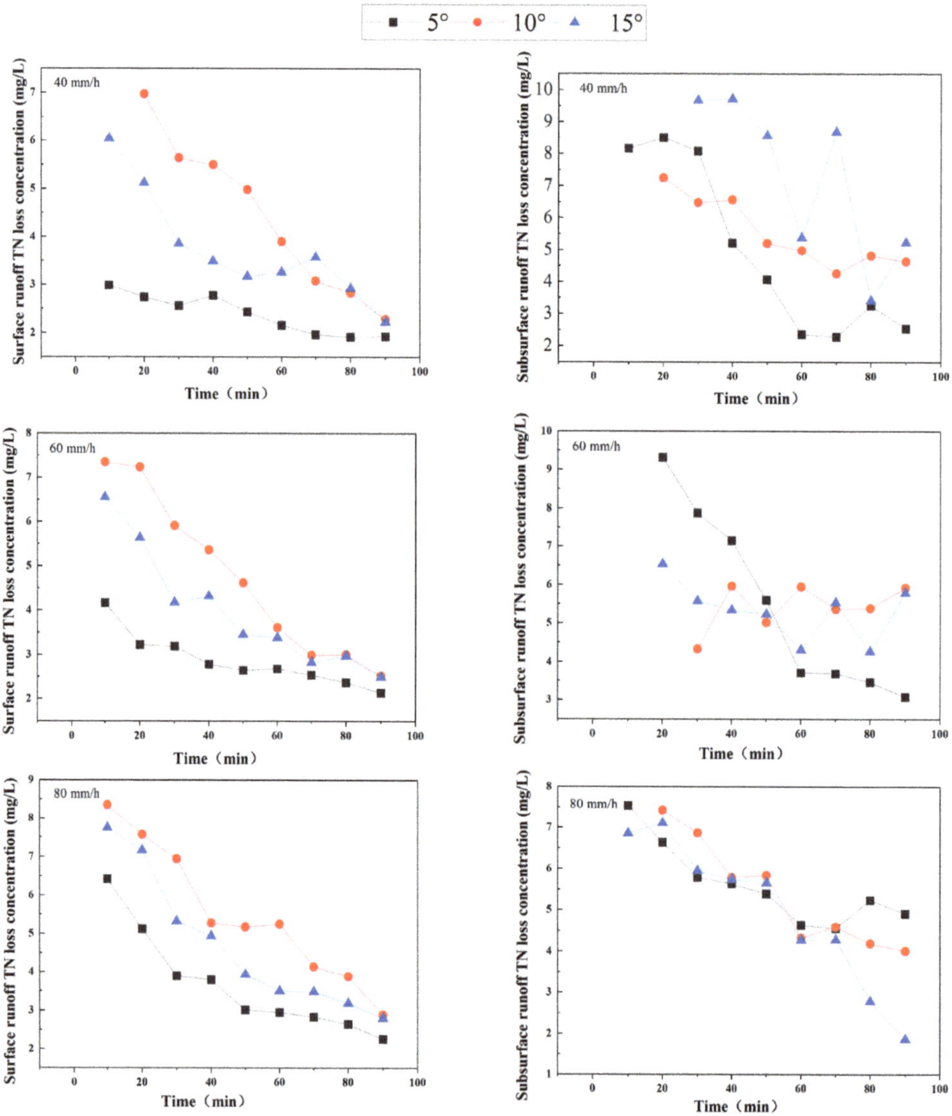

Figure 4. Surface runoff and subsurface runoff TN loss under different slope gradient and rainfall intensity.

There is a significant difference in TN loss in surface runoff under different rainfall intensities ($p < 0.05$), while there was no significant difference in TN loss in subsurface runoff under a 5° slope ($p < 0.05$). Under the same rainfall conditions, there is a significant difference in TN loss in surface runoff between different slopes ($p < 0.05$), and there is a significant difference in TN loss in subsurface runoff among different slopes under 40 rain-fall intensity ($p < 0.05$) (Table 4).

Table 4. Significant analysis of runoff TN loss under different slope gradient and rainfall intensitys and rainfall intensities.

Slope Gradient S/(°)	Surface Runoff TN Concentration (mg/L)			Subsurface Runoff TN Concentration (mg/L)		
	40 mm/h	60 mm/h	80 mm/h	40 mm/h	60 mm/h	80 mm/h
5	22.41 ± 0.93 Cc	25.14 ± 0.72 Bc	32.76 ± 0.57 Ac	22.23 ± 0.86 Cc	43.44 ± 1.06 Ba	50.29 ± 1.01 Aa
10	35.16 ± 1.07 Ca	43.35 ± 0.86 Ba	50.05 ± 0.57 Aa	36.31 ± 1.08 Ba	37.46 ± 1.07 Bb	42.81 ± 0.88 Ab
15	33.06 ± 0.53 Cb	36.01 ± 0.24 Bb	42.63 ± 0.60 Ab	32.45 ± 1.06 Bb	42.33 ± 1.03 Aa	43.41 ± 0.97 Ab

Note: value = mean ± standard error (SE). Different capital letters in the same slope gradient denote significant differences between different rainfall intensities of the same slope gradient ($p < 0.05$). Different lowercase letters in the same column denote significant differences in different slope gradients of the same rainfall intensity ($p < 0.05$).

3.3. Runoff Coefficients of Surface Runoff and Subsurface Runoff

At the same slope gradient, the surface runoff coefficient (α_s) increased with increased rainfall intensity (Figure 5) and was linearly positively correlated with it (Table 5). Conversely, under the same rainfall intensity, as the slope increases, the subsurface runoff coefficient (α_i) decreases and is linearly negatively correlated with rainfall intensity (Table 5). Under the same slope gradients, α_i decreased with increasing rainfall intensity, in the order 5 > 10 > 15°.

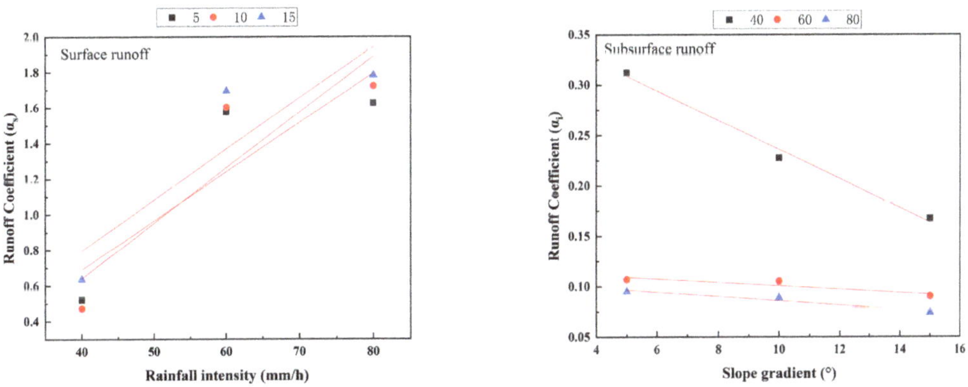

Figure 5. Runoff coefficient under different slope gradients and rainfall intensity.

Table 5. Fitting results of runoff coefficients for runoff f under different slope gradients and rainfall intensity.

Types of Runoff	Simulate Conditions		Fitted Equation	Fitting Degree R^2
Surface Runoff	Slope Gradient $S/(°)$	5	$\alpha_5 = 0.02759I - 0.41382$	0.7836
		10	$\alpha_{10} = 0.03119I - 0.60653$	0.8204
		15	$\alpha_{15} = 0.02866I - 0.34951$	0.8072
Subsurface Runoff	Raionfall Intensity $I/(mm/h)$	40	$\alpha_5 = -0.0145S + 0.38083$	0.9902
		60	$\alpha_{10} = -0.00167S + 0.11722$	0.8242
		80	$\alpha_{15} = -0.00213S + 0.10708$	0.9465

The effects of rainfall intensity and slope gradient on runoff coefficients were complementary. A correlation between αs and slope gradients S was not obvious, and the fitting effect was poor (Table 5). At 40 and 60 mm/h, α_s was 15 > 5 > 10°, but at 80 mm/h, it was 10 > 15 > 5°, and the correlation coefficient did not change. The fitting effect between α_i and slope gradient S increased with increasing slope under different rainfall intensities, the R^2 of all rainfall intensities was <0.60, and the fitting effect was poor, while the correlation coefficient decreased with increasing slope. The surface runoff coefficient α_s was mainly affected by rainfall intensity, while the subsurface runoff coefficient α_i was mainly affected by slope.

3.4. Slope Surface Runoff, Nitrogen Loss Distribution and Runoff Modulus

The slope remains unchanged, and the proportion of surface runoff in total runoff increased with increasing rainfall intensity (Table 6). As rainfall intensity increases, the effective precipitation per unit time also increases, which accelerates soil aggregate fragmentation, and the surface crust reduces the soil infiltration rate, increasing the proportion of surface runoff and decreasing the proportion of subsurface runoff. On different slope gradients, the proportion of surface runoff under the same rainfall intensity was basically the same, which is caused by low rainfall infiltration. The sum of surface runoff and subsurface runoff is the total runoff, so the proportion of subsurface runoff in total runoff is the opposite of surface runoff.

Table 6. The proportion of surface runoff in total runoff (%).

Rainfall Intensity/(mm/h)	5°	10°	15°
40	62.57	67.5	79.13
60	93.66	93.85	94.95
80	94.16	95.09	95.53

Under the same slope conditions, the proportion of nitrogen loss in subsurface runoff to total nitrogen loss decreased with increasing rainfall intensity (Table 7). This is because subsurface runoff is generated by infiltration and can dissolve and migrate nitrogen in the soil, while surface runoff can only take away nitrogen in the surface soil, which is easy to lose, but the total amount is limited. There was no obvious nitrogen loss pattern in subsurface runoff with slope at the same rainfall intensity, indicating that the critical slope gradient differed under different rainfall intensities.

Table 7. The proportion of nitrogen loss to total nitrogen loss in subsurface runoff (%).

Rainfall Intensity/(mm/h)	5°	10°	15°
40	67.49	55.67	60.09
60	63.01	51.83	54.30
80	60.46	51.14	51.38

Subsurface runoff generation lags behind that of surface runoff, but there is a period of time when they occur together, and total runoff in this period is called mixed runoff [25]. The order of different runoff modulus (runoff per unit area per unit time) types was mixed > surface > subsurface runoff (Figure 6). On different slope gradients at the same rainfall intensity, surface, subsurface, and mixed runoff moduli were basically the same. The mixed runoff modulus was the highest under different slope gradients and rain intensities, and it increased significantly with increasing rain intensity on the same slope gradients. The runoff modulus variation of surface and mixed runoff was similar. With increasing rainfall intensity, the subsurface runoff modulus initially decreased, then increased, but was lower than surface runoff.

3.5. Factors Influencing Slope Surface Runoff and Nitrogen Loss

There is a highly significant positive correlation between rainfall intensity and surface runoff ($p < 0.01$). Rainfall is the primary generator of surface runoff, and rainfall intensity affects the amount of rainfall. When the underlying surface conditions are the same, slope surface runoff is greater with higher rainfall, and the slope gradient, as a topographic factor, cannot cause surface runoff; instead, it mainly affects runoff at the outlet in terms of slope-rainfall bearing capacity and soil infiltration rate. Slope gradient and rainfall intensity promote slope runoff and nitrogen loss.

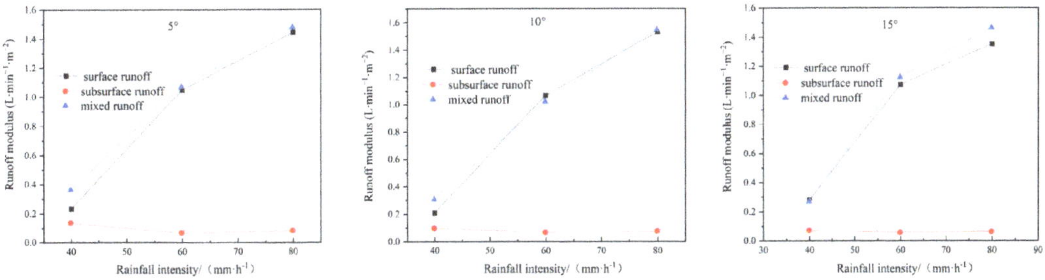

Figure 6. Variation trend of Surface runoff, subsurface runoff, and mixed runoff modulus with rainfall intensity under different slopes.

To quantify the contribution rate of various factors to sediment production in slope surface runoff, we calculated the interaction between different influencing factors on nitrogen loss (Figure 7). Rainfall intensity and slope significantly affect runoff generation and nitrogen loss. Contribution rates to surface and subsurface runoff were in the order of rainfall intensity > rainfall intensity and slope gradient interactions > slope gradient (Table 8). Among them, the contribution of rainfall intensity was the highest, with surface runoff being 99.6% and subsurface runoff being 52.8%. Contribution rates to total nitrogen loss of surface flow were in the order of slope gradient > rainfall intensity > slope gradient and rainfall intensity interactions. The contribution rate of slope gradient was the highest, at 66.8%. Contribution rates to total nitrogen loss in subsurface runoff were in the order of rainfall intensity and slope gradient interactions > slope gradient > rainfall intensity. Interactions between rainfall intensity and slope gradient contributed the most, at 44.1%. Overall, rainfall intensity significantly affected slope surface runoff, and slope gradient significantly affected nitrogen loss.

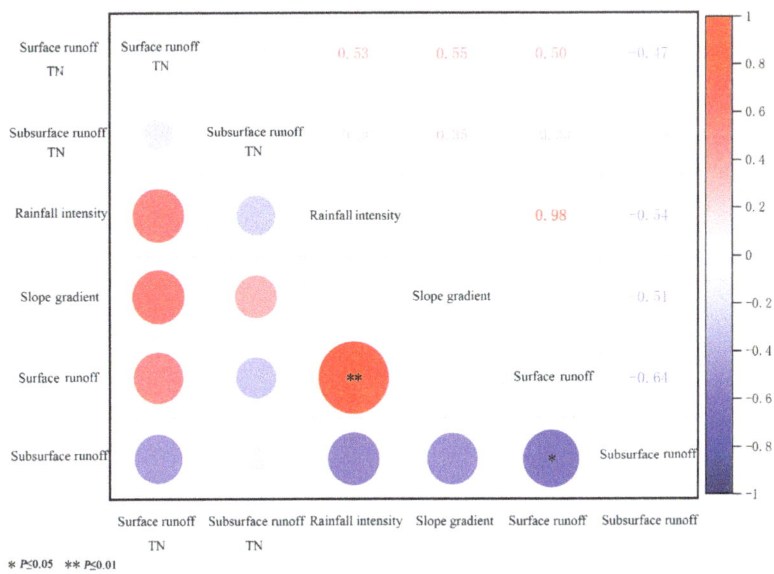

Figure 7. Correlation between rainfall intensity, slope, slope runoff, and nitrogen loss.

Table 8. The contribution rate of factors influencing slope surface runoff and nitrogen loss (%).

Factor	Surface Runoff	Subsurface Runoff	TN of Surface Runoff	TN of Subsurface Runoff
Rainfall Density	99.6	52.8	30.2	21.4
Slope Gradient	0.1	26.9	66.8	37.5
Rainfall Density × Slope Gradient	0.3	20.3	3.0	41.1

4. Discussion

The main factors influencing soil erosion on the slopes of Miyun Reservoir are rainfall intensity and slope gradient. As rainfall intensity increased from light (40 mm/h) to moderate (60 mm/h) to extreme (80 mm/h), both surface and subsurface runoff increased. Compared with a rainfall intensity of 40 mm/h at a slope of 15°, the average surface runoff rate at 60 mm/h and 80 mm/h increased 2.38 and 3.60 times, respectively. This may be because during runoff, rainfall kinetic energy acts on the soil surface. The intensity of rainfall determines the runoff magnitude [26]. Therefore, the soil erosion rate increases with the increase in rainfall intensity [27]. The main form of runoff generation was surface runoff, with a proportion of surface runoff in total runoff of 75.77 to 96.16%.

Rainfall intensity remains unchanged; the surface runoff rate increased with increasing slope gradient, in the order 5 > 15 > 10°, and tended to initially rise and then decrease, indicating that there was a critical slope between 5 and 10°. This may be due to a decreasing slope gradient significantly increasing the time to reach the maximum surface runoff rate while reducing the surface runoff. When the slope gradient increases, part of the water is more permeable and flows down the slope by gravity. Consequently, surface runoff also increases with increasing slope gradient, while subsurface runoff decreases. Compared to slopes with high gradients, those with lower slope gradients better inhibit soil particle loss with runoff during rainfall [28]. Slope gradient is also the main influencing factor of soil erosion. As the slope increases, surface runoff increases, infiltration rate decreases, and surface erosion increases [29]. In this study, surface, subsurface, and mixed runoff modulus increased with slope gradient. However, the subsurface runoff, subsurface runoff rate, and subsurface runoff modulus decrease with increasing slope gradient. Low slope gradients better inhibit the downward loss of soil particles from runoff during rainfall [30]. Compared with single influencing factors, comprehensive interactions between rainfall intensity and slope gradient had a more obvious impact on the slope surface runoff mode. Furthermore, the need to comprehensively understand the interactions between influencing factors in the slope erosion process was demonstrated.

The main influencing factors of soil nitrogen loss are rainfall and runoff. In this study, runoff TN concentration fluctuated at the beginning of surface runoff and began to stabilize after 60 min. For example, on a 5° slope, the average TN loss in surface runoff under rainfall intensity of 80 mm/h was 1.28 times that of 60 mm/h, and in subsurface runoff was 1.03 times that of 60 mm/h rainfall. Therefore, rainfall intensity had a significant effect on soil nitrogen loss. On a 5° slope, below subsurface TN loss under 80 mm/h and 60 mm/h rainfall conditions was 1.53 and 1.92 times greater than on the surface. Therefore, subsurface runoff loss was greater than surface runoff loss, potentially making pollution of groundwater sources more serious, especially in areas with shallow soils and those with underground pores and fissures. Moreover, northern soils and those in rocky mountainous areas are looser, and their water is more prone to infiltration, which washes away the soil nitrogen. Therefore, in such areas, subsurface runoff often plays a decisive role in nitrogen loss. In practical application, the occurrence of subsurface runoff should be prevented or reduced, for example, by fertilizing to increase soil water-holding capacity. The nitrogen loss in karst areas of South China is mainly carried by groundwater runoff, and this experiment is consistent with this situation [31]. Gao et al. showed that the surface

TN loss rate was greater when the rainfall intensity was ≥50 mm/h and was the main cause of TN loss [32].

Our results show that slope gradient and rainfall intensity can promote runoff, with rainfall intensity having the greatest impact on surface runoff, contributing 99.6%. The slope gradient contributed the most to total nitrogen loss from surface runoff, at 66.8%. The interaction between rainfall intensity and slope gradient contributed the most to total nitrogen loss in the subsurface runoff, at 44.1%. As rainfall intensity increases, raindrop erosion capacity increases, and soil particle dispersion increases. Increased rainfall intensity will increase runoff erosive forces and sediment-carrying capacity [33]. The slope gradient increases the slope rain-bearing area, which leads to increased runoff and flow velocity, and then water flow erosion capacity increases, which promotes slope surface erosion in the gullies. Compared with Li et al., who qualitatively studied the interaction between rainfall intensity, slope gradient, and slope length, we considered the quantitative contribution of rainfall intensity, slope gradient, and the interaction of various factors to runoff and sediment production [34]. Our approach may better reflect the relationship between the comprehensive effects of influencing factors and slope surface runoff and nitrogen loss.

The slope gradient in the Miyun reservoir area has an important impact on nitrogen loss, especially on steep slopes, which could be slowed down by reducing the slope gradient. However, we did not consider the underground migration of nutrients. Although the loss of underground nitrogen is higher than on the surface, due to its special geographical location, it is difficult to control the loss of underground nitrogen. Therefore, preventing and controlling groundwater pollution is of great significance for controlling non-point source pollution. Additionally, agricultural fertilizer use could be reduced [35], or alternatively, increasing vegetation cover would reduce nutrient loss on the slope surface [36]. This may help to mitigate the effects of rainfall on nitrogen and underground seepage in cultivated soils, thereby increasing yields and preventing pollution.

Although the effects of rainfall intensity and slope gradient on slope erosion in the Miyun Reservoir area were analyzed and a critical slope gradient determined, slope length was not considered. In the future, slope length characteristics will be added, and runoff and nutrient loss at critical slope lengths will be discussed. We only considered the contributions of rainfall intensity and slope gradient to erosion, but the contribution of one factor depended on changes in others, and these interactions need to be more fully addressed.

5. Conclusions

Initial runoff generation time gradually decreases with increasing rainfall intensity and slope gradient. Initial subsurface runoff generation time lags behind that of surface runoff but decreases with increasing rainfall intensity and slope gradient. Surface runoff rate increased with increasing rainfall intensity and slope gradient, and the subsurface runoff rate showed an upward trend with increasing rainfall intensity, initially increasing and then decreasing with increasing slope gradient. Surface runoff TN loss concentration initially decreased and then stabilized with time. Rainfall intensity has a significant effect on the change of TN loss concentration in surface runoff. The concentration of TN loss in the subsurface runoff decreased with increasing rainfall intensity.

The effects of rainfall intensity and slope gradient on surface and subsurface runoff coefficients were complementary. The main runoff loss path from slopes is surface runoff, and the main path of nitrogen loss is subsurface runoff. The order of different runoff modulus types was mixed > surface > subsurface runoff. Rainfall intensity significantly affected the slope subsurface, and slope gradient significantly affected nitrogen loss.

Author Contributions: Conceptualization, N.W. and L.W.; Methodology, J.X. and Z.Y.; Resources, D.W. and L.J.; Software, J.W. (Junqiang Wang) and M.P.; Supervision, Y.L.; Visualization, L.W.; Writing—Original Draft, N.W. and L.W.; Writing—Review and Editing, N.W., L.W., L.J., J.W. (Jiajun Wu), M.P., D.W., Y.L., J.W. (Junqiang Wang), T.X., Z.Y. and J.X. All authors have read and agreed to the published version of the manuscript.

Funding: This research was funded by the National Key R&D Plan (2022YFD1500200-1), the Science and Technology Capacity Improvement Project of Beijing Academy of Agricultural and Forestry Sciences (ZHS202304), the general program of NSFC (41977095), the Beijing Academy of Agricultural and Forestry Sciences Youth Fund (QNJJ202214), Research Business Fee Project of Heilongjiang Provincial Research Institute (CZKYF2023-1-B006), Outstanding Youth Fund Project of Heilongjiang Academy of Agricultural Sciences (2021JCQN004), Beijing Postdoctoral Fund, Postdoctoral Research Fund of Beijing Academy of Agricultural and Forestry Sciences (2020-ZZ-026), Beijing Academic Program.

Data Availability Statement: Data are contained within the article.

Acknowledgments: We appreciate and thank the anonymous reviewers for their helpful comments that led to an overall improvement of the manuscript. We also thank the Journal Editor Board for their help and patience throughout the review process.

Conflicts of Interest: The authors declare no conflict of interest.

References

1. Deng, Y.; Tang, D.S.; Li, Y.Y.; Meng, C.; Liu, H.; Li, X.; Li, Q.; Wu, J. Influences of riparian landscape pattern on river phosphorus output in typical subtropical agroforestry mixed watershed. *J. Soil Water Conserv.* **2022**, *36*, 64–70.
2. Chen, Z.F.; Gong, A.M.; Ning, D.W.; Zhang, L.; Wang, J.; Xiang, B. Characteristics of soil erosion and nutrient loss in Yunnan Province based on RUSLE model. *J. Soil Water Conserv.* **2021**, *35*, 7–14.
3. Gao, Y.L.; Li, H.B. Influence of landscape pattern change on soil erosion in Han River Basin. *Acta Ecol. Sin.* **2021**, *41*, 2248–2260.
4. Sun, R.H.; Sun, L.; Su, X.K.; Chen, L. Study on the coupling of landscape pattern and ecological process: Inheritance and innovation. *Acta Ecol. Sin.* **2021**, *41*, 415–421.
5. Wang, L.; Li, Y.; Wu, J.; An, Z.; Suo, L.; Ding, J.; Li, S.; Wei, D.; Jin, L. Effects of the Rainfall Intensity and Slope Gradient on Soil Erosion and Nitrogen Loss on the Sloping Fields of Miyun Reservoir. *Plants* **2023**, *12*, 423. [CrossRef] [PubMed]
6. Zheng, W.; Lan, R.; Zhangzhong, L.; Yang, L.; Gao, L.; Yu, J. A Hybrid Approach for Soil Total Nitrogen Anomaly Detection Integrating Machine Learning and Spatial Statistics. *Agronomy* **2023**, *13*, 2669. [CrossRef]
7. Luo, L.X.; Gao, B.; Yan, X.Y.; Jiang, X.; Dun, C. Contribution of agricultural sources to water nitrogen pollution in the Taihu Lake Lake area: A case study of Yili River basin. *J. Agro-Environ. Sci.* **2015**, *34*, 2318–2326.
8. Deng, L.; Zhang, L.; Fan, X.; Wu, Y.; Sun, T.; Fei, K. Characteristics of runoff and sediment yield under different rainfall intensities and slope gradients in erosive weathered granite area. *Trans. Chin. Soc. Agric. Eng.* **2018**, *34*, 143–150.
9. Ge, J.; Yang, J.; Zhan, M.; Dai, S.; Sheng, J. Experimental Study on the Influence of Slope and Rainfall Intensity on Slope Erosion and Sediment Production Process. *Water Resour. Power* **2015**, *33*, 101–104.
10. Zhang, X.Q.; Gu, L.B.; Zhang, K.L. Effects of slope on abortion and sediment production on slope in northwest Guizhou. *J. Soil Water Conserv.* **2015**, *29*, 18–22.
11. Tan, Z.X.; Wang, Z.L.; Wang, S.; Liu, J. The erosion process of fine gullies on loess slopes. *Sci. Soil Water Conserv.* **2012**, *10*, 1–5.
12. Wang, Y.Z.W. Improved interrill erosion prediction by considering the impact of the near-surface hydraulic gradient. *Soil Tillage Res.* **2020**, *203*, 104687. [CrossRef]
13. Bai, L.; Shi, P.; Li, Z.; Li, P.; Wang, W.; Zhao, Z.; Dong, J. The synergistic effect of terraced fields and siltation dam slope and gully control measures on runoff and sediment production in the Loess Plateau. *J. Agric. Eng.* **2023**, *39*, 96–104.
14. Shi, P.; Schulin, R. Erosion-induced losses of carbon, nitrogen, phosphorus and heavy metals from agricultural soils of contrasting organic matter management. *Sci. Total Environ.* **2018**, *618*, 210–218. [CrossRef] [PubMed]
15. Deng, L.Z.; Fei, K.; Sun, T.Y.; Zhang, L.P.; Fan, X.J.; Ni, L. Characteristics of runoff processes and nitrogen loss via surface flow and interflow from weathered granite slopes of Southeast China. *J. Mt. Sci.* **2019**, *16*, 1048–1064. [CrossRef]
16. Liu, J.; Ouyang, X.Q.; Shen, J.L.; Li, Y.; Sun, W.R.; Jiang, W.Q.; Wu, J.S. Nitrogen and phosphorus runoff losses were influenced by chemical fertilization but not by pesticide application in a double rice-cropping system in the Subtropical Hilly Region of China. *Sci. Total Environ.* **2020**, *715*, 136852. [CrossRef] [PubMed]
17. Chen, G.; Elliott, J.A.; Lobb, D.A.; Flaten, D.N.; Braul, L.; Wilson, H.F. Changes in runoff chemistry and soil fertility after multiple years of cattle winter bale feeding on annual cropland on the Canadian prairies. *Agric. Ecosyst. Environ.* **2017**, *240*, 1–13. [CrossRef]
18. Wu, Y.H.; Zhang, L.P.; Deng, L.Z.; Fan, X.J. Simulation of slope and rainfall intensity effects of slope nitrogen loss. *J. Soil Water Conserv.* **2018**, *32*, 27–33.
19. Wang, J.; Chen, J.; Jin, Z.; Guo, J.; Yang, H.; Zeng, Y.; Liu, Y. Simultaneous removal of phosphate and ammonium nitrogen from agricultural runoff by amending soil in lakeside zone of Karst area, Southern China. *Agric. Ecosyst. Environ.* **2020**, *289*, 106745. [CrossRef]
20. Wang, L.; Wu, J.; Xie, J.; Wei, D.; Li, Y.; Wang, J.; Xu, T.; Yang, Z.; Jin, L. Effects of Different Hedgerow Patterns on the Soil Physicochemical Properties, Erodibility, and Fractal Characteristics of Slope Farmland in the Miyun Reservoir Area. *Plants* **2022**, *11*, 2537. [CrossRef]
21. Bao, S. Soil agrochemical analysis. *China Agricultural Publishing*. 2000.

22. Villatoro, S.M.; Bissonnais, Y.L.; Moussa, R.; Rapidel, B. Temporal dynamics of runoff and soil loss on a plotscale under a coffee plantation on steep soil (Ultisol), Costa Rica. *J. Hydrol.* **2015**, *523*, 409–426. [CrossRef]
23. Li, Y.; Jin, L.; Wu, J.; Shi, C.; Li, S.; Xie, J.; An, Z.; Suo, L.; Ding, J.; Wei, D.; et al. Laws Governing Nitrogen Loss and Its Numerical Simulation in the Sloping Farmland of the Miyun Reservoir. *Plants* **2023**, *12*, 2042. [CrossRef] [PubMed]
24. Peng, X.; Dai, Q.; Ding, G.; Li, C. Role of underground leakage in soil, water and nutrient loss from a rock-mantled slope in the Karst rocky desertification area. *J. Hydrol.* **2019**, *578*, 124086. [CrossRef]
25. Hoffman, A.R.; Polebitski, A.S.; Penn, M.R.; Busch, D.L. Long-term Variation in Agricultural Edge-of-Field Phosphorus Transport during Snowmelt, Rain, and Mixed Runoff Events. *J. Environ. Qual.* **2019**, *48*, 931–940. [CrossRef] [PubMed]
26. Zhang, Y.; Wu, H.; Yao, M.; Zhou, J.; Wu, K.; Hu, M.; Shen, H.; Chen, D. Estimation of nitrogen runoff loss from croplands in the Yangtze River Basin: A meta-analysis. *Environ. Pollut.* **2021**, *272*, 116001. [CrossRef]
27. Pan, C.; Ma, L.; Wainwright, J. Particle selectivity of sediment deposited over grass barriers and the effect of rainfall. *Water Resour Res.* **2016**, *52*, 7963–7979. [CrossRef]
28. Yuan, Z.; Chu, Y.; Shen, Y. Simulation of surface runoff and sediment yield under different land-use in a Taihang Mountains watershed, North China. *J. Soil Tillage Res.* **2015**, *153*, 7–19. [CrossRef]
29. Wu, L.; Peng, M.; Qiao, S.; Ma, X.Y. Effects of rainfall intensity and slope gradient on runoff and sediment yield characteristics of bare loess soil. *Environ. Sci. Pollut. Res. Int.* **2018**, *25*, 3480–3487. [CrossRef]
30. Li, H.; He, R.; Hu, J.; Zhou, Y.; Xie, M.; Deng, W.; Wang, J.; Zhao, W.; Zhang, S.; Jiang, Y.; et al. Identifying conservation priority zones and their driving factors regarding regional ecosystem services. *Environ. Dev. Sustain.* **2023**, 1–23. [CrossRef]
31. Song, X.; Gao, Y.; Green, S.M.; Dungait, J.A.J.; Peng, T.; Quine, T.A.; Xiong, B.; Wen, X.; He, N. Nitrogen loss from Karst area in China in recent 50 years: An in-situ simulated rainfall experiment's assessment. *Ecol. Evol.* **2017**, *7*, 10131–10142. [CrossRef]
32. Gao, R.; Dai, Q.; Gan, Y.; Peng, X.; Yan, Y. The production processes and characteristics of nitrogen pollution in bare sloping farmland in a Karst region. *Environ. Sci. Pollut. Res.* **2019**, *26*, 26900–26911. [CrossRef]
33. Yang, F. Research on Particles Detachment and Transport Processes of the Slope in Beijing Mountainous Area. Ph.D. Thesis, Beijing Forestry University, Beijing, China, 2016.
34. Li, J.L.; Cai, Q.G.; Sun, L.Y. Analysis of interaction effects of rainfall intensity, slope degree and slope length on rill erosion. *Sci. Soil Water Conserv.* **2011**, *9*, 8–13.
35. Fan, J.; Lu, X.; Gu, S.; Guo, X. Improving nutrient and water use efficiencies using water-drip irrigation and fertilization technology in Northeast China. *Agric. Water Manag.* **2020**, *241*, 106352. [CrossRef]
36. Chengren, O.; Kaixian, W.; Bozhi, W. Characteristics of Flow Hydraulics and Soil Erosion in Maize and Potato Intercropping Systems. *J. Soil Sci. Plant Nutr.* **2023**, *23*, 5097–5108.

water

MDPI

Article

Salt Drainage Efficiency and Anti-Clogging Effects of Subsurface Pipes Wrapped with Geotextiles

Xu Wang [1,2], Yonghong Zhang [1,2,*], Liqin Fan [1,2] and Jingli Shen [1,2]

1 Institute of Agricultural Resources and Environment, Ningxia Academy of Agriculture and Forestry Sciences, Yinchuan 750002, China; wangxu640321@126.com (X.W.); fanlqnx@126.com (L.F.); lily_s90@163.com (J.S.)

2 Station of Observation and Experiment National Agricultural Environment in Yinchuan, Yinchuan 750002, China

* Correspondence: zyh8401@163.com

Abstract: Subsurface drainage pipes covered with filters and geotextiles are the key to preventing clogging and ensuring efficient drainage. To improve the salt discharge efficiency of these subsurface drainage pipes, different layers of geotextiles were set outside the pipes with the aid of uniform gravel filters. This paper reports our findings from laboratory simulation of subsurface drainage pipes and experiments. The study examined the influence of different layers of geotextiles on the drainage efficiency, salt discharge effects of subsurface drainage pipes, and the effect of superimposed geotextiles on the salt drainage efficiency as well as the anti-clogging effect of subsurface drainage pipes. The results are as follows: (1) The geotextile and filter material wrapped around the subsurface pipe facilitated the movement of water towards the subsurface pipe, which could promote the salt discharge of the subsurface pipe. However, in the single leaching experiment, the reduction in soil pH was not significant for different scenarios. (2) The salt removal rate of the geotextile-wrapped subsurface pipes was more than 95%. The salt removal rate of the double-layer geotextile scenario was the highest (96.7%), and the total salt content of soil profiles was 8.3% and 31.3% lower than those of the single-layer and triple-layer geotextile scenarios, respectively. The drainage efficiency of the double-layer geotextile scenario was the highest, and the salt distribution in the 0–60 cm profile was relatively uniform, ranging from 2.3 to 3.0 $g \cdot kg^{-1}$. (3) The clogging in the triple-layer geotextile scenario was caused by the geotextile, i.e., a dense filter cake layer formed on the surface of the geotextile. The clogging in the single-layer and double-layer geotextile scenarios was the clogging of the geotextile itself, i.e., soil particles retained in the fiber structure of geotextiles. (4) In the case of the single-layer and double-layer geotextile scenarios, the soil particles failed to completely clog the selected geotextiles, and there were still a large number of pores retained. The double-layer geotextiles integrate filtration, clogging prevention, and drainage promotion to provide the best salt drainage with the subsurface pipe. This study reveals the influence of the filter on soil water salt and salt discharge and provides a theoretical explanation and technical justification for the application of the subsurface pipes salt discharge technology in saline soil ameliorate.

Keywords: subsurface pipes drainage; salinity; geotextile; drainage rate; salt discharge rate; leaching desalination rate

1. Introduction

Soil salinization is one of the factors that causes soil degradation, reduces food production, and affects the ecological health of agriculture and forestry [1]. It is reported that more than 100 countries and regions worldwide are affected by soil salinization, with an area of 950 million hectares [2,3]. More than half of the arable land will experience varying degrees of salinization by the middle of this century [4]. Saline soils in China cover an area of nearly 100 million hectares and are mainly distributed in arid inland regions of north China, northeast China, coastal regions, and northwest China [5]. Ningxia Hui

Autonomous Region is located in the arid inland regions of northwest China. The Yellow River irrigation region in the north is obviously characterized by high evaporation and low precipitation, a flat and low-lying terrain, high groundwater levels, and salt migration, which result in severe soil salinization. The area of saline-alkali-cultivated land is about 1.4×10^6 ha, accounting for 32.5% of the total arable land area in the Yellow River irrigation area of Ningxia [6]. Rational utilization of saline-alkali land can serve as an alternative for land resources for the sustainable development of agriculture and forestry, which is of great significance for holding the red line of 1.2 billion ha of arable land. In the meantime, the utilization of saline-alkali land plays an important role in improving the ecological environment in the middle and upper reaches of the Yellow River, as well as promoting ecological protection and high-quality development in the Yellow River Basin.

An efficient irrigation and drainage system can remove solutes and act as a barrier for preventing salinity and alkalinity [7,8]. As an underground drainage facility, the subsurface drainage pipe (SDP) has the advantages of high efficiency, water conservation, and land conservation, which is conducive to mechanized operations [9–11]. SDP can reduce soil salinity while controlling the groundwater level, overcoming the drawbacks of traditional ameliorate that cannot discharge salts from the soil [12,13]. In 1980, the technology of subsurface pipes was introduced in Ningxia Hui Autonomous Region. After years of application, the area of subsurface pipes in the field has reached 1.7×10^5 ha, and the subsurface pipes have become the preferred solution for addressing the salinization hazards in the Yinbei Irrigation District [14]. Subsurface drainage systems promote salt discharge by influencing soil-saturated hydraulic conductivity and effective porosity [15]. Climate and soil conditions are the main factors that determine the spacing and depth of underground pipes [16]. Numerous scholars have proposed the depth, spacing, and diameter of subsurface pipes under different desalination standards [17,18]. However, the clogging of drainage pipes by soil particles, plant roots, and solutes during the operation of saline-alkali land can restrict efficient salt discharge by the subsurface pipes. In the process of long-term application of subsurface pipe drainage, mechanical clogging caused by soil particles, chemical clogging caused by oxidative precipitation of ions by ions being oxidized and precipitated, and biological clogging caused by microbial microorganism activities during the long-term operation of subsurface pipe drainage will cause the blockage of subsurface pipes, which is a non-negligible problem that cannot be ignored, and has become one of the main factors restricting the long-term salt discharge of subsurface pipes. It is also a problem that needs to be solved in for the promotion of subsurface pipe salt discharge technology. Coarse sand or clay loam can form a natural anti-filter layer, which is less likely to clog the filter material outside the subsurface pipes. Soils with a low clay content and a higher finer particle content are more likely to clog the drainage pipe [19]. The soil in the Yinbei region of Ningxia is mostly composed of sticky clay loam or loam clay, which is more likely to cause clogging. At present, subsurface drainage mainly relies on external geotextiles and filter materials such as sand, gravel, and straw to prevent or alleviate siltation. External filter materials can form a highly permeable layer around the drainage pipes to promote drainage and salt discharge [14]. A filter made of porous sand gravels, straw, and furnace slag installed above the subsurface pipes can restrict the movement of soil particles and screen and retain coarse particles around the filter to form a highly permeable stable zone. Geotextiles have a three-dimensional fibrous net structure, which is permeable and filtering, and during water flow, they screen the nearby soil particles and induce the formation of a highly permeable soil skeleton above the geotextile, resulting in a good filtration effect. The geotextile wrapped outside the subsurface pipes needs to satisfy three principles of good filtration, strong permeability, and the fact that it is not easy to block. The main considerations include the O_{90}/d_{90} value (O_{90} is the pore diameter of filter materials, of which 90% is smaller than this value; d_{90} is the size of the soil particles, of which 90% is smaller than this value, and should satisfy $O_{90}/d_{90} \geq 1.0$), suitable permeability coefficient, and thickness of geotextile [19]. In addition, geotextiles can be pre-wrapped, which facilitates mechanized construction and is widely used for on-site subsurface pipes [20]. Superposition treatment of geotextiles can enhance the filtration effect and achieve a high soil retention rate. How to use scientific and reasonable filter materials wrapped around the subsurface pipes to fully utilize the anti-filter effect of filter materials and to ensure efficient salt discharge of subsurface pipe are the bottlenecks that need to be overcome urgently to eliminate the salt alkali barriers.

The wrapped filter materials are the key factor affecting the efficient drainage of subsurface pipes. Field investigations were conducted to investigate the anti-clogging effects of different filter materials of subsurface pipes. There are reports on the influence of different geotextile specifications on salt discharge and clogging in subsurface pipes [21,22]. Geotextiles with a three-dimensional fibrous network structure need to be further investigated to determine whether they can increase filtration and ensure efficient salt discharge in subsurface pipes by stacking. For this purpose, this study systematically evaluated the impact of the wrapped geotextile on the salt discharge efficiency and anti-siltation effect of subsurface pipes by using a self-designed laboratory simulation device for subsurface pipe drainage. The surface and internal structures of geotextiles were observed by using an electron microscope, and the impact of geotextile stacking on salt discharge and the anti-clogging effects of subsurface pipes were revealed. The findings provide a scientific basis for optimizing the engineering parameters of subsurface drainage pipes and improving the quality of low-lying saline soil.

2. Materials and Methods

2.1. Tested Soils

The experiment was conducted from May to September 2022. The soil used for experiments was taken from salt wasteland in Haiyan Village, Yanzidun Township, Huinong District, Shizuishan City, Ningxia (39°03′ N, 106°54′ E). This region belongs to the Yinbei Irrigation Area of the Ningxia Hui Autonomous Region, with low-lying terrain, high groundwater levels, and severe soil salinization. The region has a temperate continental climate and is an arid semi-desert saline land in the middle and upper reaches of the Yellow River. The annual average precipitation and evaporation are 173 mm and 1755 mm, respectively. The precipitation is sparse and unevenly distributed and is mainly concentrated between July and September. The depth of groundwater ranges from 1.3 m to 2.0 m. Soil samples were collected from 0 to 60 cm below the ground surface, with a high salt content of more than 70 $g \cdot kg^{-1}$ and a pH below 8.43 ± 0.01 (Table 1). The 0–60 cm soil layer contained 9.0% clay particles (<0.002 mm), 38.8% silt sand particles (0.002–0.02 mm), and 52.2% sand particles (0.02–2.0 mm). The soil d_{90} values for the 0–20 cm, 20–40 cm, and 40–60 cm soil layers were 87.5 ± 4.3 μm, 95.4 ± 2.9 μm, and 99.6 ± 1.1 μm, respectively. d_{90} is the size of the soil particle, of which 90% are smaller than this value. The ratio of soil clay content to silt content was 0.23, which was lower than 0.5 [19]. The soil was loamy soil and had low cohesion. It is not easy to form a natural filter layer around the subsurface drainage pipes, leading to a high possibility of clogging. Subsurface drainage pipes that are not wrapped with geotextiles and filter materials are susceptible to mechanical clogging, which affects the normal operation of the drainage system.

Table 1. Physical and chemical properties of soil at experimental Site.

Soil Depth/(cm)	Soil Texture	pH	Soil Salinity/($g \cdot kg^{-1}$)	Dry Density/($g \cdot cm^{-3}$)	Porosity/(%)	Field Capacity/(%)
0−20	Loam soil	8.43 ± 0.01	99.86 ± 6.04	1.47 ± 0.05	44.53 ± 0.71	19.0 ± 0.50
20−40	Loam soil	8.42 ± 0.01	54.98 ± 3.13	1.53 ± 0.06	42.26 ± 0.68	19.1 ± 0.35
40−60	Loam soil	8.42 ± 0.01	71.17 ± 5.39	1.56 ± 0.04	41.13 ± 0.74	20.3 ± 0.80

2.2. Simulation Device for Subsurface Pipe Drainage

The referenced field project had a depth of 150 cm, a filter material thickness of 30 cm, a backfill soil layer thickness of 112.5 cm, and an outer diameter of subsurface pipe of

7.5 cm. With reference to the laboratory subsurface pipe drainages, simulated experiments designed by Tao et al. (2016) with a 1:3 scale ratio were adopted for experiments [22]. The designed laboratory simulation test device for subsurface pipe drainage consisted of a water supply system, a seepage box, and a collector box. The seepage box was 1.0 m long, 0.6 m wide, and 0.8 m high (Figure 1). To prevent preferential flow along the wall during the leaching process, the inner layer of the testing device was polished with sandpaper. The depth of the subsurface pipe in the experimental device was 50 cm, and the thickness of the filter material was 10 cm. The thickness of the backfill soil was 37.5 cm, and the diameter of the subsurface pipe was 2.5 cm. The thickness of the organic glass around the seepage box was 8 mm. An overflow port was installed 5 cm away from the top of the seepage box to avoid excessive water accumulation during the leaching process, and three sampling ports were installed on the box wall.

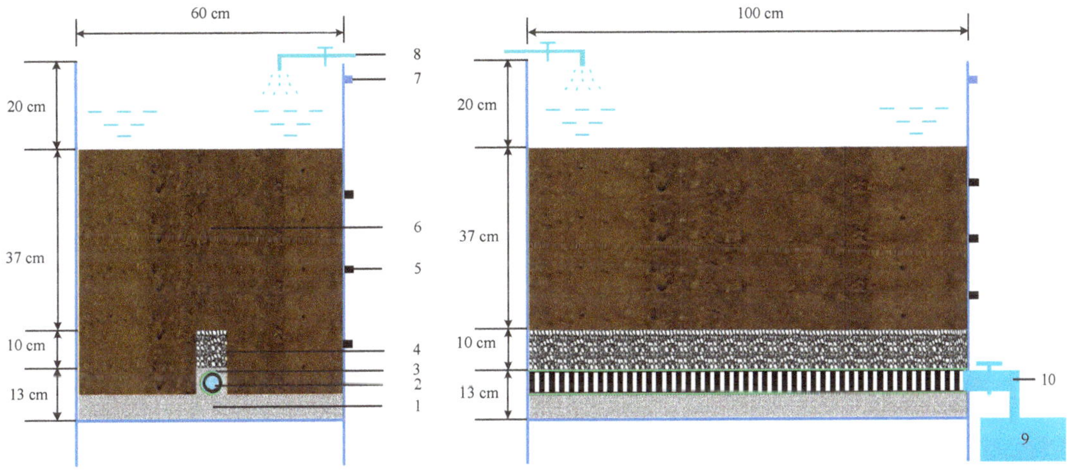

1. Quartz sands, 2. Subsurface pipe, 3. Geotextiles, 4. Gravels, 5. Sampling port, 6. Soil, 7. Overflow port, 8. Water intake port, 9. Header tank, 10. Subsurface pipe outlet

Figure 1. Diagram of simulation device.

2.3. Experiment Design and Process

Geotextiles have a three-dimensional fibrous network structure made of short fiber polypropylene material with a fluffy structure and large porosity, which makes them permeable and filterable (Figure 2). To clarify the impact of geotextile stacking on the salt discharge and anti-clogging effects of subsurface pipes, experiments were conducted in the Yanzidun Township from May to September 2022 based on a self-designed laboratory simulation device for subsurface pipe drainage. The geotextiles wrapped around the subsurface pipes were designed with three different scenarios: single-layer geotextile (T1), double-layer geotextiles (T2), and triple-layer geotextiles (T3), with three replicates for each scenario. The geotextile used in the experiment was a short fiber needle-punched nonwoven geotextile made of polypropylene, with a specification of 100 $g \cdot m^{-2}$, a density of 1.35 ± 0.12 $g \cdot cm^{-3}$, a thickness of 0.75 ± 0.04 mm, a breaking strength of 2.5 ± 0.13 $kN \cdot m^{-1}$, a tearing strength of 0.08 ± 0.01 kN, a vertical permeability coefficient of 0.06 ± 0.01 $cm \cdot s^{-1}$, and an equivalent diameter O_{90} of 0.12 ± 0.01 mm. The geotextile used in the experiment has good alkaline resistance and can be used below pH 9.0. The nonwoven geotextile is a flake with good water collection and drainage performance.

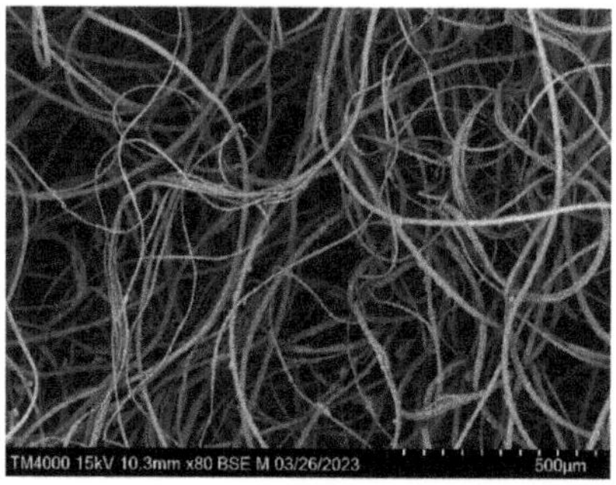

Figure 2. Initial scanning electron microscopy image of geotextile.

Plant residues and debris, such as gravel, were removed from soil samples. The samples were then air-dried, ground, and passed through a 2 mm sieve. A 5 cm thick layer of sand was placed on the bottom, and a subsurface pipe wrapped with geotextile was placed on the quartz sand filter layer. The whole area percent of the subsurface pipe was 5%. The soil was loaded into the test box layer by layer and compacted layer by layer to a thickness of 60 cm. Thereafter, it was left overnight to establish a water content equilibrium. The leaching quota was calculated based on the formula suggested by Hu et al. (2010) [23]. The calculated leaching quota was 6750 $m^3 \cdot hm^{-2}$, which was equivalent to 0.4 m^3 per seepage box. The leaching process began on 19 May 2022 at 9:00 AM. Multiple leaching was accomplished using the water injection method. The depth of the water layer during the leaching process was maintained at 10 cm, and the water source was municipal tap water. The conductivity of the leaching water was 0.55 $mS \cdot cm^{-1}$.

2.4. Sample Collection and Measurement

The anti-clogging effect of the geotextile in the subsurface pipe was demonstrated by the weight of soil lost in the seepage box, as well as the amount and rate of clogging. Three geotextile-treated subsurface pipes were weighed prior to the experiment. During the experiment, the drained water and mineralization of the subsurface pipes were regularly monitored every day, and the mineralization of the leaching water was regularly measured. Upon completion of the experiment (i.e., when the drainage of the subsurface pipe ceased), the soil profile was manually excavated, and samples were taken from both sides of the subsurface pipes to measure the moisture content and salinity. The geotextile wrapped outside the subsurface pipe was weighed, and the geotextile samples were taken from the upstream surface for observation using a scanning electron microscope. Soil profile samples were collected using an auger, and samples were taken every 20 cm from the surface to the top of the filter material directly above the subsurface pipe. The samples were taken at distances of 5 cm, 15 cm, and 25 cm from the center of the subsurface pipe, and each soil sample was repeated three times.

2.5. Measurement and Methods

The moisture content of the soil was determined by the drying method. The supernatant was fully shaken at a soil-to-water ratio of 1:2.5, and pH was measured using the Mettler Toledo S220 multi-parameter tester (METTLER TOLEDO, Switzerland. We bought it in Shanghai, China. Microtrac S3000, York, PA, USA. Hitachi TM4000plus, Tokyo, Japan). The supernatant was fully shaken at a soil-to-water ratio of 1:5 and was measured using a DDS-307A soil conductivity meter, which was converted to the total salt content [24]. The mechanical composition was measured by the laser particle size analyzer (Microtrac S3000). The water filling amount was measured by a flow meter, and mineralization was measured by a conductivity meter. The drainage duration was measured by a stopwatch, and the drainage volume of subsurface pipes was measured by measuring cylinders. The structures of geotextiles were observed using scanning electron microscopy (Hitachi TM4000plus). The quality of soil loss was the weight of soil lost through geotextiles after the completion of experiments. The clogging amount refers to the soil mass adsorbed on the upstream surface of the geotextile and retained inside the geotextile after the completion of experiments. The clogging rate was the ratio of the amount of clogging to the original quality of the geotextile.

2.6. Calculation Formulae

The formulae for calculating the drainage rate (v), drainage efficiency (R_w), salt discharge rate (R_s), desalination rate (L_R), discharge to removal ratio (R), and clogging rate (w) of subsurface pipes are as follows [25]:

$$v = \frac{Q}{TA} \tag{1}$$

where v is the drainage rate of the subsurface pipe, with a unit of cm·h^{-1}; Q is the drainage volume of the subsurface pipe within a certain period of time, with a unit of cm^3; T is the drainage duration, with a unit of h; and A is the cross-sectional area of the seepage channel, with a unit of cm^2.

$$R_w = \frac{Q_w}{L} \times 100\% \tag{2}$$

where R_w is the drainage efficiency of the subsurface pipe, with a unit of %; Q_w is the total amount of drainage from the subsurface pipe, with a unit of cm^3; and L is the total amount of leaching water, with a unit of cm^3.

$$R_s = \frac{Q_s}{S_0 + L_s} \times 100\% \tag{3}$$

where R_s represents the salt discharge rate of the subsurface pipe, with a unit of %; Q_s is the total amount of salt discharged from the subsurface pipe, with a unit of kg; S_0 is the initial total salt content of the soil, with a unit of kg; and L_s is the salt content of the leaching water, with a unit of kg.

$$L_R = \frac{S_0 - S_t}{S_0} \times 100\% \tag{4}$$

where L_R is the soil leaching desalination rate, with a unit of %; S_0 is the total salt content of the soil before leaching, with a unit of kg; and S_t is the total salt content of the soil after leaching, with a unit of kg.

$$R = \frac{R_s}{L_R} \tag{5}$$

where R is the discharge rate of the subsurface pipe; R_s is the salt discharge rate of the subsurface pipe; and L_R is the soil leaching desalination rate.

$$w = \frac{W_t - W_0}{W_0} \times 100\% \tag{6}$$

where w is the siltation rate, with a unit of %; W_t is the weight of the geotextile after the experiment, with a unit of g; and W_0 is the initial weight of the geotextile, with a unit of g.

2.7. Data Processing and Analysis

Microsoft Excel 2010 was used for data processing, while Surfer 10 was utilized to plot nephrograms. SPSS 19.0 software to perform the significance test and correlation analysis.

3. Results

3.1. Soil Moisture Distribution under Different Scenarios

The measurements showed that the soil moisture contents in the depth of 0–60 cm ranged from 15.5 ± 0.45% to 20.5 ± 0.55% for scenario T1, from 13.5 ± 0.20% to 20.0 ± 0.75%

for scenario T2, and from 19.0 ± 0.33% to 21.4 ± 0.88% for scenario T3, respectively (Figure 3). The soil moisture content of the three scenarios follows a decreasing trend as the soil depth increases. The moisture content in the region near the subsurface pipe was significantly higher than that in the upper soil layer. The geotextile and filter material wrapped around the subsurface pipe was conducive to the movement of water toward the subsurface pipe. The soil moisture content of scenario T3 in the depth of 0–60 cm was significantly higher than those of scenarios T1 and T2. There was a small difference in the soil moisture content between scenarios T1 and T2 in the depths of 20–40 cm and 40–60 cm. Scenario T2 had double-layer geotextiles, and the soil moisture content in the depth of 0–60 cm was the lowest among the three scenarios. The stacking of geotextiles allowed for a richer three-dimensional structure and enhanced the effects of permeability and filterability, especially in the triple-layer geotextile scenario.

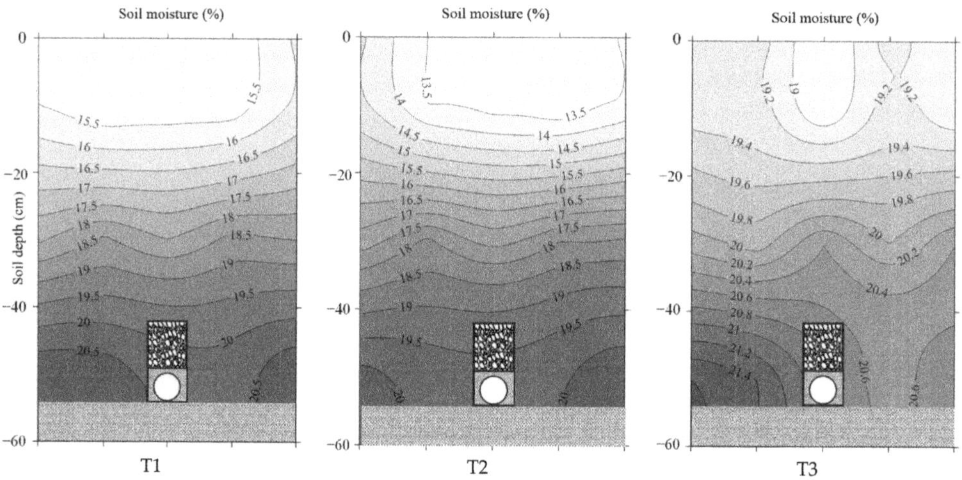

Figure 3. Distribution of soil moisture under different scenarios.

3.2. Soil Salt Distribution under Different Scenarios

The salt content of the tested soils ranged from 71.17 to 99.86 g·kg^{-1}, and the salt content in the soil profiles of the three scenarios after leaching was below 4.8 ± 0.55 g·kg^{-1} (Figure 4). The salinity of the soils in the three scenarios experienced an increasing trend as the soil depth increased. Scenario T1 had a soil profile salinity of less than 2.0 ± 0.22 g·kg^{-1} in the 0–30 cm soil layer, which was the lowest among the three scenarios. The accumulated salinity was below 40 cm in the soil layer. The soil salt content in the 0–60 cm profile of scenario T3 was higher than those of scenarios T2 and T3. The soil salt content below 40 cm was higher than 3.9 ± 0.15 g·kg^{-1}. The salt distribution in the 0–60 cm profile of scenario T2 was relatively uniform, ranging from 2.3 to 3.0 g·kg^{-1}. Compared with the scenarios of the single-layer geotextile and the triple-layer geotextile, the salt leaching of scenario double-layer geotextile was more uniform. Therefore, moderate geotextile stacking has better effects on permeability and filterability, which could promote uniform salt leaching in the soil profile.

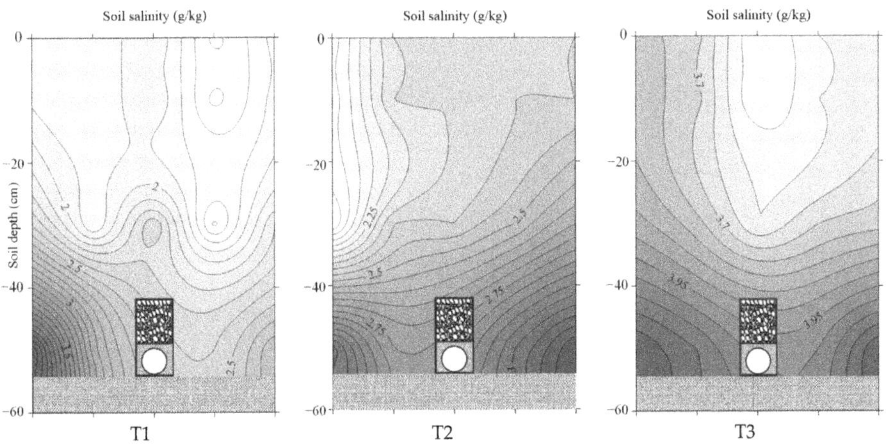

Figure 4. Distribution of soil salinity under different scenarios.

3.3. *Analysis of the Effect of Soil Salt Leaching and Alkali Reduction under Different Scenarios*

Before the experiment, the pH values of the tested soils at 0–20 cm, 20–40 cm, and 40–60 cm were 8.43 ± 0.01, 8.42 ± 0.01, and 8.42 ± 0.01, respectively, which indicated that the soil pH distributions were relatively uniform. The pH values in the depth of 0–60 cm varied from 8.29 to 8.36, from 8.30 to 8.36, and from 8.35 to 8.39 for scenarios T1, T2, and T3, respectively. In the single leaching experiment, the reduction in soil pH was not significant for different scenarios, and the difference among the three scenarios was not significant (Table 2).

Table 2. Effects of different scenarios on soil salinity and pH.

Soil Depth	Original Soil		T1		T2		T3	
	Salinity ($g \cdot kg^{-1}$)	pH	Salinity ($g \cdot kg^{-1}$)	pH	Salinity ($g \cdot kg^{-1}$)	pH	Salinity ($g \cdot kg^{-1}$)	pH
0–20 cm	99.86 ± 6.04	8.43 ± 0.01	1.8 ± 0.16 b	8.36 ± 0.02 a	2.3 ± 0.14 b	8.35 ± 0.02 a	3.6 ± 0.11 a	8.35 ± 0.01 a
20–40 cm	54.98 ± 3.13	8.42 ± 0.01	2.1 ± 0.37 b	8.29 ± 0.08 a	2.4 ± 0.24 b	8.30 ± 0.02 a	3.7 ± 0.10 a	8.35 ± 0.01 a
40–60 cm	71.17 ± 5.39	8.42 ± 0.01	3.1 ± 0.83 b	8.36 ± 0.01 a	3.0 ± 0.19 b	8.36 ± 0.03 a	4.1 ± 0.10 a	8.39 ± 0.02 a
Total salinity (kg)	33.22 ± 10.02	/	1.2 ± 0.06 b	/	1.1 ± 0.06 b	/	1.6 ± 0.02 a	/

Note: Different lowercase letters indicate differences between different scenarios in the same soil layer ($p < 0.05$).

Before the experiment, the salt content of the tested soils exceeded 70 $g \cdot kg^{-1}$, and the total salt content in the 0–60 cm depth was 33.22 ± 10.02 kg. After the leaching experiments, the soil salt contents of the three scenarios were all less than 4.1 ± 0.10 $g \cdot kg^{-1}$. The soil salt contents of scenarios T1, T2, and T3 were between 1.8–3.1 $g \cdot kg^{-1}$, 2.3–3.0 $g \cdot kg^{-1}$, and 3.6–4.1 $g \cdot kg^{-1}$, respectively. After the leaching experiments, the total soil salt contents in the seepage boxes for scenarios T1, T2, and T3 were 1.2 ± 0.06 kg, 1.1 ± 0.06 kg, and 1.6 ± 0.02 kg, respectively, which were 96.4%, 96.7%, and 95.2% lower than those before the experiment. After the leaching experiments, the soil salinity at 0–20 cm of scenario T1 was 1.8 ± 0.17 $g \cdot kg^{-1}$, which was the lowest among the three scenarios. The highest soil salinity was observed in scenario T3, which was significantly higher than those in scenarios T2 and T3 ($p < 0.05$). The soil salinity of scenarios T1 and T2 at 20–40 cm was lower than 2.4 $g \cdot kg^{-1}$, which was significantly lower than that of scenario T3 ($p < 0.05$). The soil salinity at 40–60 cm of scenarios T1 and T2 was significantly lower than that of scenario T3 ($p < 0.05$). The total soil salt content of scenario T2 was the lowest, at 8.3% and 31.3% lower than that of scenarios T1 and T3, respectively. The double-layer geotextile scenario had the best salt leaching effect because of the reasonable geotextile stacking.

3.4. Analysis of Salt Leaching Efficiency under Different Scenarios

The observations from the experiments are listed in Table 3. It can be seen that scenario T2 has the highest values for the majority of the parameters except for the drainage rate (v). The drainage rate (v) was the highest in scenario T1, and the drainage rates of scenarios T1 and T2 were significantly higher than that of scenario T3 ($p < 0.05$). Scenario T2 had the highest drainage efficiency (R_w) and salt discharge rate (R_S), with a subsurface pipe drainage ratio of 0.98, which offered the best salt discharge effect. Scenario T1 had a higher initial drainage rate (v), but the drainage efficiency (R_w) was lower than the other scenarios. Scenario T3 had the lowest drainage rate (v) and salt discharge rate (Rs) among the three scenarios. This was because the thickness of the triple-layer geotextile wrapped around the subsurface pipe was large, which increased the difficulty of water passage and affected the drainage effect, leading to a poor salt removal effect of the subsurface pipe. Scenario T2 promoted the drainage of subsurface pipe and had a removal rate of 0.98, which provided the best salt removal effect.

Table 3. The drainage rate and salt discharge rate of subsurface pipe under different scenarios.

Scenario	Drainage Rate/v ($cm{\cdot}h^{-1}$)	Drainage Efficiency/R_w (%)	Salt Discharge Rate/Rs (%)	Desalination Rate/L_R (%)	(Rs/L_R)
T1	0.037 ± 0.002 a	49.78 ± 0.80 b	83.40 ± 2.65 b	96.39 ± 4.39 a	0.87 ± 0.02 b
T2	0.034 ± 0.002 a	54.92 ± 1.74 a	95.00 ± 2.34 a	96.69 ± 3.94 a	0.98 ± 0.07 a
T3	0.023 ± 0.003 b	54.01 ± 4.23 a	74.38 ± 1.49 c	95.18 ± 2.99 a	0.78 ± 0.04 c

Note: Different lowercase letters indicate differences in the same indicator between different scenarios ($p < 0.05$).

3.5. Geotextile Clogging and Soil Retention Effect

Particles in soils were entrained by moving water, with some particles passing through the geotextile into the drainage system while others intercepted by the geotextile. The soil particles were generally intercepted through clogging and siltation. Scenarios T1 and T2 resulted in siltation inside the geotextile, where fine soil particles were trapped in the fiber structure of the geotextile, leading to a decrease in its permeable area. Scenario T3 resulted in clogging of the upstream surface of the geotextile, where soil particles were collected to form a layer of low-permeability filter cakes on the upstream surface of the geotextile (Figure 5). Observations of the surface and internal structures of geotextile using scanning electron microscopy revealed that the pores of the geotextile wrapped around the subsurface pipe were clogged or blocked by the clusters composed of fine soil particles, resulting in a decrease in the water surface of the geotextile. Scenario T3 resulted in the formation of a clear and dense filter cake layer on the upstream surface. In scenarios T1 and T2, soil particles did not completely block the geotextile, and a large number of pores were present.

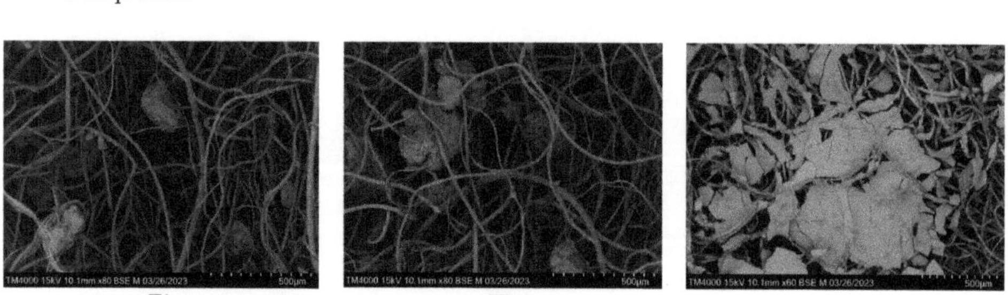

T1 T2 T3

Figure 5. Electron microscope scanning images of geotextile.

The results from the experiments are listed in Table 4, which demonstrates the differences between the scenarios and, more importantly, the order of the clogging rates was

T3 > T2 > T1. After one drainage experiment, the soil loss of all three scenarios exceeded 110 g, with the highest soil loss in scenario T1 and the lowest in scenario T3. Soil loss in scenario T3 was 35.3% and 17.0% lower than those in scenarios T1 and T2, respectively, with significant pairwise differences among the three scenarios ($p < 0.05$). All three scenarios exhibited different forms of clogging. Scenario T1 had the lowest clogging rate, which was 7.0% and 18.5% lower than those of scenarios T2 and T3, respectively. The triple-layer geotextile had a strong interception effect with less soil loss. However, it failed to form a permeable coarse particle skeleton above the subsurface pipe, and more particles would accumulate above the geotextile, making it easier to form filter cakes. The soil retention and anti-clogging effects of geotextile-wrapped subsurface pipe were closely related to the particle content of the tested soil and the thickness of the geotextiles. A smaller thickness would cause some soil particles to form a natural filter layer, pass through the geotextile, and enter the drainage pipes, resulting in soil losses. In practice, the soil particles lost during the drainage process would lead to sedimentation within the drainage pipe. The single-layer geotextile in scenario T1 had a small thickness, and the fine soil particles were prone to enter the drainage pipes through the geotextile, leading to soil losses. The triple-layer geotextile in scenario T3 had a larger thickness, which had a good soil retention effect. However, it was prone to forming a dense filter cake layer on the upstream surface, resulting in a high siltation rate. The double-layer geotextile in scenario T2 considered both soil retention and anti-clogging effects.

Table 4. Soil loss and siltation on the subsurface pipe geotextile material.

Scenario	Weight of Soil Loss/(g)	Initial Weight of Geotextile/(g)	Weight of Clogged Geotextile/(g)	Clogging Rate/(%)
T1	182.73 ± 6.62 a	9.16 ± 0.11 c	12.58 ± 0.68 c	37.33 ± 6.54 c
T2	142.37 ± 3.70 b	18.41 ± 0.58 b	25.80 ± 1.16 b	40.12 ± 5.24 b
T3	118.17 ± 5.60 c	27.96 ± 1.44 a	40.78 ± 2.42 a	45.81 ± 1.65 a

Note: Different lowercase letters indicate differences in the same indicator between different scenarios ($p < 0.05$).

4. Discussion

Within a short period of time after drainage onset, soil particles undergo directional migration. Fine particles in the soil gradually move downwards to fill pores by the drag of water flow. The redistribution of particles leads to continuous compaction of the soil, and the drainage flow rate of the subsurface pipe gradually decreases. In severe cases, it can lead to the loss of drainage function [22].

The interior of geotextile is a fibrous network with many small pores that can form flow channels and block the passage of soil particles. The larger the equivalent pore size of geotextiles, the greater the soil loss. The smaller the thickness, the stronger the interception ability of soil. The stacking of geotextiles can improve the interception and particle-screening ability of soil. The geotextile wrapped around the subsurface pipe is used to screen nearby soil particles under the action of water flow. Large particles are intercepted and gradually accumulate outside to form a soil-permeable skeleton with high permeability, inducing the formation of a natural filter layer in the soil above [26,27].

The research found that geotextiles with a thickness of 0.41 mm had a better permeability and anti-clogging ability than those with a thickness of 0.35 mm [28]. The clogging rate of thick geotextiles alone was 16% lower than that of thin geotextiles. Thin geotextiles not only experience significant clogging on their own but also have a tighter filter cake formed on the surface of the geotextile. Reasonable filter materials could promote the soil above the pipe to form the bridge, filter cake, and natural soil areas from bottom to top, forming a good filter structure. Geotextiles can increase the characteristic particle size (d_{90}) value of the soil by more than 20% through particle screening, inducing the formation of a highly permeable soil skeleton on its surface [19]. However, the available study pointed out that the surface of spun-bonded polypropylene geotextiles had smooth surfaces and was prone to producing thin particles called "pancakes" after contact with the soil [29]. After long-term operation, the small particles in the soil accumulate in the filter layer, causing mechanical clogging and affecting the drainage effect of subsurface pipes. In this experiment, the scanning electron microscopy results showed that the utilization of triple-layer geotextile formed a significant filter cake layer on the upstream surface, and the presence of the filter cake layer continued to absorb smaller soil particles that moved with water. The thickness of the filter cake gradually increased over time, leading to a decrease in permeability. For the scenarios of single-layer and double-layer geotextile, the soil particles were trapped in the fiber structure of the geotextile. The soil particles did not completely block the geotextile, and there were still a large number of effective pores.

Subsurface pipes had a more significant effect on desalination of the upper soil layer in the field, which could reduce the spatial heterogeneity of soil salinity and promote the transformation of soil salinity from "high salinity heterogeneity" to "low salinity homogeneity" [12]. The results of this experiment also verified the conclusion that the soil desalination rate of the wrapped subsurface pipes exceeded 95%, and the desalination effect was significant. The soil salt content changed from surface aggregation to desalination. The pre-leaching water flow rate of the thin geotextile in this experiment was fast, and the salt leaching of the upper soil was adequate. However, the interception effect of the thin geotextile on soil particles was limited.

During water flow, soil particles were easily trapped in the geotextile or through the geotextile into the subsurface pipe. The distribution of soil particles in the fiber structure of the geotextile was not uniform. The thickness of the triple-layer geotextile scenario increased the difficulty for soil particles to pass through the geotextile, making it easier to form a filter cake layer on the upstream surface. This is also the main reason for the different soil moisture movements in the three scenarios. In the early stage of the single-layer geotextile scenario, the leaching water flow was faster due to the large pores of the single-layer geotextile and the high drainage rate in the early stage. Over time, a large amount of water carried sediment into the filter material, and soil particles could easily stay in the geotextile fiber structure and block it, reducing drainage and salt discharge rates in the later stage. The drainage rate of the triple-layer geotextile scenario was low, and the salt removal effect was poor. The double-layer geotextile integrated filtration, anti-clogging, and drainage promotion. The soil profile salt leaching was uniform, and the salt removal effect was the best.

The long-term operation of the subsurface pipe drainage in the field has obvious effects on controlling the groundwater level, reducing the salinity of groundwater, preventing the occurrence of soil secondary salinization, and improving the surface ecological environment. Small subsurface pipe spacing can weaken the spatial heterogeneity of soil salinity and increase the dissolved oxygen content in the soil water, but it can also lead to the loss of soil nutrients, especially nitrogen. Nitrogen loss in farmland is mainly total nitrogen, and the loss of ammonium nitrogen is small [30]. The study pointed out that a reasonable increase in the leaching rate could help enhance the salt removal effect of the subsurface pipes [25]. Since the simulation experiment was conducted in the open-air environment in summer, and the evaporation was large in summer in the northwest arid region, the water loss caused by evaporation during the waterlogging leaching process would lead to insufficient leaching water volume. In this experiment, the salt leaching effect of 0–20 cm soil was satisfactory, but the salt content of some soil layers in the profile exceeded 4 $g \cdot kg^{-1}$, and the salt content did not leach to the expected value. Therefore, it is necessary to fully consider the water loss caused by evaporation during the summer waterlogging leaching process and then adjust the amount of leaching water to obtain the expected results.

To deal with the problem of soil salt easily migrating upward in arid areas, it is particularly important to increase the surface coverage. Dense planting of crops or surface covering with film can help to reduce surface evaporation [31]. With the advancement of high-standard farmland construction projects, drip irrigation combined with a subsurface pipe drainage system has been widely applied, and a subsurface pipe has played an important role in controlling groundwater levels. However, the amount of drip irrigation washing water is limited, especially in arid areas where precipitation is scarce and evaporation is large. Appropriate increase in the vertical porous flow pipes filled with straw or sand can contribute to the salt drainage effect of subsurface pipes [32].

5. Conclusions

The desalination rate of the outside geotextile-wrapped subsurface pipes exceeded 95%, and the desalination rate of the stacked geotextile double-layer scenario was the highest as the moderate geotextile stacking yielded better effects of permeability and filterability, which could promote salt leaching in soil profile. The total salt content of the soil profile was 8.3% and 31.3% lower than those of the single-layer and triple-layer geotextile scenarios, respectively. The removal ratio with the double-layer geotextile scenario was the highest, and the soil profile salt leaching was uniform. The clogging caused by the triple-layer geotextile scenario was caused by the geotextile because the large thickness increased the difficulty of water passage, which formed a dense filter cake layer on the upstream surface of the geotextile. The siltation caused by the single-layer and double-layer geotextile scenarios was due to the clogging of the geotextile itself. That is, soil particles were retained in the fiber structure of the geotextile, the soil particles did not completely block the pores of the geotextile, and there were still a large number of effective pores. The thickness of the double-layer geotextile was moderate enough to provide a suitable three-dimensional structure. It integrates the advantages of filtration, anti-clogging, and drainage promotion to provide the best salt discharge effect for the subsurface pipe. The study revealed the influence of wrapped geotextiles on the salt discharge efficiency and anti-clogging effect of subsurface pipes, which provides evidence for the application of subsurface pipe salt discharge technology to ameliorate saline soils.

Author Contributions: All authors contributed to the conception and design. X.W. and Y.Z. carried out the experimental design, investigation, formal analysis, and writing. L.F. and J.S. contributed to the review and editing. All authors have read and agreed to the published version of the manuscript.

Funding: This research was funded by the National Natural Science Foundation Project (42367043); the Chinese Academy of Sciences 'Western Light' talent training program 'Western Young Scholars' project; the National Key Research and Development Plan Project (2021YFD1900605-05); Agricultural Science and Technology Independent Innovation Project of Ningxia Hui Autonomous Region (NGSB-2021-11-03); and the Observation and Monitoring of Basic Long-Term Scientific and Technological Work in Agriculture (NAES091AE18).

Data Availability Statement: Data are contained within the article.

Conflicts of Interest: The authors declare no conflicts of interest.

References

1. Tarolli, P.; Luo, J.; Park, E.; Barcaccia, G.; Masin, R. Soil salinization in agriculture: Mitigation and adaptation strategies combining nature-based solutions and bioengineering. *iScience* **2024**, *27*, 108830. [CrossRef] [PubMed]
2. Cuevas, J.; Daliakopoulos, I.; Moral, F.; Hueso, J.; Tsanis, I. A review of soil-improving cropping systems for soil salinization. *Agronomy* **2019**, *9*, 295. [CrossRef]
3. Minhas, P.; Ramos, T.; Ben, A.; Pereira, L. Coping with salinity in irrigated agriculture: Crop evapotranspiration and water management issues. *Agric. Water Manag.* **2020**, *227*, 105832. [CrossRef]
4. James, O. History of the roles of gypsum in soil reclamation and establishment of SAR/EC water quality guidelines. In Proceedings of the First IUSS Conference on Sodic Soil Reclamation, Changchun, China, 10–12 May 2021.
5. Yang, J.; Yao, R.; Wang, X.; Xie, W.; Zhang, X.; Zhu, W.; Zhang, L.; Sun, R. Research on Salt-affected Soils in China: History, Status Quo and Prospect. *Acta Pedol. Sin.* **2022**, *59*, 10–27. (In Chinese)
6. Liu, T.; Wang, B.; Xiao, H.; Wang, R.; Yang, B.; Cao, Q.; Cao, Y. Differentially improved soil microenvironment and seedling growth of Amorpha fruticosa by plastic, sand and straw mulching in a saline wasteland in northwest China. *Ecol. Eng.* **2018**, *122*, 126–134. [CrossRef]
7. Askri, B.; Khodmi, S.; Bouhlila, R. Impact of subsurface drainage system on waterlogged and saline soils in a Saharan palm grove. *Catena* **2022**, *212*, 106070. [CrossRef]

8. Askar, M.; Youssef, M.; Chescheir, G.; Negm, L.; King, K.; Hesterberg, D.; Amoozegar, A.; Skaggs, R. DRAINMOD simulation of macropore flow at subsurface drained agricultural fields: Model modification and field testing. *Agric. Water Manag.* **2020**, *242*, 106401. [CrossRef]
9. Ritzema, H.; Abdel-Dayem, S.; El-Atfy, H.; Nasralla, M.; Shaheen, H. Challenges in modernizing the subsurface drainage systems in Egypt. *Agric. Water Manag.* **2023**, *288*, 108484. [CrossRef]
10. Soe, Y.; Shinogi, Y.; Taniguchi, T. Changes in Certain Paddy Soil Properties under Perforated Sheet Pipe as Subsurface Shallow Drainage. *Jpn. Agric. Res. Q. JARQ* **2022**, *56*, 59–66. [CrossRef]
11. Heng, T.; Feng, G.; Yang, L.; He, X.; Yang, G.; Li, F.; Xu, X.; Feng, Y. Soil salt balance in a cotton field under drip irrigation and subsurface pipe drainage systems. *Agron. J.* **2021**, *113*, 4875–4888. [CrossRef]
12. Acharya, U.; Chatterjee, A.; Daigh, A. Effect of Subsurface Drainage Spacing and Depth on Crop Yield. *Agron. J.* **2019**, *111*, 1675–1681. [CrossRef]
13. Chen, G.; Wei, Z.; Liu, H. Study on Soil Desalination Process of Saline-Alkaline Grassland along the Yellow River in Western Inner Mongolia under Subsurface Drainage. *Sustainability* **2022**, *14*, 14494. [CrossRef]
14. Rong, Z.; Wang, S.; Hao, R.; Hao, R.; Tao, Y. Permeability and anti-clogging performance of geotextile envelope material around subsurface drainage pipe in Yinbei Irrigation District in Ningxia. *Trans. Chin. Soc. Agric. Eng.* **2021**, *37*, 68–75. (In Chinese)
15. Talukolaee, M.; Naftchali, A.; Parvariji, L.; Ahmadi, M. Investigating long-term effects of subsurface drainage on soil structure in paddy fields. *Soil Tillage Res.* **2018**, *177*, 155–160. [CrossRef]
16. Mante, A.; Ranjan, R.; Bullock, P. Subsurface drainage for promoting soil strength for field operations in southern Manitoba. *Soil Tillage Res.* **2018**, *184*, 261–268. [CrossRef]
17. Han, D.; Chen, C.; Wang, F.; Li, W.; Peng, H.; Jin, Q.; Bi, B.; Shaghaleh, H.; Hamoud, Y.A. Effects of Subsurface Pipe Drainage Spacing on Soil Salinity Movement in Jiangsu Coastal Reclamation Area. *Sustainability* **2023**, *15*, 13932. [CrossRef]
18. Tian, F.; Miao, Q.; Shi, H.; Li, R.; Dou, X.; Duan, J.; Liu, J.; Feng, W. Study on Water and Salt Transport under Different Subsurface Pipe Arrangement Conditions in Severe Saline–Alkali Land in Hetao Irrigation District with DRAINMOD Model. *Water* **2023**, *15*, 3001. [CrossRef]
19. Liu, W.; Luo, W.; Jia, Z.; Pan, Y.; Tang, S.; Yuan, H.; Li, S. Experimental study on geotextile envelope for subsurface drainage in Yellow River Delta. *Trans. Chin. Soc. Agric. Eng.* **2013**, *29*, 109–116. (In Chinese)
20. Tian, F.; Shi, H.; Miao, Q.; Li, R.; Duan, J.; Dou, X.; Feng, W. Soil Water and Salt Transport in Severe Saline–Alkali Soil after Ditching under Subsurface Pipe Drainage Conditions. *Agriculture* **2023**, *13*, 2196. [CrossRef]
21. Guo, C.; Wu, J.; Zhu, Y.; Lin, Z.; He, S.; Qian, Y.; Yang, H.; Li, H.; Wei, M. Influence of clogging substances on pore characteristics and permeability of geotextile envelopes of subsurface drainage pipes in arid areas. *Geotext. Geomembr.* **2020**, *48*, 735–746. [CrossRef]
22. Tao, Y.; Wang, S.; Xu, D.; Zhai, X. Effect of Structure-type on Improved Subsurface Drainage Performance. *Trans. Chin. Soc. Agric. Mach.* **2016**, *47*, 113–118. (In Chinese)
23. Hu, S.; Tian, C.; Song, Y. Calculation of salinity leaching quota based on saturated infiltration theroy. *Acta Pedol. Sin.* **2010**, *47*, 563–567. (In Chinese)
24. Wang, X.; Sun, Z.; Liu, X.; Bao, Z.; Jiao, B.; Li, X.; Zeng, Y.; Sameh, E. Amelioration of Alkalized Solonchak Soils by Subsurface Gravel Blind Ditches and Desulfurized Gypsum. *Appl. Ecol. Environ. Res.* **2019**, *17*, 7865–7879. [CrossRef]
25. Chen, M.; Huang, J.; Zeng, W.; Ao, C.; Liu, D.; Liu, Y. Water-salt transport law of subsurface pipes with geotextiles under the condition of salt discharge. *Trans. Chin. Soc. Agric. Eng.* **2020**, *36*, 130–139. (In Chinese)
26. Alavi, S.; Naseri, A.; Bazaz, A.; Ritzema, H.; Hellegers, P. Performance evaluation of the Hydroluis drainpipe-envelope system in a saline-sodic soil. *Agric. Water Manag.* **2021**, *243*, 106486. [CrossRef]
27. Lu, P.; Yang, Y.; Luo, W.; Zhang, Y.; Jia, Z. Numerical Simulation of Soil Water-Salt Dynamics and Agricultural Production in Reclaiming Coastal Areas Using Subsurface Pipe Drainage. *Agronomy* **2023**, *13*, 588. [CrossRef]
28. Zhang, D.; Jia, Z.; He, Y.; Zhang, Y.; Luo, K. Experimental study on hydraulic performance of subsurface drainage geotextile envelope in Yangzhou Yanyun Irrigation District. *J. Drain. Irrig. Mach. Eng.* **2023**, *41*, 38–43. (In Chinese)
29. Elzoghby, M.; Jia, Z.; Luo, W. Experimental study on the hydraulic performance of nonwoven geotextile as subsurface drain filter in a silty loam area. *Ain Shams Eng. J.* **2021**, *12*, 3461–3469. [CrossRef]
30. Tan, P.; Wang, S.; Fu, T.; Liu, J.; Han, L. Development history, present situation and prospect of the subsurface drainage technology in China. *Chin. J. Eco-Agric.* **2021**, *29*, 633–639. (In Chinese)
31. Xu, Y.; Liu, H.; Gong, P.; Li, P.; Li, L.; Xu, Q.; Xue, B.; Guo, Y.; Zhang, Y.; Tian, R. Model-Based Optimization of Design Parameters of Subsurface Drain in Cotton Field under Mulch Drip Irrigation. *Water* **2022**, *14*, 3369. [CrossRef]
32. Heng, T.; He, X.; Yang, L.; Xu, X.; Feng, Y. Mechanism of Saline–Alkali land improvement using subsurface pipe and vertical well drainage measures and its response to agricultural soil ecosystem. *Environ. Pollut.* **2022**, *293*, 118583. [CrossRef]

water

MDPI

Article

Research on Salt Drainage Efficiency and Anti-Siltation Effect of Subsurface Drainage Pipes with Different Filter Materials

Xu Wang [1,2], Jingli Shen [1,2], Liqin Fan [1,2] and Yonghong Zhang [1,2,*]

1 Institute of Agricultural Resources and Environment, Ningxia Academy of Agriculture and Forestry Sciences, Yinchuan 750002, China; wangxu640321@126.com (X.W.); lily_s90@163.com (J.S.)
2 Station of Observation and Experiment National Agricultural Environment in Yinchuan, Yinchuan 750002, China
* Correspondence: zyh8401@163.com

Abstract: Subsurface pipes covered with geotextiles and filters are essential for preventing clogging and ensuring efficient drainage. To address low salt discharge efficiency due to subsurface drainage pipes (SDPs) clogging easily, sand gravel, straw, and combined sand gravel–straw were set above SDPs, respectively, within a setting of uniform geotextiles. The influences of different filter materials on the drainage efficiency and salt discharge effect of the SDPs, as well as the effects of different filter materials on the salt drainage efficiency and anti-siltation effect of the SDPs were studied by performing simulation experiments in a laboratory. The results confirmed the following: (1) The salt removal rates of the SDPs externally wrapped with materials exceeded 95%. The subsurface pipe treated with the sand gravel filter material had the highest desalting rate (93.69%) and soil profiles with total salt contents that were 17.7% and 20.5% lower than those treated with the straw and combined sand gravel–straw materials, respectively. (2) The soil salinity of the sand gravel filter material around the SDPs was between 1.57 and 3.6 g/kg, and the drainage rate (R) was 0.97, so its salt-leaching effect was the best. (3) The sand gravel filter material increased the characteristic particle size of the soil above the SDP by 8.4%. It could effectively intercept coarse particles, release fine particles, and facilitate the formation of a highly permeable soil skeleton consisting of coarse particles, such as sand particles surrounding the soil. (4) The use of the straw filter material produced dense filter cake layers on the upstream surfaces of the geotextiles. When the sand gravel and combined sand gravel–straw filter materials were used, soil particles remained in the geotextile fiber structure, and a large number of pores were still retained. Therefore, the sand gravel filter material was the most suitable for the treatment of Yinbei saline–alkali soil in Ningxia Hui Autonomous Region.

Keywords: subsurface pipe drainage; salinity; filter material; drainage rate; salt discharge rate; leaching desalination rate

1. Introduction

Soil salinization is one of the factors that causes soil degradation, reduces food production, and affects the ecological health of agriculture and forestry [1]. It has been reported that more than 100 countries and regions worldwide are affected by soil salinization with an area of 950 million hectares (ha) [2,3]. By the middle of the twenty-first century, more than half of arable lands will experience different degrees of salinization [4]. Saline soils in China cover an area of about 100 million ha and are mainly distributed in arid inland regions of Northern, Northeastern, Northwestern, and coastal China [5]. Ningxia Hui Autonomous Region is located in the arid inland region of Northwest China. The Yellow River irrigation region in the north is clearly characterized by high evaporation and low precipitation levels, a flat and low-lying terrain, high groundwater levels, and salt migration, resulting in severe soil salinization. In the Yellow River irrigation region of Ningxia Hui Autonomous Region, the area of saline–alkali cultivated land is about 1.4×10^6 ha, accounting for 32.5% of the total arable land area [6]. The rational utilization of saline–alkali land can be considered as an alternative for land resources for the sustainable development of agriculture and forestry, which is of great importance for holding the red line of 1.2 billion ha of arable land. The amelioration and utilization of saline–alkali land can reduce groundwater levels and prevent secondary salinization. Planting plants can improve the soil structure, increase the level of coverage, and improve the field microclimate [7]. Especially with the intensification of the greenhouse effect, arid areas are facing both salinization and water shortages, which undoubtedly have an adverse impact on the economy, society, and ecological environment. Therefore, utilization of saline–alkali land plays a key role in the improvement of the ecological environment in the middle and upper reaches of the Yellow River as well as the promotion of ecological protection and high-quality development in the Yellow River Basin.

An efficient irrigation and drainage system can eliminate solutes and serve as a barrier to prevent salinity and alkalinity [8,9]. As part of underground drainage facilities, SDPs are highly efficient and contribute to water as well as land conservation, giving rise to mechanized operations [10–12]. SDPs can decrease soil salinity while controlling groundwater levels and overcome the drawbacks of traditional amendment methods that cannot discharge salts from soil [13,14]. A subsurface pipe drainage technique was introduced in Ningxia Hui Autonomous Region in 1980. After years of its application, the total area of SDPs in the field has reached 1.7×10^5 ha, and subsurface pipes have become the preferred solution for solving salinization hazards in Yinbei Irrigation District [15]. Subsurface drainage systems promote salt discharge by affecting the saturated hydraulic conductivity and effective porosity of the soil [16]. Climatic and soil conditions are the main factors determining the spacing and depth of SDPs [17]. Several researchers have proposed measurements for the depth, spacing, and diameter of SDPs based on different desalination standards [18,19]. However, the clogging of SDPs by soil particles, plant roots, and solutes in saline–alkali land can restrict their efficient salt discharge. Coarse sand or clay loam can form natural anti-filter layers, which are less likely to clog the filter material around SDPs. Soils with low clay contents and high finer particle contents are more likely to clog SDPs [20]. The soil in the Yinbei region of Ningxia Hui Autonomous Region mostly consists of sticky clay loam or loam clay, which are more likely to cause clogging.

Currently, the main components of the subsurface drainage system are external geotextiles and filter materials such as sands, gravels, and straw to prevent or alleviate siltation. Filter materials can effectively prevent or reduce siltation caused by soil that enters the SDPs with the water flow. The SDPs must be equipped with external filter materials to form a high permeable layer around them to reduce the water flow resistance and head loss [21]. Sand gravel filter materials are laid around the SDPs. Since the filter materials wrapped around the SDPs are porous media of the same texture, a continuous soil–water potential gradient field is formed around the SDPs. The water in the soil has a significant hysteresis effect, while the water and salt in the soil accumulate at the bottom of the SDPs, weakening the drainage effect. Sand–gravel filter materials are laid on top of the SDP. The water conductivity difference between the fine sand mat and clay loam is employed to slow down the soil water flow rate around the bottom of the SDP, so that a local saturated zone is generated in the upper part of the SDP, promoting salt drainage through the SDP [22]. Reasonable filter materials allow the soil above the SDP to form a bridge, filter cake, and natural soil areas from bottom to top with a good filter structure. The main considerations when setting up the filter material for the SDP include the following: the O_{90}/d_{90} value (O_{90} is the pore diameter of the filter materials, of which 90% are smaller than this value; d_{90} is the size of the soil particles, of which 90% are smaller than this value; and $O_{90}/d_{90} \geq 1.0$ should be satisfied), suitable permeability coefficient, and thickness of the filter material [20]. Research on the anti-siltation of SDP filter material mainly adopts laboratory simulation experiments [23], involving seepage boxes, soil columns, soil boxes, and so on. Research has also been conducted by sampling outdoor field surveys [24]. The application of filter materials is greatly influenced by soil characteristics, and their reverse filtration effect is closely related to the composition of soil particles and the pore size of filter materials [25]. The clogging type of geotextiles wrapped with sand and straw filter materials as well as the formation of soil-permeable skeletons around filter materials are bottlenecks that need to be investigated and resolved.

Filter materials are the key factor affecting the drainage efficiency of SDPs. Field experiments have been performed to explore the anti-clogging effects of various SDP filter materials. Laboratory simulation experiments have confirmed an improvement in drainage and desalination by external adsorption materials [24,26]. However, it is worthwhile to conduct further research on the effects of different filter materials, such as sand and straw, on the salt drainage efficiency and anti-silting effect of SDPs by using typical saline–alkali soil in the Yinbei region of Ningxia Hui Autonomous Region and a laboratory simulation device.

Therefore, we designed a laboratory simulation device to systematically evaluate the effects of sand gravel and straw filter materials on the salt discharge efficiency and anti-silting effect of SDPs. The soil particle size composition above the filter materials was analyzed. The surface and internal structures of the geotextiles were evaluated using an electron microscope to reveal the clogging types of SDPs externally wrapped with different filter materials, so as to guide the scientific development of SDP filter materials. Our findings provide a scientific basis for the long-term drainage and desalination of SDPs as well as an improvement in the quality of low-lying saline soil.

2. Materials and Methods

2.1. Tested Soils

Experiments were performed from May to October 2023. The soil samples used for experiments were taken from Haiyan Village, Yanzidun Township, Huinong District, Shizuishan City, Ningxia Hui Autonomous Region (39°03′ N, 106°54′ E, Figure 1). This region belongs to the Yinbei Irrigation Area of Ningxia Hui Autonomous Region, with low-lying terrain and high groundwater levels. Under the influence of the arid climate, the soil salt in this region has risen significantly, and soil salinization is severe. The region is an arid semi-desert saline land in the middle and upper reaches of the Yellow River with a temperate continental climate. Annual average precipitation and evaporation rates are 173 mm and 1755 mm, respectively. Precipitation in this region is sparse, unevenly distributed, and mainly concentrated from July and September. The depth of groundwater ranges from 1.3 to 2.0 m. Soil samples were collected from 0 to 60 cm below ground surface, with high salt content up to 50 g·kg^{-1} and pH below 8.93 (Table 1). The 0–60 cm soil layer contained 10.2% clay particles, 36.4% silt sand particles, and 53.4% sand particles, while the soil texture was loam soil [27]. The d_{90} values of soil layers at depths of 0–20, 20–40, and 40–60 cm were 86.8, 94.7, and 99.4 μm, respectively. The ratio of soil clay content to silt content was 0.28, which was lower than 0.5 [20]. The soil type was loamy soil with a low level of cohesion. It is not easy to form a natural filter layer around subsurface drainage pipes, leading to a high possibility of clogging. SDPs that are not wrapped with geotextiles and filter materials are susceptible to mechanical clogging, which affects the normal operation of drainage systems.

Table 1. Physical and chemical properties of soil at the experimental site.

Soil Depth /(cm)	Soil Texture	pH	Soil Salinity/(g/kg)	Dry Density/(g/cm^3)	Porosity/ (%)	Field Capacity/ (%)
0~20	Loam soil	8.83 ± 0.02	87.5 ± 4.14	1.45 ± 0.03	45.28 ± 0.52	18.7 ± 0.42
20~40	Loam soi	8.93 ± 0.05	59.6 ± 3.47	1.52 ± 0.04	42.64 ± 0.46	19.0 ± 0.37
40~60	Loam soi	8.79 ± 0.03	58.4 ± 3.85	1.58 ± 0.02	40.38 ± 0.38	20.1 ± 0.64

Note: Field capacity is water content by mass.

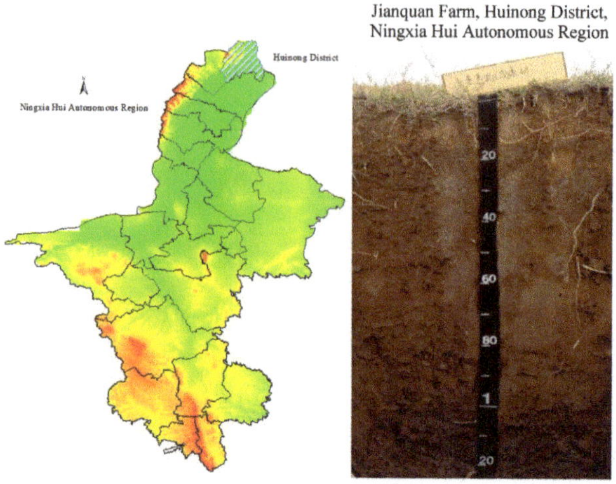

Figure 1. Geographical position of the study area and photo of soil type [27] (Red represents mountains in Ningxia, yellow represents hills in Ningxia, and green represents plains in Ningxia).

2.2. Simulation Device for Subsurface Pipe Drainage

The depth of the object at the studied site was 150 cm, the thickness of filter material was 30 cm, the thickness of backfill soil layer was 112.5 cm, and the outer diameter of SDP was 7.5 cm. With reference to laboratory simulations of subsurface pipe drainage, simulation experiments designed by Tao et al. (2016) with a 1:3 scale ratio were adopted for our experiments [28]. The designed laboratory simulation test device for subsurface pipe drainage was composed of a water supply system, a seepage box, and a collector box. Seepage box was 1.0 m long, 0.6 m wide, and 0.8 m high (Figure 2). To prevent preferential flow along the wall during leaching process, the inner surface of testing device was polished with sandpaper. In the experimental device, the depth of SDP was 50 cm, and the thickness of filter material was 10 cm. The thickness of backfill soil was 37.5 cm, and diameter of SDP was 2.5 cm. The thickness of the organic glass around the seepage box was 8 mm. An overflow port was installed 5 cm away from the top of the seepage box to prevent excessive water accumulation during leaching process, and three sampling ports were installed on box wall.

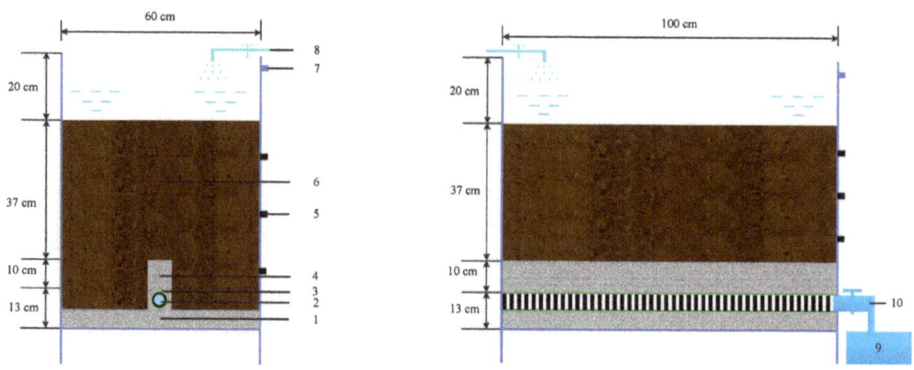

(**a**) Front view (**b**) Side view

Figure 2. Schematic diagram of simulation device. 1. Quartz sands, 2. SDP, 3. Geotextiles, 4. Gravels, 5. Sampling port, 6. Soil, 7. Overflow port, 8. Water intake port, 9. Header tank, and 10. Outlet of SDP.

2.3. Experiment Design and Process

To determine the effects of sand gravel, straw, and combined sand gravel–straw filter materials on salt drainage and anti-silting effects of SDPs, the laboratory simulation experiments were conducted in Yanzidun Township from May to October 2023 based on a self-designed laboratory simulation device for subsurface pipe drainage. Using unified covering of geotextiles for subsurface drainage pipes, three scenarios with sand gravel (T1), straw (T2), and combined sand gravel–straw (T3) as wrapped filter materials were designed, with three tests repeated for each scenario. Average values of the data from three tests were used to analyze the test results.

Plant residues and debris such as gravel were removed from soil samples. The prepared samples were then air-dried, grinded, and passed through a 2 mm sieve. A 5 cm thick sand layer was placed on the bottom, and a subsurface pipe wrapped with geotextiles was placed on quartz sand filter layer. The hole area percentage of subsurface pipe was 5%. The soil was loaded into the test soil box and was compacted layer by layer to a thickness of 60 cm. Thereafter, it was left overnight to establish a water content equilibrium. Leaching quota was calculated according to the equation derived by Hu et al. (2010) [29]. The calculated leaching quota was 6750 $m^3 \cdot hm^{-2}$, which was equivalent to 0.4 m^3 per seepage box. Leaching process began at 9:00 AM on 7 June 2023. The leaching process was repeated multiple times using water injection method. The depth of water level during leaching process was maintained at 10 cm, and municipal tap water was used. Leaching water salinity was 0.44 $g \cdot L^{-1}$. Photos of simulation experiment are presented in Figure 3.

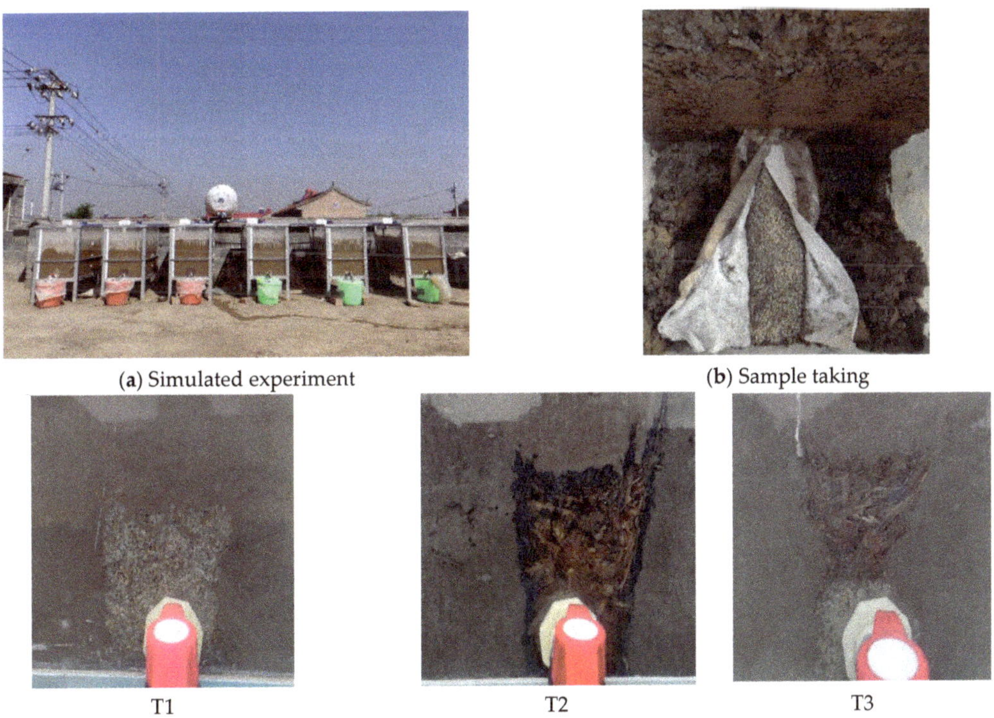

Figure 3. Photos of simulated experiment.

2.4. Sample Collection and Measurement

The anti-clogging effect of geotextiles for SDPs was demonstrated by the weight of the soil lost in seepage box as well as the amount and rate of clogging. Three SDPs treated with geotextiles were weighed before experiments. During the experiments, drained water and mineralization of SDPs were regularly monitored every day, while the mineralization of leaching water was regularly measured. Upon the completion of experiments (i.e., when subsurface pipe drainage ceased), soil profile was manually excavated, and samples were collected at distances 5, 15, and 25 cm away from the center of the SDP to measure water content, salt content, and soil particle size. Each soil sample was collected three times. The soil discharged through the SDP was collected and dried for weighing, the geotextile wrapped around the SDP was removed and weighed, and geotextile samples were taken from face water surface for observation by scanning electron microscopy.

2.5. Measurement and Methods

The water content of soil was determined by drying method. The supernatant was fully shaken at a soil-to-water ratio of 1:2.5, and the pH was measured using a Mettler Toledo S220 (METTLER TOLEDO, Greifensee, Switzerland) multi-parameter tester. The obtained supernatant was fully shaken at a soil-to-water ratio of 1:5 and was measured using a DDS-307A (INESA, Shanghai, China) soil conductivity meter, which was converted to the total salt content [30]. The mechanical composition of the soil was determined by a laser particle size analyzer (Microtrac S3000, Microtrac, York, PA, USA). Water filling amount was measured by a flow meter, and mineralization was measured using a conductivity meter. Drainage duration was measured using a stopwatch, and drainage volume of the SDP was obtained using measuring cylinders. Geotextile structures were observed using scanning electron microscopy (Hitachi TM4000plus, Hitachi, Tokyo, Japan). Loss of soil mass was calculated as the weight of soil lost through geotextiles after the completion of experiments. Clogging amount was considered as the soil mass adsorbed on the upstream surface of the geotextile and retained inside it after the completion of experiments. Clogging rate was defined as the ratio of the clogging amount to the original weight of the geotextile.

2.6. Calculation Equations

The equations for calculating drainage rate (R_w), salt discharge rate (R_s), desalination rate (L_R), discharge-to-removal ratio (R), and clogging rate (w) of SDPs are expressed as follows [31]:

$$R_w = \frac{Q_w}{L} * 100\% \quad (1)$$

where R_w is the drainage rate of SDP, in %; Q_w is total drainage amount from SDP, in cm^3; and L is total leaching water amount, in cm^3.

$$R_s = \frac{Q_s}{S_0 + L_s} * 100\% \quad (2)$$

where R_s is the salt discharge rate of the subsurface pipe, in %; Q_s is the total amount of salt discharged from the subsurface pipe, in kg; S_0 is the initial total salt content of soil, in kg; and L_s is the salt content of leaching water, in kg.

$$L_R = \frac{S_0 - S_t}{S_0} * 100\% \quad (3)$$

where L_R is the soil leaching desalination rate, in %; S_0 is the total salt content of soil before leaching, in kg; and S_t is the total salt content of soil after leaching, in kg.

$$R = \frac{R_s}{L_R} \quad (4)$$

where R is the discharge rate of SDP; R_s is the salt discharge rate of SDP; and L_R is the soil leaching desalination rate.

$$w = \frac{W_t - W_0}{W_0} * 100\% \quad (5)$$

where w is the siltation rate, in %; W_t is the weight of the geotextile after experiment, in g; and W_0 is the initial weight of the geotextile, in g.

2.7. Data Processing and Analysis

Microsoft Excel 2010 was employed for data processing while Surfer 10 was used to plot nephograms. SPSS 19.0 software was applied to perform significance tests and correlation analyses.

3. Results

3.1. Soil Moisture Distribution under Different Scenarios

Our measurements showed that the water contents of the soil at a depth of 0–60 cm ranged from 12.6 ± 0.23% to 19.9 ± 0.32% for scenario T1, from 15.5 ± 0.15% to 19.8 ± 0.28% for scenario T2, and from 13.7 ± 0.24% to 19.6 ± 0.36% for scenario T3 (Figure 4). The water content of the soil layer at a depth of 0–20 cm under scenario T1 was the lowest, while that under scenario T2 was the highest. The water content of the soil layer at a depth of 0–60 cm under scenario T2 was higher than those under scenarios T1 and T3, and the difference in the water contents of the soil layers at depths of 20–40 cm and 40–60 cm between scenarios T1 and T3 was small. The water contents of the soil under the three scenarios increased with the increase in soil depth, and the water content of the area near the SDPs was remarkably higher than that of the upper soil. The SDPs were coated with sand, straw, and other filter materials, which may have resulted in the movement of water towards them.

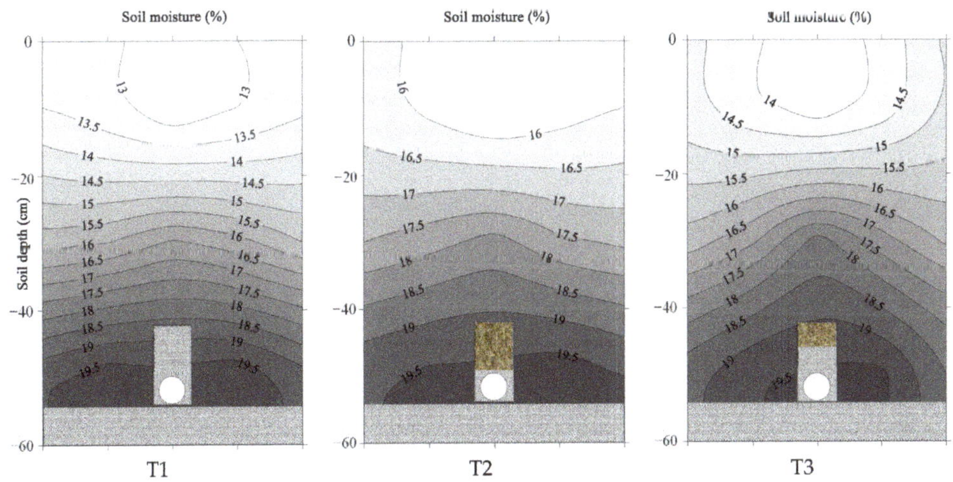

Figure 4. Soil moisture distributions under different scenarios.

3.2. Soil Salt Distributions under Different Scenarios

The salt contents of the tested soils ranged from 58.4 to 87.5 g·kg^{-1}, and the salt contents in the soil profiles under the three scenarios after leaching were below 3.9 g·kg^{-1} (Figure 5). The salt content of the soil profile under scenario T1 was between 1.57 ± 0.15 and 3.60 ± 0.17 g/kg, while those under scenarios T2 and T3 were between 2.70 ± 0.25 and 3.92 ± 0.08 g/kg and between 2.35 ± 0.15 and 3.75 ± 0.30 g/kg, respectively. The salt contents of the soil profiles of the three scenarios increased with the increase in soil depth. The salt content of the soil immediately above the SDPs was remarkably lower than that on both sides of the SDPs, with the lowest value. The salinity of the soil profile at a depth of 0–20 cm under scenario T1 was lower than 2.4 g/kg, which was the lowest among the three scenarios. The salt accumulated in the soil layer at a depth below 40 cm. The salt distribution of the soil layer at a depth of 0–60 cm under scenario T3 was uniform, and the salinity of the soil profile ranged from 2.4 to 3.8 g/kg.

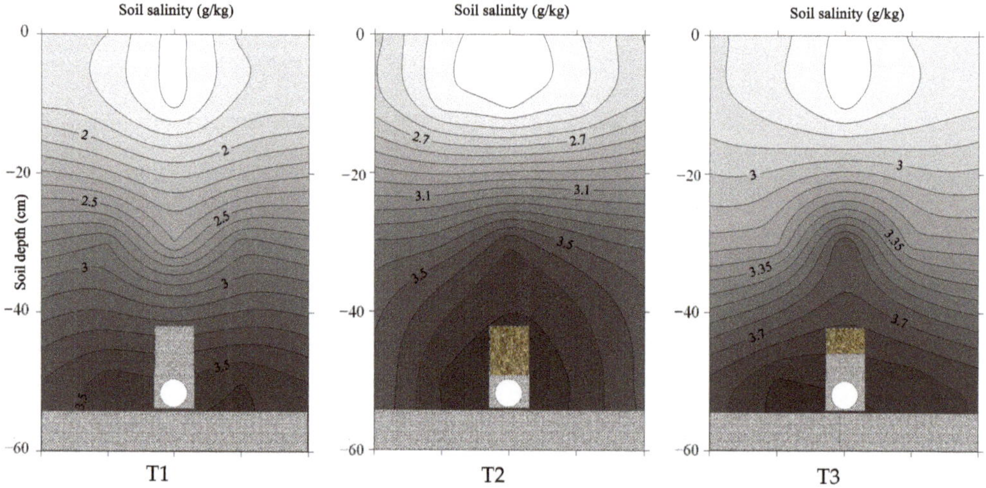

Figure 5. Distributions of soil salinity under different scenarios.

3.3. Analysis of the Effects of Soil Salt Leaching and Alkali Reduction under Different Scenarios

Before the experiment, the pH values of the samples collected from the soil layers at depths of 0–20, 20–40, and 40–60 cm were 8.83 ± 0.02, 8.93 ± 0.05, and 8.79 ± 0.03, respectively, which indicated that the pH distributions of the soil were relatively uniform. The pH values at depths of 0–60 cm under scenarios T1, T2, and T3 varied from 8.52 to 8.76, from 8.46 to 8.77, and from 8.50 to 8.76, respectively. In the single leaching experiment, the reduction in soil pH was not significant under the different scenarios, and no significant differences were found among the three scenarios (Table 2).

Table 2. Effects of different scenarios on soil salinity and pH.

Soil Depth/(cm)	Original Soil		T1		T2		T3	
	Salinity (g/kg)	pH	Salinity (g/kg)	pH	Salinity (g/kg)	pH	Salinity (g/kg)	pH
0~20	87.5 ± 4.14	8.83 ± 0.02	1.75 ± 0.16	8.52 ± 0.02	2.43 ± 0.14	8.53 ± 0.02	2.78 ± 0.07	8.53 ± 0.03
20~40	59.6 ± 3.47	8.93 ± 0.05	2.70 ± 0.26	8.76 ± 0.03	3.42 ± 0.08	8.77 ± 0.04	3.36 ± 0.30	8.76 ± 0.03
40~60	58.4 ± 3.85	8.79 ± 0.03	3.45 ± 0.09	8.53 ± 0.01	3.75 ± 0.25	8.46 ± 0.01	3.78 ± 0.16	8.50 ± 0.02
Total salinity (kg)	30.03 ± 0.17	/	1.16 ± 0.07	/	1.41 ± 0.07	/	1.46 ± 0.03	/

Before the experiments, the salt content of the tested soils exceeded 58 $g{\cdot}kg^{-1}$, and the total salt content in the soil layer at a depth of 0–60 cm was 30.03 kg. After the leaching experiments, the soil salt contents under the three scenarios were all less than 3.8 $g{\cdot}kg^{-1}$. The soil salt contents under scenarios T1, T2, and T3 were between 1.75 and 3.45, 2.43 and 3.75, and 2.78 and 3.78 $g{\cdot}kg^{-1}$, respectively. After the leaching experiments, the total soil salt contents in the seepage boxes under scenarios T1, T2, and T3 were 1.16 ± 0.07, 1.41 ± 0.07, and 1.46 ± 0.03 kg, which were 96.1%, 95.3%, and 95.1% lower than those before the experiments, respectively. After the leaching experiments, the soil salinity at a depth of 0–20 cm under scenario T1 was 1.75 ± 0.16 $g{\cdot}kg^{-1}$, which was significantly lower than those under scenarios T2 and T3. The soil salt content of the soil layer at a depth of 20–40 cm under scenario T1 was 2.70 ± 0.26 g/kg, which was much lower than those under scenarios T2 and T3. The soil salt content of the soil layer at a depth of 40–60 cm under scenario T1 was lower than those under scenarios T2 and T3, and no significant differences were observed among the three scenarios ($p > 0.05$). The total soil salt under scenario T1 was the lowest, which was 17.7% and 20.5% lower than those under scenarios T2 and T3, respectively. The sand–gravel scenario had the best salt-leaching effect.

3.4. Analysis of Salt-Leaching Efficiency under Different Scenarios

Table 3 summarizes our experimental observations. It was seen that the drainage rate (R_w) was the largest (34.42%) under scenario T1 and the lowest under scenario T2 and that the R_w values of the three scenarios were significantly different ($p < 0.05$). Scenario T1 had the largest R_w, salt discharge rate (Rs), and ratio of drainage rate (R) among the three scenarios, and there were significant differences in the R_w and Rs indexes among the three scenarios ($p < 0.05$). The desalination rate (L_R) under scenario T1 was 96.16%, which was significantly higher than those under scenarios T2 and T3 ($p < 0.05$). Scenario T2, in which straw filter material was used, had a higher initial R_w value but lower R_w and Rs values than those under the other two scenarios, which was due to the large pores of the straw filter material and its high drainage rate in the early stage. With the passage of time, a large amount of water carried sediment into the filter materials, the soil particles were prone to stay in the straw filter material, and clogging occurred, causing a decrease in R_w and Rs values in the later stage. The drainage ratio (R) under scenario T1 was 0.97, which was the best for salt drainage.

Table 3. Drainage rate and salt discharge rate of SDPs under different scenarios.

Scenario	Drainage Rate/R_w (%)	Salt Discharge Rate/Rs (%)	Desalination Rate/L_R (%)	R (Rs/L_R)
T1	34.42 ± 0.17 a	93.69 ± 0.14 a	96.16 ± 0.15 a	0.97 ± 0.0002 b
T2	32.52 ± 0.27 c	87.72 ± 0.17 c	95.31 ± 0.14 b	0.92 ± 0.0004 a
T3	33.43 ± 0.19 b	89.72 ± 0.18 b	95.16 ± 0.16 b	0.94 ± 0.0004 c

Note: Different letters indicate significant differences among different scenarios at 0.05 level.

3.5. Geotextile Clogging and Soil Retention Effect

After one leaching test cycle, the soil losses under the three scenarios exceeded 120 g (Table 4). The soil loss under scenario T1 was the highest, while that under scenario T2 was the lowest. The soil loss under scenario T1 was 34.9% and 10.2% lower than those under scenarios T2 and T3, respectively, and significant differences were found among the three scenarios ($p < 0.05$). All three scenarios exhibited different clogging degrees, and the clogging rate under scenario T1 was the lowest, which was 22.2% and 6.3% lower than those under scenarios T2 and T3, respectively. Compared with the original soil, after leaching, the clay and silt contents of the soil above the filter material under scenario T1 were decreased, while the sand content was increased. However, the clay and silt contents of the soil above the filter material under scenario T2 were increased, while the sand content was decreased (Table 5). The d_{90} value of the soil above the filter material under scenario T1 was 8.4% higher than that of the original soil. Under scenario T1, the clogging rate was decreased, and the d_{90} value of the soil above the filter material was significantly higher than those under scenarios T2 and T3 ($p < 0.05$), which presented a good filtration performance. The straw filter material scenario had a strong interception effect, in which more particles were gathered above the geotextiles, a filter cake was more likely to form, and the soil preservation effect was good. However, a dense filter cake layer could be easily formed on the upstream surface, resulting in high clogging rates. The pores of the sand–gravel filter material were more uniform, which could intercept coarse soil particles and release fine particles, forming a good filter body structure and improving the filtration effect.

Table 4. Soil loss and siltation on geotextile materials around SDPs.

Scenario	Weight of Soil Loss/(g)	Initial Weight of Geotextile/(g)	Weight of Clogged Geotextile/(g)	Clogging Ratio/(%)
T1	198.78 ± 2.33 a	9.12 ± 0.13 b	12.57 ± 0.09 c	37.75 ± 0.01 c
T2	129.43 ± 5.32 c	9.30 ± 0.06 ab	13.81 ± 0.03 a	48.51 ± 0.01 a
T3	144.13 ± 5.94 b	9.35 ± 0.10 a	13.12 ± 0.13 b	40.28 ± 0.02 b

Note: Different letters indicate significant differences among scenarios at the same soil depth at 0.05 level.

Table 5. Analysis results of soil particle size above the filter material.

Scenario	Mass Fraction/(%)			d_{90}/(μm)
	Clay	Silt	Sand	
Original soil	10.22 ± 1.24	36.41 ± 1.45	53.41 ± 1.29	94.70 ± 4.10
T1	8.02 ± 0.31 c	24.06 ± 7.28 b	67.92 ± 7.43 a	108.25 ± 5.32 a
T2	18.20 ± 1.54 a	42.93 ± 4.25 a	38.87 ± 5.78 b	74.25 ± 2.71 c
T3	13.98 ± 1.35 b	27.23 ± 2.68 b	58.79 ± 3.81 a	89.83 ± 4.58 b

Note: Different letters indicate significant differences among treatments at the same soil depth at 0.05 level.

3.6. Clogging Types of Geotextiles under Different Scenarios

The particles in the soils were entrained by flowing water; some particles passed through the geotextiles into the drainage system, while others were intercepted by the geotextiles. The soil particles were generally intercepted through clogging and siltation. Under scenarios T1 and T3, siltation occurred inside the geotextiles, in which fine soil particles were trapped in the geotextile fiber structure, decreasing its permeable area. Under scenario T2, clogging occurred in the upstream surface of the geotextile, where soil particles were accumulated to form a low-permeability filter cake layer (Figure 6). Our observations of the surface and internal structures of the geotextile using scanning electron microscopy showed that the pores of the geotextile wrapped around the SDPs were clogged or blocked by clusters of fine soil particles. Under scenario T2, a clear and dense filter cake layer was formed on the geotextile upstream surface. Under scenarios T1 and T3, the soil particles did not completely block the geotextile, and a large number of pores were present.

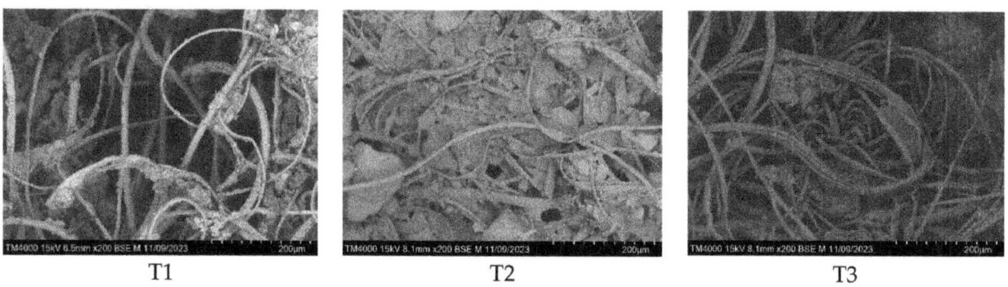

Figure 6. Electron microscopy scanning images of geotextiles.

4. Discussion

The SDPs had stronger effects on the desalination of the upper soil layer in the field, which could decrease the spatial heterogeneity of the soil salinity and transform the soil salinity from "high-salinity heterogeneity" to "low-salinity homogeneity" [13]. The results also verified that the soil desalination rates of the wrapped SDPs exceeded 95% and that the desalination effect of the soil immediately above the SDP was obvious. The pores of SDP filter material were closely related to the migration laws of the water and salt in the soil. Tao et al. (2016) showed that a filter layer with a good particle gradation was one of the key factors in the reverse filtration effect [28]. The clogging degree of the filter material was found to affect the drainage efficiency and the total displacement of SDPs, resulting in differences in the distributions of water and salt in the soil profile. In this experiment, the sand gravel filter material yielded the lowest clogging rate, the highest drainage and salt discharge rates, the largest total displacement, and the best salt-leaching effect for the soil layer at a depth of 0–60 cm. The straw filter material presented a high silting rate, leading to low drainage and salt discharge rates as well as a low total displacement. The salt-leaching effect of the profile under the straw filter material was not as good as those under other filter materials. The filter materials could reasonably facilitate the soil above the SDP to form a bridge, filter cake, and natural soil areas from bottom to top, with a good filter structure. Some researchers have improved subsurface pipes' drainage capacity by improving the structure and material of the filter layer around the SDPs. However, some of these techniques have only been at the stage of simulation tests and have not been applied on a large scale in the field [25,26]. The medium porosity of the SDP sand filter material mixed with fine sand and fine gravel could form a stable filter layer, resulting in water stratification in the soil profile, improving the water retention rate in the subsurface soil and preventing the salt from returning [32]. The experiments also confirmed that the sand filter materials could form a good filter layer and promote the salt discharge of the SDPs. In the straw filter material scenario, the leaching water flow was faster in the early stage, which was due to the large pores of the straw filter material and high drainage rate in the early stage. With the passage of time, a large amount of water carried the sediment into the filter material, the soil particles could easily stay in the straw filter material, and clogging occurred, which decreased the drainage and salt discharge rates in the later stage.

The directional migration of the soil particles occurred within a short period of time after the drainage began. Fine particles in the soil gradually moved downwards by the drag of the water flow to fill pores. This particle redistribution led to the continuous compaction of the soil, and the drainage flow rate of the SDPs gradually decreased. In severe cases, this could result in the loss of drainage function [25]. The interior of the geotextile is a network of fibers with many small pores that can create flow channels and block the passage of soil particles. Filter materials such as sand and straw can increase the soil's likelihood of interception and particle-screening ability. The filter materials wrapped around the SDPs were used to screen nearby soil particles under the action of the water flow. Large particles were intercepted and gradually accumulated outside to form a soil-permeable skeleton with a high level of permeability, inducing the formation of a natural filter layer in the soil above [21,33]. The geotextiles increased the characteristic particle size d_{90} values of the soils by more than 20% through particle screening, resulting in the formation of a highly permeable soil skeleton on their surfaces [24]. In our research, the sand–gravel filter material increased the characteristic particle size d_{90} value of the soil above the SDPs by 8.4%, which effectively played the role of a filter material in intercepting coarse particles and releasing fine particles and induced the surrounding soil to form a soil-permeable skeleton consisting of coarse particles, such as sand particles.

Existing research has shown that spunbonded polypropylene geotextiles have smooth surfaces and are prone to produce thin particles called "pancakes" after contact with the soil [34]. After the long-term operation of SDPs, small particles in the soil are accumulated in the filter layer, causing mechanical clogging and affecting their drainage effects. In this study, the geotextiles' surface was clogged under the scenario of the straw filter material, which might be due to the fact that the filter material did not have a significant screening effect on the soil particles above it due to the non-uniformity of the pores of the straw, which failed to form a highly permeable soil skeleton structure. Soil particles of different sizes entered the straw filter material and eventually remained in the straw and geotextile. In this study, our scanning electron microscopy results revealed that the straw filter material scenario formed a dense filter cake layer on the upstream surface, and this filter cake layer continued to absorb smaller soil particles that moved with the water. The thickness of the filter cake gradually increased over time, decreasing the permeability. For the scenarios of the sand gravel filter and combined sand gravel–straw filter materials, soil particles were trapped in the geotextiles' fiber structure. The soil particles did not completely block the geotextiles, and there were still a large number of effective pores.

5. Conclusions

In order to solve the technical bottleneck of low salt discharge efficiency due to SDPs clogging easily, this study systematically evaluated the influences of three kinds of filter materials, namely sand gravel, straw, and combined sand–gravel straw, on the salt discharge efficiency and anti-silting effect of SDPs by adopting a self-designed laboratory simulation experiment device. The following conclusions were drawn:

(1) The salt removal rates of the SDPs externally wrapped with the filter materials exceeded 95%. The sand gravel filter material scenario had the highest desalting rate, and its soil profiles had total salt contents that were 17.7% and 20.5% lower than those of the straw and combined sand gravel–straw scenarios, respectively. The soil salinity of the sand gravel filter material around the SDP was between 1.57 and 3.6 g/kg, and its salt-leaching effect was the best.
(2) Under the straw filter material scenario, dense filter cake layers were formed on the upstream surfaces of the geotextiles. Under the sand gravel and combined sand gravel–straw scenarios, the soil particles remained in the geotextiles' fiber structure, and a large number of pores were retained.
(3) The sand gravel filter materials increased the characteristic particle size of the soil above the SDP by 8.4%, which effectively intercepted coarse particles, released fine particles, and induced the formation of a highly permeable soil skeleton consisting of coarse particles, such as sand particles, surrounding the soil. Therefore, the sand gravel filter material was the most suitable for the treatment of Yinbei saline–alkali soil in Ningxia Hui Autonomous Region.

Author Contributions: Investigation, X.W. and Y.Z.; Data curation, J.S. and Y.Z.; Writing—original draft, J.S. and L.F.; Writing—review & editing, X.W.; Visualization, L.F. All authors have read and agreed to the published version of the manuscript.

Funding: This research was funded by National Natural Science Foundation (42367043); National Key Research and Development Plan (2021YFD1900605-05); Chinese Academy of Sciences 'Western Light' Talent Training Program 'Western Young Scholars'; Agricultural Science and Technology Independent Innovation of Ningxia Hui Autonomous Region (NGSB-2021-11-03); and Observation and Monitoring of Basic Long-Term Scientific and Technological Work in Agriculture (NAES091AE18).

Data Availability Statement: Data are contained within the article.

Conflicts of Interest: The authors declare no conflicts of interest.

References

1. Tarolli, P.; Luo, J.; Park, E.; Barcaccia, G.; Masin, R. Soil salinization in agriculture: Mitigation and adaptation strategies combining nature-based solutions and bioengineering. *iScience* **2024**, *27*, 108830. [CrossRef] [PubMed]
2. Cuevas, J.; Daliakopoulos, I.; Moral, F.; Hueso, J.; Tsanis, I. A review of soil-improving cropping systems for soil salinization. *Agronomy* **2019**, *9*, 295. [CrossRef]
3. Minhas, P.; Ramos, T.; Ben, A.; Pereira, L. Coping with salinity in irrigated agriculture: Crop evapotranspiration and water management issues. *Agric. Water Manag.* **2020**, *227*, 105832. [CrossRef]
4. James, O. History of the roles of gypsum in soil reclamation and establishment of SAR/EC water quality guidelines. In Proceedings of the First IUSS Conference on Sodic Soil Reclamation, Changchun, China, 30 July–2 August 2021.
5. Yang, J.; Yao, R.; Wang, X.; Xie, W.; Zhang, X.; Zhu, W.; Zhang, L.; Sun, R. Research on Salt-affected Soils in China: History, Status Quo and Prospect. *Acta Pedol. Sin.* **2022**, *59*, 10–27. (In Chinese)
6. Liu, T.; Wang, B.; Xiao, H.; Wang, R.; Yang, B.; Cao, Q.; Cao, Y. Differentially improved soil microenvironment and seedling growth of Amorpha fruticosa by plastic, sand and straw mulching in a saline wasteland in northwest China. *Ecol. Eng.* **2018**, *122*, 126–134. [CrossRef]
7. Wang, Z.; Yin, D.; Wang, H.; Zhao, C.; Li, Z. Effects of biochar on waterlogging and the associated change in micro-ecological environment of maize rhizosphere soil in saline-alkali land. *Bioresources* **2020**, *15*, 9303–9323. [CrossRef]
8. Askri, B.; Khodmi, S.; Bouhlila, R. Impact of subsurface drainage system on waterlogged and saline soils in a Saharan palm grove. *Catena* **2022**, *212*, 106070. [CrossRef]
9. Askar, M.; Youssef, M.; Chescheir, G.; Negm, L.; King, K.; Hesterberg, D.; Amoozegar, A.; Skaggs, R. DRAINMOD simulation of macropore flow at subsurface drained agricultural fields: Model modification and field testing. *Agric. Water Manag.* **2020**, *242*, 106401. [CrossRef]
10. Ritzema, H.; Abdel-Dayem, S.; El-Atfy, H.; Nasralla, M.; Shaheen, H. Challenges in modernizing the subsurface drainage systems in Egypt. *Agric. Water Manag.* **2023**, *288*, 108484. [CrossRef]
11. Soe, Y.; Shinogi, Y.; Taniguchi, T. Changes in Certain Paddy Soil Properties under Perforated Sheet Pipe as Subsurface Shallow Drainage. *Jpn. Agric. Res. Q. JARQ* **2022**, *56*, 59–66. [CrossRef]
12. Heng, T.; Feng, G.; Yang, L.; He, X.; Yang, G.; Li, F.; Xu, X.; Feng, Y. Soil salt balance in a cotton field under drip irrigation and subsurface pipe drainage systems. *Agron. J.* **2021**, *113*, 4875–4888. [CrossRef]

13. Acharya, U.; Chatterjee, A.; Daigh, A. Effect of Subsurface Drainage Spacing and Depth on Crop Yield. *Agron. J.* **2019**, *111*, 1675–1681. [CrossRef]
14. Chen, G.; Wei, Z.; Liu, H. Study on Soil Desalination Process of Saline-Alkaline Grassland along the Yellow River in Western Inner Mongolia under Subsurface Drainage. *Sustainability* **2022**, *14*, 14494. [CrossRef]
15. Rong, Z.; Wang, S.; Hao, R.; Hao, R.; Tao, Y. Permeability and anti-clogging performance of geotextile envelope material around subsurface drainage pipe in Yinbei Irrigation District in Ningxia. *Trans. Chin. Soc. Agric. Eng.* **2021**, *37*, 68–75. (In Chinese)
16. Talukolaee, M.; Naftchali, A.; Parvariji, L.; Ahmadi, M. Investigating long-term effects of subsurface drainage on soil structure in paddy fields. *Soil Tillage Res.* **2018**, *177*, 155–160. [CrossRef]
17. Mante, A.; Ranjan, R.; Bullock, P. Subsurface drainage for promoting soil strength for field operations in southern Manitoba. *Soil Tillage Res.* **2018**, *184*, 261–268. [CrossRef]
18. Han, D.; Chen, C.; Wang, F.; Li, W.; Peng, H.; Jin, Q.; Bi, B.; Shaghaleh, H.; Hamoud, Y.A. Effects of Subsurface Pipe Drainage Spacing on Soil Salinity Movement in Jiangsu Coastal Reclamation Area. *Sustainability* **2023**, *15*, 13932. [CrossRef]
19. Tian, F.; Miao, Q.; Shi, H.; Li, R.; Dou, X.; Duan, J.; Liu, J.; Feng, W. Study on Water and Salt Transport under Different Subsurface Pipe Arrangement Conditions in Severe Saline–Alkali Land in Hetao Irrigation District with DRAINMOD Model. *Water* **2023**, *15*, 3001. [CrossRef]
20. Liu, W.; Luo, W.; Jia, Z.; Pan, Y.; Tang, S.; Yuan, H.; Li, S. Experimental study on geotextile envelope for subsurface drainage in Yellow River Delta. *Trans. Chin. Soc. Agric. Eng.* **2013**, *29*, 109–116. (In Chinese)
21. Hu, L.; Yang, S.; Liang, Z.; Zhang, W.; Wang, W. An Experimental Study on Wrapping Materials of Subsurface Drain for Farmland in the Downstream Hetao Irrigation District. *J. Irrig. Drain.* **2022**, *41*, 141–148. (In Chinese)
22. Qin, W.; Li, M.; Li, Y.; Liu, H. Proposed Gravel Filters for Pipe-Drain to Improve the Efficacy of the Drainage System under Drip Irrigation. *J. Irrig. Drain.* **2017**, *36*, 80–85. (In Chinese)
23. Ghane, E.; Dialamel, B.; Abdalaal, Y.; Ghane, M. Knitted-sock geotextile envelopes increase drain inflow in subsurface drainage systems. *Agric. Water Manag.* **2022**, *274*, 107939. [CrossRef]
24. Bao, Z.; Tong, B.; Zhang, Z. Application Effects of Wraped Materials Outside Drainage Pipe in Ningxia Irrigation Area of Yellow River. *J. Irrig. Drain.* **2007**, *26*, 47–50. (In Chinese)
25. Zhang, Y.; Wang, S.; Hao, R.; Rong, Z. Comparative test on the anti-filtration effect of geotextile envelope material around subsurface drainage pipe using two kinds of soil. *Trans. Chin. Soc. Agric. Eng.* **2023**, *39*, 270–276. (In Chinese)
26. Li, J.; Wang, H.; Ma, J.; Guo, L.; Jin, M. Improving the desalination of subsurface drain pipe with envelope adsorption filter using seepage tank test. *Trans. Chin. Soc. Agric. Eng.* **2023**, *39*, 121–130. (In Chinese)
27. Zhang, G. *Soil Series of China*; Science Press: Beijing, China, 2020. (In Chinese)
28. Tao, Y.; Wang, S.; Xu, D.; Zhai, X. Effect of Structure-type on Improved Subsurface Drainage Performance. *Trans. Chin. Soc. Agric. Mach.* **2016**, *47*, 113–118. (In Chinese)
29. Hu, S.; Tian, C.; Song, Y. Calculation of salinity leaching quota based on saturated infiltration theroy. *Acta Pedol. Sin.* **2010**, *47*, 563–567. (In Chinese)
30. Wang, X.; Sun, Z.; Liu, X.; Bao, Z.; Jiao, B.; Li, X.; Zeng, Y.; Sameh, E. Amelioration of Alkalized Solonchak Soils by Subsurface Gravel Blind Ditches and Desulfurized Gypsum. *Appl. Ecol. Environ. Res.* **2019**, *17*, 7865–7879. [CrossRef]
31. Chen, M.; Huang, J.; Zeng, W.; Ao, C.; Liu, D.; Liu, Y. Water-salt transport law of subsurface pipes with geotextiles under the condition of salt discharge. *Trans. Chin. Soc. Agric. Eng.* **2020**, *36*, 130–139. (In Chinese)
32. Liu, J.; Tian, J.; Zhang, M.; Wang, J.; Yang, Z.; Zhao, G. Effect of Sand Filter in Subsurface Drain on Water and Salt Movement in Paddy Soil. *J. Irrig. Drain.* **2022**, *41*, 114–121. (In Chinese)
33. Tao, Y.; Wang, S.; Xu, D.; Guan, X.; Ji, M.; Liu, J. Theoretical analysis and experimental verification of the improved subsurface drainage discharge with ponded water. *Agric. Water Manag.* **2019**, *213*, 546–553. [CrossRef]
34. Elzoghby, M.; Jia, Z.; Luo, W. Experimental study on the hydraulic performance of nonwoven geotextile as subsurface drain filter in a silty loam area. *Ain Shams Eng. J.* **2021**, *12*, 3461–3469. [CrossRef]

water

MDPI

Article

Study on the Effects of Irrigation Quotas and Amendments on Salinized Soil and Maize Growth

Liang Chen [1], Shaoli Yue [1], Lifeng Sun [1], Ming Gao [2] and Rui Wang [2,*]

[1] Ningxia Agricultural Comprehensive Development Center, Yinchuan 750002, China
[2] College of Agriculture, Ningxia University, Yinchuan 750021, China
* Correspondence: amwangrui@126.com

Abstract: Salt damage affects crop yields and wastes limited water resources. Implementing water-saving and salt-controlling strategies along with amendments can enhance crop productivity and support the development of salinized soils towards. In this study, we used "Jia Liang 0987" maize as the test material, and a two-factor split block design was executed to investigate the effects of synergistic management of irrigation volume (W1: 360 mm, W2: 450 mm, and W3: 540 mm) and amendments (T1: microbial agent 816.33 kg·hm^{-2}, T2: humic acid 6122.45 kg·hm^{-2}, T3: microsilica powder 612.25 kg·hm^{-2}) on water, salt and soil indices, and growth characteristics. The combination of 450 mm of irrigation with humic acid (W2T2) or with microsilica powder (W2T3) significantly lowered the groundwater level by 0.24 m and 0.19 m, respectively. The soil mineralization was significantly reduced by 2.60 g/L and 1.75 g/L with W2T2 and 540 mm of irrigation combined with humic acid (W3T2), respectively. The soil moisture content increased with depth and over time, showing the greatest improvement with W2T2. This combination also showed optimal results for pH and total salt, organic matter, available phosphorus, quick-acting potassium, Cl^-, and SO_4^{2-} contents. W2T2 and W3T2 improved soil field capacity and HCO_3^- contents, and significantly increased total nitrogen and phosphorus content, improving the soil nutrient grade. W2T2 showed the greatest maize plant height (323.67 cm) and stem thickness (21.54 mm for diameter), enhancing above-ground dry biomass (72,985.49 kg·hm^{-2}) and grain yield (14,646.57 kg·hm^{-2}). Implementing water-saving and salt-controlling strategies with amendments effectively improved soil fertility and crop yield in salinized soils, and the amendments factor played a major role. In saline–alkali soils in the northwest of China, 450 mm of irrigation combined with humic acid is especially helpful for enhancing soil fertility and maize productivity.

Keywords: irrigation; amendments; salinized soil; maize; soil fertility

1. Introduction

Salinization and water resource shortages are the two major obstacles to sustainable agriculture in arid and semi-arid regions [1], and they are also global challenges [2]. The high salinity and sodium levels in the soil cause the soil to inflate, disperse, and lose its structural integrity. This has a detrimental effect on the permeability coefficient, water infiltration, and porosity of the soil [3]. Thus, this causes the disintegration of the soil structure, loss of organic matter, and a nutrient deficit, which results in decreased soil fertility [4]. It also hinders soil water conductivity and air permeability. Furthermore, the high pH and electrical conductivity and low permeability potential of saline–sodic soil have a direct impact on the rate at which soil enzymes participate in biochemical processes and on microbial activity. This has a detrimental effect on crop growth, because it further lowers the nutrient content of the soil and prevents element cycling within the soil system [5]. According to incomplete statistics from the United Nations Educational, Scientific, and Cultural Organization and the Food and Agriculture Organization of the United Nations (FAO), the global area of saline–alkali soils is approximately 0.833 billion hm^2 [6]. The structural qualities of the soil are harmed by excessive soil salinity and alkalization, which also hinder biological activity and microbial diversity and lower fertilizer application efficiency [7]. These elements have a negative impact on soil-dwelling plants, reducing corn and wheat crop yields. This severely restricts the sustainable development of irrigated agriculture, which in turn, threatens human food security [8].

Salinized and secondary salinized soils are important land resources. Rational development and improvement of these soils into arable land meet current requirements and agricultural demands, providing researchers with a clear direction for exploration [9]. The essence of salinized soil improvement is to use technical measures to enhance soil fertility, improve the soil, and make it suitable for crop growth and development, thereby achieving high crop yields [10]. This will be an important strategy for the sustainable development of irrigated agriculture in China, and alleviate the status quo of "more people and less land, and tight land resources" [11]. Such improvements will transform adverse factors into favorable conditions; improve the ecological environment; and play a significant role in promoting the sustainable development of regional economies, society, and ecology.

In recent years, strategies for reclaiming saline–alkali land have been studied and developed [12]. Such strategies mainly involve the cultivation of salt-tolerant crops [13], water-saving irrigation [10], back filling of guest soil [11], modifying agent application [14], and water conservation measures [15]. Particularly, water-saving irrigation (WSI) is widely used in arid and semi-arid regions. In China, a number of WSI regimes have been implemented, including controlled irrigation, drip irrigation, and rain-gathering irrigation [16–18]. It is a highly efficient technology that modifies the physical state, microbial activity, and crop growth of soil, resulting in reduced soil salinization risk and water consumption [19,20]. By optimizing the amount of irrigation, WSI has been shown to be effective at saving water and increasing corn yield. It is also thought that WSI practice helps lower runoff loads from non-point nitrogen and phosphorus contamination [21]. Therefore, it is crucial to determine the optimal WSI quota for maximum crop yield and minimum water consumption. Applying organic amendments is a useful strategy for enhancing saline soil structure since they can provide sufficient organic matter [22]. Organic amendments, which contain organic acids, act as soil aggregate binders and give microorganisms a source of carbon [23]. Past research has confirmed that the availability and accessibility of organic matter to microorganisms is strongly correlated with the particulate organic matter and the C/N ratio of the organic amendment [24]. Applying nutrients can decrease the stability of soil aggregates and speed up the process through which microorganisms break down organic amendments. The second most common mineral element in the crust of the earth, silicon (Si), improves a plant's ability to respond to both biotic and abiotic stressors, hence reducing salinity [25]. Crop plants generate resilience to stress in the presence of microsilica by scavenging the harmful effects of reactive oxygen species, lowering excessive Na ion absorption, and reducing electrolyte leakage and malondialdehyde concentration [26]. However, there are few studies showing that a combination of improvement methods is more conducive to saline soils.

Mildly salinized soil is present in the northwest irrigation area of the Hetao Plain in the upper reaches of the Yellow River, located in Liuzhong Village, Pingluo County, in the northern part of the Yinchuan Plain, Ningxia. This area experiences drought and low rainfall throughout the year, with high inherent soil salinity [27]. In the early 21st century, local policies promoting ecological migration and irrigation from the Yellow River led to extensive land reclamation for farming. However, the lack of rainfall to wash away salts, combined with irrigation only reaching the topsoil, caused frequent water and salt movement, resulting in salinization [28]. Coupled with drought and water scarcity, there remains an urgent need to conserve water and control salt while ensuring crop yields, an important way to promote high-quality development in the Yellow River Basin [29]. Using soil salt-barrier-targeting technology and directional regulation technology, in combination with agronomic practices and water and fertilizer management techniques, it is possible to develop a comprehensive improvement plan for mildly salinized land that enhances both yield and quality. Therefore, this study compared different irrigation quotas and types of amendments to analyze the interactions within the salinized soil–water–salt system. It aimed to precisely select and assemble irrigation techniques and integrate soil fertility improvement methods to enhance soil productivity. This research sought to improve salinized soils and increase crop yields, providing technical insights and data support for water-saving and soil fertility enhancement in mildly salinized soils in Ningxia. It holds significant importance for promoting the sustainable economic, social, and ecological development of the Yellow River irrigation region.

2. Materials and Methods

2.1. Overview of the Experimental Site

The study area is located in Liuzhong Village, Pingluo County, in the northern part of Yinchuan Plain, Ningxia. It lies downstream of the Qingtongxia Irrigation District, with Helan Mountain to the west and the Yellow River to the east, with the geographic coordinates, 38°9′04″ N, 106°5′45″ E. The region has a temperate arid continental climate, at an altitude of approximately 1090 m. It experiences drought and low rainfall throughout the year, with an annual average precipitation of 173.2 mm, mostly occurring from June to August. The annual average temperature is 9.0 °C, with a large diurnal temperature variation. The region receives more than 2387.9 h of sunshine annually, with a frost-free period of 171 d. The total annual amount of sunshine exceeds 2800 h, and the frost-free period ranges from 150 to 170 d. Evaporation is intense, and the surface layer consists of salinized irrigated alluvial soil, with a thickness of approximately 0.92 m and a relatively clayey texture [30]. The characteristics of the soil before amendments are listed in Table 1.

Table 1. Basic properties of the soil in the study area.

Soil Layer	pH	Total Salt	SOM	AH-N	AP	AK	SO_4^{2-}	HCO_3^-	Cl^-
cm		g/kg	g/kg	mg/kg	mg/kg	mg/kg	g/kg	g/kg	g/kg
0~20	8.52	2.25	13.63	36.4	15.85	149.44	0.05	0.04	0.03
20~40	8.47	2	12.87	30.1	8.7	133.49	0.04	0.04	0.02
40~60	8.45	1.09	8.15	21.12	0.29	120.37	0.08	0.04	0.02
60~80	8.46	1.14	7.23	21.12	0.87	123.16	0.09	0.04	0.02
80~100	8.47	1.13	6.85	21.12	0.87	121.7	0.11	0.04	0.02

2.2. Experimental Materials and Design

The test crop was "Jia Liang 0987" maize, and the study period was from March to November 2022. This study used a two-factor split block design, as detailed in Table 1. The main plots were designated for irrigation quotas (W), using tap water drip irrigation with three irrigation levels: W1: 360 mm, W2: 450 mm, and W3: 540 mm. The subplots were designated for soil amendments (T), using three types of amendments: no treatment (T0), microbial agent (T1), humic acid (T2), and microsilica powder (T3). Each treatment had three replicates, totaling 36 treatment plots. The specific application rates of the amendments are detailed in Table 2. Each treatment plot was enclosed with 50 cm wooden stakes and lines, with labeled tags for differentiation. Each plot was 7 m long and 7 m wide, with an effective planting area of 49 m^2. The field trial arrangement is detailed in Table 3.

Microbial agent: The formulation was primarily composed of *Bacillus subtilis*, and contained lactic acid bacteria and yeast, with a concentration of 2.00×10^7 colony-forming units (CFU) per gram. It was produced by Ningxia Qiyuan Biotechnology Co., Ltd., Yinchuan, China.

Humic acid: The total humic acid content was ≥50%, the N content was ≥3%, and the moisture content was ≤2%. It was produced by Ningxia Qiyuan Biotechnology Co.

Microsilica powder: The microsilica powder was a porous material with a particle size of 0.1–10 μm and a silica content of more than 90%. It was produced by Ningxia Qiyuan Biotechnology Co.

Table 2. Field trial arrangement.

Main Plot/Irrigation Quota	Subplot/Amendments			
W1	T0	T1	T2	T3
	T2	T3	T0	T1
	T0	T1	T2	T3
W2	T2	T3	T0	T1
	T0	T1	T2	T3
	T2	T3	T0	T1
W3	T0	T1	T2	T3
	T2	T3	T0	T1
	T0	T1	T2	T3

Table 3. The dosage of the amendments.

Amendments	Dosage/kg·hm^{-2}
microbial agents	816.33
humic acid	6122.45
microsilica	612.25

Base fertilizer: urea, total nitrogen (TN) content of ≥46.0%, particle size range: (Φ) 0.85–2.80 mm, produced by Inner Mongolia Ordos United Chemical Co. All treatments used conventional base fertilizer (914.29 kg·hm^{-2} of urea applied), consistent with other field management practices.

2.3. Measurement Items and Methods

2.3.1. Physical and Chemical Properties of Soils and Their Measurement Methods

Soil samples were collected before (1 April), during (1 July), and after (1 October) soil improvement at the study site in 2022 to determine various physical and chemical properties. Physical properties included soil water content and field capacity. Chemical properties included the following soil fertility indicators: pH and total salts, organic matter, total nitrogen, alkaline hydrolyzable nitrogen, available phosphorus, and available potassium content. The soil salt ions measured included HCO_3^-, Cl^-, and SO_4^{2-}.

Soil samples were collected from 0–20 cm, 20–40 cm, 40–60 cm, 60–80 cm, and 80–100 cm depths using the ring knife method before soil improvement (1 April). Soil samples from a depth of 0–20 cm were collected using the ring knife method during soil improvement (1 July) and after soil improvement (1 October). A portion of the fresh soil sample was placed in a covered aluminum box, weighed, and then dried in an oven at 105 °C until it reached a constant weight, after which it was weighed again to calculate the soil moisture content. The soil field capacity was measured using the ring knife method. Other collected soil samples were dried naturally, ground finely, and sieved through 1 mm and 0.25 mm sieves. With a 5:1 water-to-soil ratio to extract the soil water solution, soil salt content was measured by a conductivity meter [31]. The pH value was measured using a PHS-25 precision acidity meter, and electrical conductivity was measured using a conductivity meter, which was then converted to total soil salt content. Soil organic matter (SOM) content was measured using potassium dichromate with external heating; total nitrogen (TN) content was measured using the Kjeldahl method; total phosphorus (TP) content was measured using the molybdenum-antimony colorimetric method; and alkaline hydrolysable nitrogen (AH-N) content was measured using the alkali diffusion method. Available phosphorus (AP) content was measured using the sodium bicarbonate extraction-molybdenum-antimony colorimetric method. Available potassium (AK) content was measured using the ammonium acetate-flame photometry method [32]. Soil salt ion content was measured as follows: SO_4^{2-} content was measured using the EDTA indirect chromogenic titration method, HCO_3^- contents were measured using the double indicator titration method, and Cl^- content was measured using the $AgNO_3$ titration method [33].

2.3.2. Groundwater Level and Mineralization, and Measurement Methods

In 2022, soil samples were collected before soil improvement (1 April), during soil improvement (1 July), and after soil improvement (1 October). A sectional rotary soil auger with a scale was used to drill from the surface downward until water was reached. The depth at which the auger submerged was recorded as the groundwater level. The muddy water extracted was collected using an automatic water suction pump and stored in 500 mL polyethylene bottles, which were then taken to the laboratory for mineralization measurements [34].

2.3.3. Maize Growth Characteristics and Their Measurement

Plant height, stem thickness, aboveground dry matter mass, and maize kernel yield were determined at the time of maize harvest. After maturity, three representative maize plants with uniform growth were selected. Plant height was measured using a telescopic 5 m tape measure, and stem thickness was measured using a vernier caliper. In the study plots, six representative maize plants with uniform growth (three from the inner rows and three from the outer rows) were selected. Yield measurement was conducted using the high-yield creation method of the Ministry of Agriculture and Rural Affairs, measuring grain yield for each treatment. Six representative plants (three plants in each inner and outer row) with uniform growth were selected. Samples of the whole plant above ground were taken back to the laboratory for weighing and rinsing with distilled water. They were then placed in an oven to inactivate at 105 °C for 30 min, followed by drying at 70 °C until a constant weight was achieved. The final weight measured was recorded as the above-ground dry biomass.

2.4. Data Analysis

The study data were organized using Excel 2010 software, plotted using Origin2021 (OriginLab Corporation, Northampton, MA, USA). ANOVA, correlation analysis and principal component analysis were executed with SPSS 25.0 software (IBM, Armonk, NY, USA). All data were expressed as the mean ± standard deviation (SD) of three replicates and were compared using the least significant difference (LSD) method ($p < 0.05$). Using the method of factor analysis with SPSS software, total explained variance and the core coefficient matrix of each principal component was determined. The comprehensive score of each treatment was calculated as follows:

$$F = \sum_{i=1}^{n} N_i \times F_i$$

$$N_i = \frac{\beta_i}{\sum_{i=1}^{n} i}$$

where F is the comprehensive score of each treatment, N_i is the weight of each principal component, F_i is the average core of each component, and β_i is the variance contribution rate of each component.

The soil nutrient level was determined by the 'Technical specifications for assessment and rating criteria of cultivated land quality' (DB 33/T 895-2013) [35].

3. Results

3.1. Characteristics of Groundwater Level Movement in Salinized Soils

After applying different irrigation quotas and amendments, the overall groundwater level showed a decreasing trend, with an annual reduction of 0.06 m to 0.24 m (Figure 1). Under the same amendment, the W2 irrigation quota showed a greater reduction in the groundwater level, while W1 had the least effect. At the same irrigation quota, the effect of different amendments on soil groundwater level reduction was T2 > T3 > T1 > T0. After amendment, the W2T2 treatment had the most significant effect on groundwater level reduction, with a 0.24 m reduction, followed by W2T3, with a 0.19 m reduction.

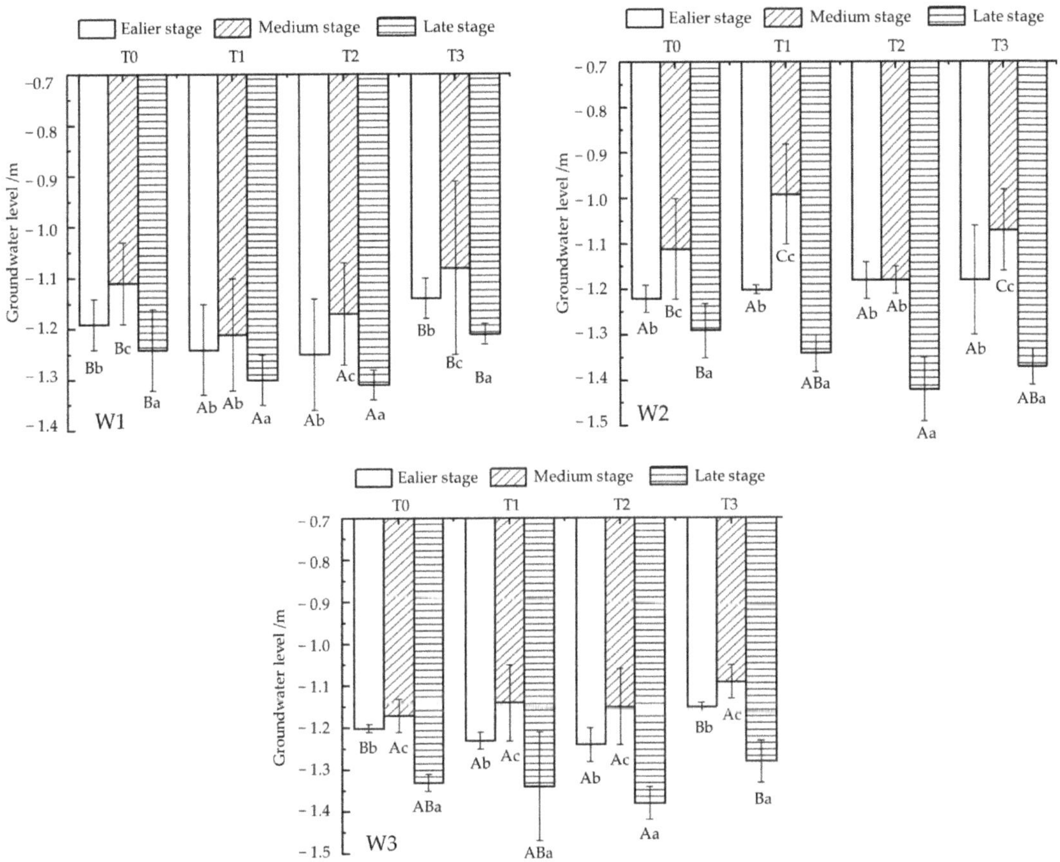

Figure 1. Migration characteristics of groundwater level in salinized soil. Note: W1: 360 mm irrigation quota, W2: 450 mm irrigation quota, W3: 540 mm irrigation quota; T0: no conditioning, T1: microbial agent, T2: humic acid, T3: microsilica powder. Uppercases indicate significant differences among treatments with same stage at the 0.05 level and lowercases indicate significant differences among stages with same treatment at the 0.05 level. Earlier stage: 1 April was before the improvement period. Medium stage: 1 July was during the improvement period. Late stage: 1 October was after the improvement period.

3.2. Variation Patterns of Groundwater Salinity in Salinized Soil

The groundwater salinity in the study area ranged from 5.88 to 11.43 g/L. Irrigation leaching mitigated the degree of soil salinization to a certain extent, and the groundwater salinity showed an initial increase followed by a decrease (Figure 2). W2 had the greatest effect on groundwater mineralization reduction, W3 had a slightly smaller effect than W2, and W1 had the smallest effect. At the same irrigation quota, the reduction in groundwater mineralization was more significant with T2. As a result, W2T2 had the most significant reduction in groundwater mineralization, with a reduction of 2.60 g/L.

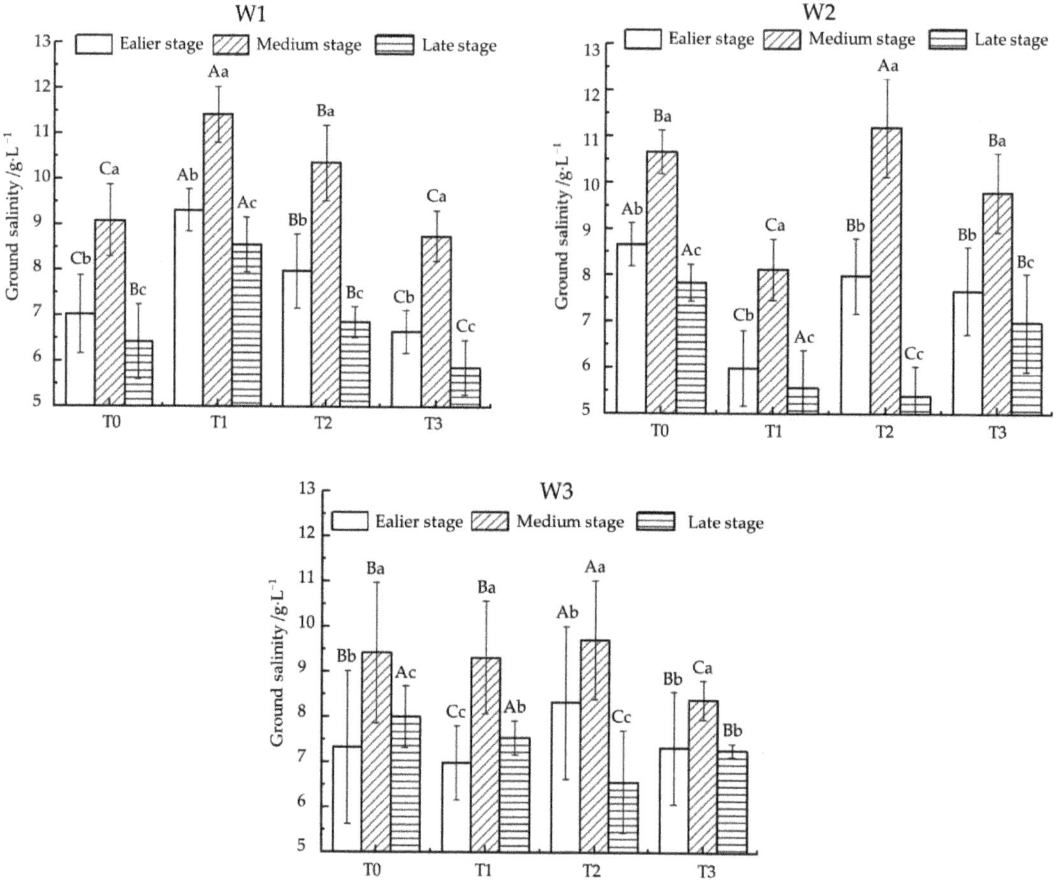

Figure 2. Migration law of groundwater salinity in salinized soil. Note: W1: 360 mm irrigation quota, W2: 450 mm irrigation quota, W3: 540 mm irrigation quota; T0: no conditioning, T1: microbial agent, T2: humic acid, T3: microsilica powder. Uppercase indicates significant differences among treatments with same stage at the 0.05 level and lowercase indicates significant differences among stages with same treatment at the 0.05 level. Earlier stage: 1 April was before the improvement period. Medium stage: 1 July was during the improvement period. Late stage: 1 October was after the improvement period.

3.3. Characteristics of pH and Salinity Changes in Salinized Soils

During and after the improvement, both soil pH and total salt content showed a decreasing trend (Table 4). During the mid-improvement period, the W2 irrigation quota combined with the T2 amendment showed the best results for reducing soil pH, achieving reductions of 0.26 to 0.35. This effect continued into the late-improvement period. The pH of W2T2 was reduced by 0.98 from the pre-improvement level and significantly reduced by 0.10 to 0.71 from all other treatments. The soil total salt content was reduced from 8.00% to 28.00% and from 33.33% to 52.44% during and after the amendment, respectively, compared to the pre-amendment period. Under the combined influence of irrigation quotas and amendments, the W2T2 treatment showed the greatest reduction, significantly lowering the total salt content by 21.74% and 28.67% compared to that of W1T0. The direct improvement effect of the amendments was more significant than irrigation and the interactions.

Table 4. Variation characteristics of soil pH and total salt content in salinized soil.

Treatment	pH			Total Salt (g/kg)		
	Earlier Stage	Medium Stage	Late Stage	Earlier Stage	Medium Stage	Late Stage
W1T0	8.52	8.43 a	8.14 b	2.25	2.07 a	1.50 a
W1T1	8.52	8.23 ef	7.92 c	2.25	2.04 a	1.36 bc
W1T2	8.52	8.20 ef	7.63 ef	2.25	1.73 abc	1.21 de
W1T3	8.52	8.37 ab	7.96 c	2.25	1.93 efg	1.36 bc
W2T0	8.52	8.33 bc	8.16 ab	2.25	1.97 ab	1.48 a
W2T1	8.52	8.27 cde	7.75 d	2.25	1.89 bcd	1.27 cd
W2T2	8.52	8.17 f	**7.54 f**	2.25	1.62 efg	**1.07 f**
W2T3	8.52	8.35 ab	7.72 de	2.25	1.73 g	1.22 de
W3T0	8.52	8.33 bcd	8.25 a	2.25	1.77 cde	1.43 ab
W3T1	8.52	8.25 ef	7.81 d	2.25	1.76 def	1.21 de
W3T2	8.52	8.17 f	7.64 e	2.25	1.67 def	1.12 ef
W3T3	8.52	8.25 de	7.93 c	2.25	1.82 fg	1.25 d
Irrigation quota	-	15.37 **	48.47 **	-	54.92 **	32.77 **
Amendments	-	86.23 **	458.35 **	-	55.61 **	131.98 **
Interaction	-	6.82 **	11.17 **	-	8.40 **	3.19 *

Notes: W1: 360 mm irrigation quota, W2: 450 mm irrigation quota, W3: 540 mm irrigation quota; T0: no conditioning, T1: microbial agent, T2: humic acid, T3: microsilica powder. Lowercase indicates significant difference in soil properties among treatments at the 0.05 level. * indicates significant impact at the 0.05 level and ** indicates significant impact at the 0.01 level. Earlier stage: 1 April was before the improvement period. Medium stage: 1 July was during the improvement period. Late stage: 1 October was after the improvement period.

Irrigation quotas combined with amendments effectively reduced the content of various soil anions. The reduction effect on anions was more pronounced in the late-improvement period compared to the mid-improvement period (Table 5). After the improvement, the concentrations of SO_4^{2-}, HCO_3^-, and Cl^- decreased by 64.54% to 87.32%, 12.53% to 72.40%, and 20.11% to 62.86%, respectively, compared to before the improvement. In contrast, in the mid-improvement period, the decrease was only by 4.40% to 33.19%. At the same irrigation quota, the effect of each amendment on the reduction in soil anions was as follows: T2 > T1 > T3 > T0. Under the same amendment, the W2 irrigation quota had the greatest effect, followed by W3. Overall, the W2T2 treatment showed the most significant reduction in SO_4^{2-}, HCO_3^-, and Cl^-. Amendments and their interaction with irrigation quotas reduced soil anion content more significantly than irrigation factors alone.

Table 5. Anion migration in salinized soil.

Treatment	SO_4^{2-} (mg/kg)			HCO_3^- (mg/kg)			Cl^- (g/kg)		
	Earlier Stage	Medium Stage	Late Stage	Earlier Stage	Medium Stage	Late Stage	Earlier Stage	Medium Stage	Late Stage
W1T0	54.40	52.36 a	26.56 ab	39.35	35.78 ab	21.99 b	35.76	33.52 a	26.87 ab
W1T1	54.40	50.39 a	35.41 ab	39.35	27.90 bc	20.73 bcd	35.76	28.99 bcd	24.89 bc
W1T2	54.40	46.20 abc	32.32 ab	39.35	18.49 d	16.40 cdef	35.76	**23.89 e**	15.39 gh
W1T3	54.40	49.51 a	19.52 b	39.35	32.43 abc	21.01 bc	35.76	28.22 bcd	20.16 de
W2T0	54.40	52.45 a	20.11 b	39.35	26.65 c	11.93 fg	35.76	31.09 ab	16.45 fg
W2T1	54.40	50.60 a	31.41 ab	39.35	17.02 d	12.45 efg	35.76	29.49 b	22.62 cd
W2T2	54.40	**41.79 c**	**19.29 b**	39.35	**14.92 d**	**10.86 g**	35.76	25.03 de	**13.28 h**
W2T3	54.40	48.81 ab	40.67 ab	39.35	17.37 d	12.48 efg	35.76	28.80 bcd	26.45 ab
W3T0	54.40	52.07 a	49.65 a	39.35	37.62 a	34.42 a	35.76	29.19 bc	28.57 a
W3T1	54.40	46.80 abc	38.83 ab	39.35	29.99 abc	17.67 bcde	35.76	27.79 bcde	23.17 c
W3T2	54.40	42.48 bc	39.76 ab	39.35	29.70 bc	11.11 fg	35.76	25.43 cde	18.66 ef
W3T3	54.40	46.85 abc	43.65 ab	39.35	32.10 abc	15.41 defg	35.76	30.40 ab	22.20 cd
Irrigation quota	-	3.65 *	12.09 **	-	1.29	1.29	-	0.40	37.33 **
Amendments	-	22.18 **	0.61		78.07 **	78.07 **	-	36.43 **	138.94 **
Interaction	-	1.02	3.34 *	-	21.53 **	21.53 **	-	3.96 **	51.36 **

Notes: W1: 360 mm irrigation quota, W2: 450 mm irrigation quota, W3: 540 mm irrigation quota; T0: no conditioning, T1: microbial agent, T2: humic acid, T3: microsilica powder. Lowercase indicates significant difference in soil properties among treatments at the 0.05 level. * indicates significant impact at the 0.05 level and ** indicates significant impact at the 0.01 level. Earlier stage: 1 April was before the improvement period. Medium stage: 1 July was during the improvement period. Late stage: 1 October was after the improvement period.

3.4. Effects of Water and Salt Transport on Soil Fertility

Soil organic matter, alkaline hydrolysable nitrogen, available phosphorus, and available potassium contents showed an overall increasing trend during the improvement period. The effect of soil fertility enhancement was more pronounced in the late-improvement period (Table 6). During the improvement process, the optimal irrigation quota for improving each soil fertility indicator varied. In contrast, the effect of the W2 irrigation rating on soil fertility enhancement was pronounced in the late-improvement period. Additionally, the T2 amendment consistently outperformed other amendments at enhancing soil fertility throughout the entire improvement process. Under the W2T2 treatment, the increases in soil organic matter, alkaline hydrolysable nitrogen, available phosphorus, and available potassium contents were 77.48%, 86.90%, 173.31%, and 31.10%, respectively. Other treatments showed significant increases ranging from 0.46% to 32.08%. The direct effects of the amendments were more significant than those of the irrigation factors and their interactions with the amendments.

3.5. Effects of Water and Salt Transport on Maize Growth Characteristics

The plant height and stem thickness of maize plants were highly significantly affected by irrigation rates, amendments and the interaction of irrigation rates and amendments (Table 7). As the irrigation quota increased, the plant height and stem thickness of maize initially increased and then decreased. The W1 and W2 irrigation quotas resulted in higher values, with the W2 quota being the highest. Under the same irrigation quota, the T2 amendment had the most significant effect on increasing maize plant height. Therefore, the W2T2 treatment achieved the highest values, with plant height and stem thickness reaching 323.67 cm and 21.54 mm, respectively, which was significantly higher than the other treatments by 1.36% to 24.38% and 4.82% to 23.44%, respectively. The trends in above-ground dry biomass and grain yield were similar to those observed for plant height and stem thickness. The yield increase effect of different amendments was T2 > T3 > T1 > T0, with the W2 irrigation quota showing the best results. Consequently, the W2T2 treatment achieved above-ground dry biomass and grain yields of 72,985.49 kg·hm^{-2} and 14,646.57 kg·hm^{-2}, respectively, representing increases of 3.93% to 55.36% and 4.85% to 62.63% compared to the other treatments. The singular effect of the amendments was still higher than that of the irrigation quota and their interaction.

Table 6. Effects of water and salt migration on soil organic matter, alkaline hydrolysable nitrogen, available phosphorus contents and available potassium content.

Treatment	SOM (g/kg)			AH-N (mg/kg)			AP (mg/kg)			AK (mg/kg)		
	Earlier Stage	Medium Stage	Late Stage	Earlier Stage	Medium Stage	Late Stage	Earlier Stage	Medium Stage	Late Stage	Earlier Stage	Medium Stage	Late Stage
W1T0	13.63	13.84 de	17.54 d	149.44	149.35 efgh	173.38 cde	15.85	10.20 e	14.20 h	149.44	149.35 efgh	173.38 cde
W1T1	13.63	14.56 bcde	21.58 bc	149.44	152.84 efg	161.58 def	15.85	25.16 c	41.46 d	149.44	152.84 efg	161.58 def
W1T2	13.63	15.41 abc	22.77 ab	149.44	168.89 abc	192.40 abc	15.85	**43.40 a**	45.22 cd	149.44	168.89 abc	192.40 abc
W1T3	13.63	14.62 bcde	20.64 c	149.44	145.56 fgh	155.16 ef	15.85	21.29 cd	23.86 fg	149.44	145.56 fgh	155.16 ef
W2T0	13.63	13.30 e	18.32 d	149.44	141.05 gh	148.34 f	15.85	16.90 de	31.75 e	149.44	141.05 gh	148.34 f
W2T1	13.63	14.49 bcde	21.85 bc	149.44	158.63 cde	176.43 abcd	15.85	34.08 b	54.89 ab	149.44	158.63 cde	176.43 abcd
W2T2	13.63	**16.50 a**	**24.19 a**	149.44	**171.41 ab**	**195.92 a**	15.85	25.46 c	**57.32 bc**	149.44	171.41 ab	**195.92 a**
W2T3	13.63	15.52 abc	20.88 c	149.44	139.38 h	168.36 def	15.85	16.53 de	29.34 a	149.44	139.38 h	168.36 def
W3T0	13.63	13.73 de	22.09 bc	149.44	146.12 fgh	171.41 cde	15.85	15.34 de	17.69 ef	149.44	146.12 fgh	171.41 cde
W3T1	13.63	14.15 cde	21.84 bc	149.44	155.90 def	190.85 abc	15.85	27.25 bc	48.95 gh	149.44	155.90 def	190.85 abc
W3T2	13.63	15.94 ab	22.63 b	149.44	179.79 a	195.03 ab	15.85	26.95 bc	44.64 cd	149.44	**179.79 a**	195.03 ab
W3T3	13.63	15.06 abcd	20.75 c	149.44	167.23 bcd	174.06 bcde	15.85	26.84 bc	19.16 gh	149.44	167.23 bcd	174.06 bcde
Irrigation quota	-	1.41	17.30 **	-	61.99 **	34.78 **	-	1.29	115.57 **	-	19.19 **	10.21 **
Amendments	-	33.65 **	96.66 **	-	70.58 **	182.41 **	-	78.07 **	449.05 **	-	78.46 **	33.48 **
Interaction	-	1.94	21.15 **	-	9.11 **	115.08 **	-	21.53 **	7.14 **	-	9.84 **	6.39 **

Notes: W1: 360 mm irrigation quota, W2: 450 mm irrigation quota, W3: 540 mm irrigation quota; T0: no conditioning, T1: microbial agent, T2: humic acid, T3: microsilica powder. Lowercase indicates significant difference in soil properties among treatments at the 0.05 level. ** indicates significant impact at the 0.01 level. Earlier stage: 1 April was before the improvement period. Medium stage: 1 July was during the improvement period. Late stage: 1 October was after the improvement period.

Table 7. Effects of water and salt regulation on maize growth characteristics.

Treatment	Plant Height /cm	Stem Diameter /mm	Aboveground Dry Matter Mass/kg·hm^{-2}	Grain Yield /kg·hm^{-2}
W1T0	294.11 ± 4.67 c	17.45 ± 0.35 e	49,697.35 ± 2550.86 e	9620.59 ± 191.95 d
W1T1	315.22 ± 2.71 abc	19.40 ± 0.37 bcd	69,816.50 ± 4128.54 ab	13,393.24 ± 901.72 abc
W1T2	319.33 ± 9.02 ab	20.55 ± 0.30 ab	70,228.57 ± 1012.36 ab	13,964.46 ± 765.25 ab
W1T3	306.00 ± 3.48 abc	20.32 ± 0.60 ab	58,497.13 ± 1933.73 cd	11,987.76 ± 369.51 c
W2T0	260.22 ± 4.91 d	18.11 ± 0.67 de	47,274.97 ± 1037.14 e	10,220.06 ± 357.93 d
W2T1	297.42 ± 8.98 bc	19.85 ± 0.88 bc	62,423.95 ± 2013.26 c	10,204.58 ± 464.11 d
W2T2	**323.67 ± 2.85 a**	**21.54 ± 0.24 a**	**72,985.49 ± 632.73 a**	**14,646.57 ± 165.15 a**
W2T3	309.00 ± 4.18 abc	20.75 ± 0.13 ab	64,378.98 ± 1221.55 bc	13,462.67 ± 368.84 abc
W3T0	257.33 ± 19.88 d	18.46 ± 0.35 cde	46,977.16 ± 5067.40 e	9005.85 ± 454.86 d
W3T1	301.00 ± 2.00 abc	20.40 ± 1.19 ab	52,012.99 ± 1473.69 de	12,874.92 ± 697.46 bc
W3T2	293.78 ± 9.16 c	20.15 ± 0.27 ab	64,986.51 ± 566.29 bc	12,037.73 ± 139.68 c
W3T3	300.78 ± 1.58 abc	19.81 ± 0.18 bc	61,136.75 ± 2025.60 c	12,198.71 ± 776.92 c
Irrigation quota	20.42 **	3.97 *	21.73 **	2.38 *
Amendments	51.28 **	42.73 **	125.96 **	89.17 **
Interaction	5.95 **	2.77 *	11.82 **	19.58 **

Notes: W1: 360 mm irrigation quota, W2: 450 mm irrigation quota, W3: 540 mm irrigation quota; T0: no conditioning, T1: microbial agent, T2: humic acid, T3: microsilica powder. Lowercase indicates significant difference in soil properties among treatments at the 0.05 level. * indicates significant impact at the 0.05 level and ** indicates significant impact at the 0.01 level. Earlier stage: 1 April was before the improvement period. Medium stage: 1 July was during the improvement period. Late stage: 1 October was after the improvement period.

3.6. Influence of Water, Salt, and Soil Fertility on Crop Yield

Based on the correlation analysis of water and salt content and fertility indicators in salinized soil with the above-ground dry biomass and grain yield of maize under different irrigation quotas and amendments, it was found that an increase in pH and total salt contents significantly inhibited the above-ground dry biomass and grain yield of maize. In contrast, an increase in the water content, field capacity, and soil organic matter and alkaline hydrolyzable nitrogen contents significantly increased the above-ground dry biomass and grain yield of maize ($p < 0.05$).

To further explore the impact of different irrigation modes on the above-ground dry biomass and grain yield of maize, principal component analysis was conducted on related factors. The eigenvalue of the first principal component (PC1) was 12.67, with a variance contribution rate of 60.34%. The eigenvalue of the second principal component (PC2) was 2.70, with a variance contribution rate of 12.88%. Together, they explained 73.22% of the variance. This indicates that the two principal components largely represent the influence of various indicators on the above-ground dry biomass and grain yield of maize. As shown in Figure 3 (right), available phosphorus content, available potassium content, and field capacity had the most significant positive effects on promoting the above-ground dry biomass of maize. The soil moisture content and available phosphorus content had the most significant positive effects on promoting maize grain yield. The pH and total salt content had the greatest negative impact on the formation of maize above-ground dry biomass and grain yield.

After a comprehensive evaluation of crop yield indicators, the highest score was achieved by W2T2, followed by W3T2, with W1T0 having the least effective outcome (Table 8). Under the same conditions, the application of humic acid improved salinized soil more effectively than the application of microbial agents, microsilica powder, or no amendments. The 450 mm irrigation quota had a much greater effect on water-saving and salt control in saline soil compared to the 540 mm and 360 mm irrigation quotas. That is, the 450 mm irrigation quota with humic acid was most conducive to water conservation and salt control, improvement of salinized soils, and regulation to enhance the productivity of maize.

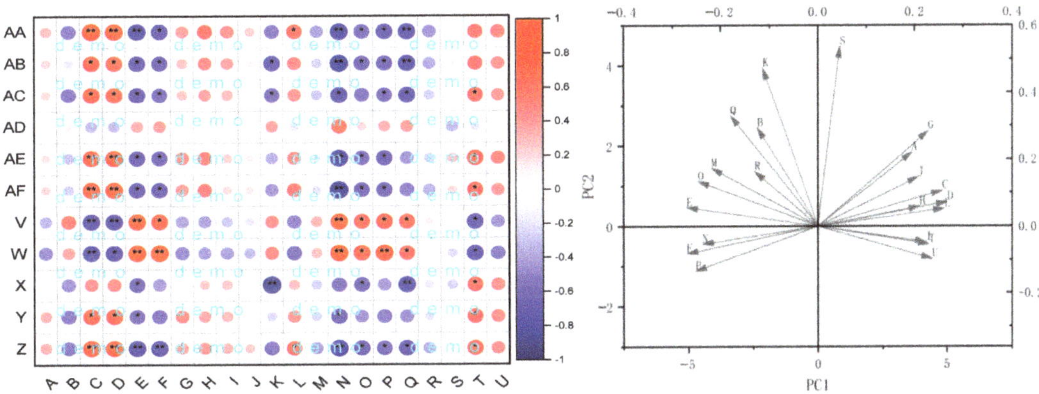

Figure 3. Correlation analysis (**left**) and principal component analysis (**right**) of water–salt–fertility and the interaction between fertility and crop yield. Note: * $p < 0.05$; ** $p < 0.01$. A—groundwater level; B—salinity; C—moisture content; D—field capacity; E—pH value; F—total salt; G—organic matter; H—alkaline hydrolysable nitrogen; I—available phosphorus; J—available potassium; K—total nitrogen; L—total phosphorus; M—K^+; N—Na^+; O—Ca^{2+}; P—Mg^{2+}; Q—HCO_3^-; R—Cl^-; S—SO_4^{2-}; T—above-ground dry biomass; U—grain yield. V—pH value; W—total salt; X—mineralization, Y—organic matter; Z—alkaline hydrolysable nitrogen; AA—available phosphorus; AB—available potassium; AC—total nitrogen; AD—total phosphorus; AE—above-ground dry biomass; AF—grain yield.

Table 8. Evaluation value and ranking of each component.

Treatment	Score	Ranking
W1T0	−4.13	12
W1T1	−0.69	8
W1T2	2.51	3
W1T3	−1.58	9
W2T0	−3.06	11
W2T1	0.94	4
W2T2	4.49	1
W2T3	0.38	6
W3T0	−2.55	10
W3T1	0.88	5
W3T2	2.72	2
W3T3	0.09	7

Note: W1: 360 mm irrigation quota, W2: 450 mm irrigation quota, W3: 540 mm irrigation quota; T0: no conditioning, T1: microbial agent, T2: humic acid, T3: microsilica powder.

4. Discussion

In saline–alkali soil improvement, soil water acts as both a solvent and carrier for soil salts, making the study of water–salt balance and precise irrigation essential. Accurate irrigation based on water–salt balance and the application of amendments are key to water conservation, reducing salinity, improving soil texture, and enhancing crop yields [36]. This has become an important method for improving saline–alkali soils. Drip irrigation can inhibit the upward movement of groundwater salts, reduce salinity near the crop root system by leaching, loosen the soil to reduce evaporation and provide sufficient water for good crop growth [37]. The findings of the current study support a similar view that drip irrigation with different irrigation quotas can increase soil moisture content, effectively reduce total soil salt content and the content of other salt ions, and decrease the sodium adsorption ratio and alkalinity. These effects are more significant as the irrigation quota increases. A previous study [38] found a negative correlation between groundwater level and mineralization, where an increase in salinity driven by a rise in groundwater level triggered an increase in mineralization, and vice versa. This study supports a similar view, as dynamic monitoring of water and salt revealed that changes in salinity in saline–alkali soils corresponded to groundwater level changes. When the groundwater level rose (from April to July), salinity increased, and when the groundwater level fell (from July to October), salinity significantly decreased. The rise in groundwater levels was due to the high rainfall in June and July. The combined effect of rainfall and drip irrigation caused the groundwater level to rise, resulting in the leaching of salts into deeper soil layers and the accumulation of salts in the surface soil, which increased the salinity of the soil [39]. As rainfall later decreased, salinity also decreased relatively. Drip irrigation effectively promotes deep soil water infiltration, increasing moisture content in the 0–40 cm soil layer. Previous studies found that field capacity decreases with increasing soil depth [40]. In this study, a slight decrease in field capacity was observed over time and depth, possibly due to the reduction in rainfall.

The soil plays a crucial role in determining the suitability of water-saving irrigation techniques and amendments. In China, sandy soils account for only 14.05% of paddy fields, which are partly distributed throughout Ningxia Province [17]. Given the low soil fertility and high penetration rate of sandy soils, it is inappropriate to design irrigation regimes that solely conserve water [41]. In the soil aggregate-forming and stabilizing process, soil carbon serves a crucially important role. By promoting fungal growth, organic amendments with a high C/N ratio have been shown to be crucial in the formation of water-stable aggregates [42,43]. Along with the use of soil amendments, it loosens the soil's aggregate structure, enhancing its water retention capacity and directing moisture to accumulate around the maize root zone. Previous findings showed that the main elements supporting the quicker formation and longer stability of coastal saline–alkali soil aggregates are the characteristics of organic amendments, which can alter microbial communities by raising soil C/N ratio and effective chemical compositions of solid-state SOM. Additionally, the inclusion of inorganic amendments can enhance the stability and development of microaggregates as opposed to macroaggregates [44]. Our results revealed that both amendments and different drip irrigation quotas effectively reduced the total salt content and the contents of Cl^-, SO_4^{2-}, and HCO_3^-. The 540 mm irrigation quota showed the best results, with humic acid being the most effective amendment. Combining irrigation with amendments can reduce soil salinity and salt ions [45], effectively mitigating soil barrier factors [36], and ensuring a favorable water–salt environment for crop growth. This study also found that the main salt ions were Cl^-, while K^+, SO_4^{2-}, and HCO_3^- were present in smaller amounts, and CO_3^{2-} was almost nonexistent. The levels of Cl^- decreased after the application of amendments. This may be due to the ions being leached by drip irrigation and the amendments conditioning the soil, as well as interactions between ions leading to their precipitation and removal.

Soil fertility nutrient indicators are crucial factors for assessing soil fertility [46]. According to the national soil nutrient-grading standards (DB 33/T 895-2013), soil is classified into six levels (as shown in Table 9). Salinized soil was improved through irrigation leaching and amendment conditioning, resulting in enhanced soil nutrient content, as detailed in Table 10. Initially, the soil's nutrient level was at level four. After improvement, the nutrient levels in treatments other than W1T0, W2T0, and W3T0 increased by one level, reaching level three.

Under the same amendment, the 450 mm and 540 mm irrigation quotas were more effective at enhancing soil nutrients and reducing soil pH. This may be because higher irrigation quotas can accelerate the dissolution of amendments and fertilizer, leading to quicker reactions, improved soil texture, and enhanced soil fertility.

Water is essential for crop growth. Amendments such as humic acid, microbial agents, and microsilica powder can improve soil texture, thereby promoting healthy crop growth and increasing crop yield. This study found that drip irrigation with a 450 mm quota combined with humic acid effectively increased maize plant height, stem thickness, above-ground dry biomass, and grain yield. The conclusion of this study is that humic acid significantly improved maize physiology and increased grain yield compared to the application of chemical fertilizers alone, similar to the findings of Ali et al. [47].

Table 9. Soil nutrient-grading standards.

Grade	SOM (g/kg)	AH-N (mg/kg)	AP (mg/kg)	AK (mg/kg)	TN (g/kg)	TP (g/kg)
1st	>40	>150	>40	>200	>2	>1
2nd	30~40	120~150	20~40	150~200	1.5~2	0.8~1
3rd	20~30	90~120	10~20	100~150	1~1.5	0.6~0.8
4th	10~20	60~90	5~10	50~100	0.75~1	0.4~0.6
5th	6~10	30~60	3~5	30~50	0.5~0.75	0.2~0.4
6th	<6	<30	<3	<30	<6	<6

Note: SOM: soil organic matter, AH-N: alkali hydrolysable nitrogen, AP: available phosphorus, AK: available potassium, TN: total nitrogen, TP: total phosphorus.

Table 10. Ratings of improved soil nutrients.

Treatment	SOM (g/kg)	AH-N (mg/kg)	AP (mg/kg)	AK (mg/kg)	TN (g/kg)	TP (g/kg)	Soil Grade
CK	13.63	36.4	15.85	149.44	0.85	0.84	4th
W1T0	17.54	33.40	14.20	173.38	1.00	1.11	4th
W1T1	21.58	48.10	41.46	161.58	1.08	1.27	3rd
W1T2	22.77	62.47	45.22	192.40	1.08	1.80	3rd
W1T3	20.64	42.40	23.86	155.16	0.88	1.40	3rd
W2T0	18.32	45.57	31.75	148.34	0.91	1.11	4th
W2T1	21.85	48.60	54.89	176.43	1.00	1.54	3rd
W2T2	24.19	68.03	57.32	195.92	1.16	2.09	3rd
W2T3	20.88	45.27	29.34	168.36	0.97	1.39	3rd
W3T0	22.09	42.90	17.69	171.41	0.75	1.27	4th
W3T1	21.84	52.40	48.95	190.85	1.07	1.69	3rd
W3T2	22.63	43.73	44.64	195.03	1.17	2.03	3rd
W3T3	20.75	61.77	19.16	174.06	0.90	1.59	3rd

Note: W1: 360 mm irrigation quota, W2: 450 mm irrigation quota, W3: 540 mm irrigation quota; T0: no conditioning, T1: microbial agent, T2: humic acid, T3: microsilica powder. SOM: soil organic matter, AH-N: alkali hydrolysable nitrogen, AP: available phosphorus, AK: available potassium, TN: total nitrogen, TP: total phosphorus.

5. Conclusions

The application of humic acid or microsilica powder at a 450 mm irrigation quota reduced the water table by 0.24 m and 0.19 m, respectively. The reduction in soil mineralization and HCO_3^- at 450 mm and 540 mm irrigation quotas with humic acid treatment was greatest. Soil pH and total salt, Cl^-, and SO_4^{2-} content was most significantly reduced with the 450 mm irrigation quota combined with humic acid. At the same time, this combination maximized the increase in soil nutrients. Maize growth reached its maximum values under the 450 mm irrigation quota combined with humic acid treatment. Moreover, the improvement effect of the amendments on the soil was significantly higher than that of the irrigation quotas and their interactions. Considering the comprehensive water and salt indicators, total salts and salt ions, soil fertility indicators, and maize growth characteristics, the 450 mm irrigation quota combined with humic acid showed the best results for water-saving, salt control, and enhancing soil fertility and productivity. More efficient amendments' usage investigation is needed on water-saving irrigation in saline–alkali land to achieve better productivity.

Author Contributions: Conceptualization and supervision, R.W.; methodology, writing—review and editing, S.Y.; software and investigation, L.S.; formal analysis, validation and data curation, L.C.; writing—original draft preparation, M.G. All authors have read and agreed to the published version of the manuscript.

Funding: This research was funded by the Ningxia Agricultural Core Technology Research Project of 2023, the National Key Research and Development Program of China (No. 2021YFD1900600) and the Ningxia Science and Technology Leading Talent Project (2023GKLRLX13).

Data Availability Statement: Data are contained within the article.

Conflicts of Interest: The authors declare no conflicts of interest.

References

1. Yan, F.; Zhang, F.; Fan, J.; Hou, X.; Bai, W.; Liu, X.; Wang, Y.; Pan, X. Optimization of Irrigation and Nitrogen Fertilization Increases Ash Salt Accumulation and Ions Absorption of Drip-Fertigated Sugar Beet in Saline-Alkali Soils. *Field Crop. Res.* **2021**, *271*, 108247. [CrossRef]
2. Maleki Tirabadi, M.; Banihabib, M.; Randhir, T. An Integrated Framework for Simultaneously Modeling Primary and Secondary Salinity at a Watershed Scale. *J. Hydrol.* **2022**, *612*, 128171. [CrossRef]
3. Zahedifar, M. Effect of biochar on cadmium fractions in some polluted saline and sodic soils. *Environ. Manag.* **2020**, *66*, 1133–1141. [CrossRef] [PubMed]
4. Tian, Y.; Xia, R.; Ying, Y.; Lu, S. Desulfurization steel slag improves the saline-sodic soil quality by replacing sodium ions and affecting soil pore structure. *J. Environ. Manag.* **2023**, *345*, 118874. [CrossRef] [PubMed]
5. Xu, Z.; Shao, T.; Lv, Z.; Yue, Y.; Liu, A.; Long, X.; Zhou, Z.; Gao, X.; Rengel, Z. The mechanisms of improving coastal saline soils by planting rice. *Sci. Total Env.* **2020**, *703*, 135529. [CrossRef] [PubMed]
6. Food and Agriculture Organization of the United Nations. 'World Soil Day: Food and Agriculture Organization of the United Nations Emphasizes that Soil Salinization Poses a Threat to Global Food Security' [EB/OL]. 3 December 2021. Available online: https://www.fao.org/newsroom/detail/world-soil-day-fao-highlights-threat-of-soil-salinization-to-food-security-031221/en (accessed on 5 April 2023).
7. Chen, L.; Zhou, G.; Feng, B.; Wang, C.; Luo, Y.; Li, F.; Shen, C.; Ma, D.; Zhang, C.; Zhang, J. Saline–Alkali Land Reclamation Boosts Topsoil Carbon Storage by Preferentially Accumulating Plant-Derived Carbon. *Sci. Bull.* **2024**, *69*, 1212–1216. [CrossRef] [PubMed]
8. Fan, P.; Chen, D.; He, Y.; Zhou, Q.; Tian, Y.; Gao, L. Alleviating salt stress in tomato seedlings using Arthrobacter and Bacillus megaterium isolated from the rhizosphere of wild plants grown on saline-alkaline lands. *Int. J. Phytoremediat.* **2016**, *18*, 1113–1121. [CrossRef] [PubMed]
9. Negacz, K.; Malek, Z.; De Vos, A.; Vellinga, P. Saline soils worldwide: Identifying the most promising areas for saline agriculture. *J. Arid. Environ.* **2022**, *203*, 104775. [CrossRef]
10. Du, Y.; Liu, X.; Zhang, L.; Zhou, W. Drip Irrigation in Agricultural Saline-Alkali Land Controls Soil Salinity and Improves Crop Yield: Evidence from a Global Meta-Analysis. *Sci. Total Environ.* **2023**, *880*, 163226. [CrossRef] [PubMed]
11. Jat Baloch, M.Y.; Zhang, W.; Sultana, T.; Akram, M.; Shoumik, B.A.A.; Khan, M.Z.; Farooq, M.A. Utilization of Sewage Sludge to Manage Saline–Alkali Soil and Increase Crop Production: Is It Safe or Not? *Environ. Technol. Innov.* **2023**, *32*, 103266. [CrossRef]
12. Stavi, I.; Thevs, N.; Priori, S. Soil salinity and sodicity in drylands: A review of causes, effects, monitoring, and restoration measures. *Front. Environ. Sci.* **2021**, *9*, 712831. [CrossRef]
13. Ma, C.; Lei, C.-Y.; Zhu, X.-L.; Ren, C.-G.; Liu, N.; Liu, Z.-Y.; Du, H.; Tang, T.; Li, R.-Z.; Cui, H.-L. Saline-Alkali Land Amendment and Value Development: Microalgal Biofertilizer for Efficient Production of a Halophytic Crop—Chenopodium Quinoa. *Land Degrad. Dev.* **2023**, *34*, 956–968. [CrossRef]
14. Liu, C.; Shang, H.; Han, L.; Sun, X. Effect of Alkali Residue and Humic Acid on Aggregate Structure of Saline-Alkali Soil. *Soil Sci. Soc. Am. J.* **2024**, *88*, 291–303. [CrossRef]
15. Heng, T.; He, X.L.; Yang, L.L.; Xu, X.; Feng, Y. Mechanism of Saline–Alkali Land Improvement Using Subsurface Pipe and Vertical Well Drainage Measures and Its Response to Agricultural Soil Ecosystem. *Environ. Pollut.* **2022**, *293*, 118583. [CrossRef] [PubMed]
16. Dong, W.; Guo, J.; Xu, L.; Song, Z.; Zhang, J.; Tang, A.; Zhang, X.; Leng, C.; Liu, Y.; Wang, L.; et al. Water regime-nitrogen fertilizer incorporation interaction: Field study on methane and nitrous oxide emissions from a rice agroecosystem in Harbin, China. *J. Environ. Sci.* **2018**, *64*, 289–297. [CrossRef] [PubMed]
17. Zhuang, Y.; Zhang, L.; Li, S.; Liu, H.; Zhai, L.; Zhou, F.; Ye, Y.; Ruan, S.; Wen, W. Effects and Potential of Water-Saving Irrigation for Rice Production in China. *Agric. Water Manag.* **2019**, *217*, 374–382. [CrossRef]
18. Zhang, S.; Zamanian, K.; Raza, S.; Raheb, A.; Feng, Y.; Zhao, X. A Balance among Irrigation and Fertilization Regimes to Reduce Greenhouse Gases Emissions from Saline and Alkaline Soils. *Land Degrad. Dev.* **2024**, *35*, 168–182. [CrossRef]
19. Dai, J.; Cui, Z.; Zhang, Y.; Zhan, L.; Nie, J.; Cui, J.; Zhang, D.; Xu, S.; Sun, L.; Chen, B.; et al. Enhancing stand establishment and yield formation of cotton with multiple drip irrigation during emergence in saline fields of Southern Xinjiang. *Field Crop. Res.* **2024**, *315*, 109482. [CrossRef]
20. Ding, B.; Cao, H.; Zhang, J.; Bai, Y.; He, Z.; Guo, S.; Wang, B.; Jia, Z.; Liu, H. Biofertilizer application improved cotton growth, nitrogen use efficiency, and yield in saline water drip-irrigated cotton fields in Xinjiang, China. *Ind. Crop. Prod.* **2023**, *205*, 117553. [CrossRef]

21. Tao, W.; Li, J.; Li, W.; Wen, C.; Gao, S.; Wang, Y.; Liu, D.; Xu, L.; Jiang, Y.; Liu, Z.; et al. Higher rice productivity and lower paddy nitrogen loss with optimized irrigation and fertilization practices in a rice-upland system. *Agric. Ecosyst. Environ.* **2024**, *374*, 109176. [CrossRef]
22. Xing, J.; Li, X.; Li, Z.; Wang, X.; Hou, N.; Li, D. Remediation of soda-saline-alkali soil through soil amendments: Microbially mediated carbon and nitrogen cycles and remediation mechanisms. *Sci. Total Environ.* **2024**, *924*, 171641. [CrossRef] [PubMed]
23. Dai, X.; Guo, Q.; Song, D.; Zhou, W.; Liu, G.; Liang, G.; He, P.; Sun, G.; Yuan, F.; Liu, Z. Long-Term Mineral Fertilizer Substitution by Organic Fertilizer and the Effect on the Abundance and Community Structure of Ammonia-Oxidizing Archaea and Bacteria in Paddy Soil of South China. *Eur. J. Soil. Biol.* **2021**, *103*, 103288. [CrossRef]
24. Fang, Y.; Singh, B.P.; Van Zwieten, L.; Collins, D.; Pitt, W.; Armstrong, R.; Tavakkoli, E. Additive effects of organic and inorganic amendments can significantly improve structural stability of a sodic dispersive subsoil. *Geoderma* **2021**, *404*, 115281. [CrossRef]
25. Dhakate, P.; Kandhol, N.; Raturi, G.; Ray, P.; Bhardwaj, A.; Srivastava, A.; Kaushal, L.; Singh, A.; Pandey, S.; Chauhan, D.K.; et al. Silicon Nanoforms in Crop Improvement and Stress Management. *Chemosphere* **2022**, *305*, 135165. [CrossRef] [PubMed]
26. Kim, Y.H.; Khan, A.L.; Waqas, M.; Lee, I.J. Silicon regulates antioxidant activities of crop plants under abiotic-induced oxidative stress: A review. *Front. Plant Sci.* **2017**, *8*, 510. [CrossRef] [PubMed]
27. Fan, G.; Zhang, D.; Zhang, J.; Li, Z.; Sang, W.; Zhao, L.; Xu, M. Ecological Environmental Effects of Yellow River Irrigation Revealed by Isotope and Ion Hydrochemistry in the Yinchuan Plain, Northwest China. *Ecol. Indic.* **2022**, *135*, 108574. [CrossRef]
28. Li, J.; Chen, J.; He, P.; Chen, D.; Dai, X.; Jin, Q.; Su, X. The Optimal Irrigation Water Salinity and Salt Component for High-Yield and Good-Quality of Tomato in Ningxia. *Agric. Water Manag.* **2022**, *274*, 107940. [CrossRef]
29. Wu, Z.; Li, Y.; Wang, R.; Xu, X.; Ren, D.; Huang, Q.; Xiong, Y.; Huang, G. Evaluation of Irrigation Water Saving and Salinity Control Practices of Maize and Sunflower in the Upper Yellow River Basin with an Agro-Hydrological Model Based Method. *Agric. Water Manag.* **2023**, *278*, 108157. [CrossRef]
30. Gu, X.; Zhang, F.; Xie, X.; Cheng, Y.; Xu, X. Effects of N and P Addition on Nutrient and Stoichiometry of Rhizosphere and Non-Rhizosphere Soils of Alfalfa in Alkaline Soil. *Sci. Rep.* **2023**, *13*, 12119. [CrossRef]
31. Endo, T.; Abdalla, M.; Elkarim, A.; Toyoda, M.; Yamamoto, S.; Yamanaka, N. Simplified evaluation of salt affected soils using 1:5 soil-water extract. *Commun. Soil. Sci. Plant Anal.* **2021**, *52*, 2533–2549. [CrossRef]
32. Li, G.; Ma, Z.; Wei, L.; Wu, C.; Chen, H.; Guo, B.; Ge, T.; Wang, J.; Li, J. Long-term agricultural cultivation decreases microbial nutrient limitation in coastal saline soils. *Sci. Total Environ.* **2024**, *949*, 175005. [CrossRef]
33. Heydari, B.; Khayat, N.; Nazarpour, A. Assessment of the immediate impact of anions and cations on clay soils resistance parameters case study (garmsiri project iran). *Case Stud. Constr. Mater.* **2024**, *20*, e03105. [CrossRef]
34. Khasanov, S.; Li, F.; Kulmatov, R.; Zhang, Q.; Qiao, Y.; Odilov, S.; Yu, P.; Leng, P.; Hirwa, H.; Tian, C.; et al. Evaluation of the Perennial Spatio-Temporal Changes in the Groundwater Level and Mineralization, and Soil Salinity in Irrigated Lands of Arid Zone: As an Example of Syrdarya Province, Uzbekistan. *Agric. Water Manag.* **2022**, *263*, 107444. [CrossRef]
35. *DB 33/T 895-2013*; Technical Specification for Assessment and Rating Criteria of Cultivated Land Quality. Quality and Technical Supervision Bureau of Zhejiang Province: Hangzhou, China, 2013.
36. Yan, S.; Gao, Y.; Tian, M.; Tian, Y.; Li, J. Comprehensive Evaluation of Effects of Various Carbon-Rich Amendments on Tomato Production under Continuous Saline Water Irrigation: Overall Soil Quality, Plant Nutrient Uptake, Crop Yields and Fruit Quality. *Agric. Water Manag.* **2021**, *255*, 106995. [CrossRef]
37. Liu, S.; Wang, N.; Li, D.; Tian, C.; Zhang, K.; Hu, M.; Niu, H.; Zhao, Z. Effects of Halophyte Suaeda Salsa Continuous Cropping on Physical and Chemical Properties of Saline Soil under Drip Irrigation in Arid Regions. *Agric. Ecosyst. Environ.* **2024**, *371*, 109076. [CrossRef]
38. Ghasempour, R.; Aalami, M.T.; Kirca, V.S.O.; Roushangar, K. Assessing the Soil Salinity Vulnerability and Groundwater Quality Variations Due to Drying up of the Lake. *Environ. Sci. Pollut. Res.* **2023**, *30*, 115611–115627. [CrossRef] [PubMed]
39. Xu, X.; Xin, P.; Yu, X. Interactions of macropores with tides, evaporation and rainfall and their effects on pore-water salinity in salt marshes. *J. Hydrol.* **2024**, *630*, 130740. [CrossRef]
40. Lena, B.; Bondesan, L.; Pinheiro, E.; Ortiz, B.; Morata, G.; Kumar, H. Determination of irrigation scheduling thresholds based on HYDRUS-1D simulations of field capacity for multilayered agronomic soils in Alabama, USA. *Agric. Water Manag.* **2022**, *259*, 107234. [CrossRef]
41. Zheng, j.; Chen, T.; Xia, G.; Chen, W.; Liu, G.; Chi, D. Effects of zeolite applicationon grain yield, water use and nitrogen uptake of rice under alternate wetting and drying irrigation. *Int. J. Agric. Biol. Eng.* **2018**, *11*, 157–164. [CrossRef]
42. Luo, R.; Kuzyakov, Y.; Liu, D.; Fan, J.; Luo, J.; Lindsey, S.; He, J.; Ding, W. Nutrient addition reduces carbon sequestration in a Tibetan grassland soil: Disentangling microbial and physical controls. *Soil. Biol. Biochem.* **2020**, *144*, 107764. [CrossRef]
43. Zhang, Y.; Liu, X.; Li, P.; Xiao, L.; Wang, X. Temporal effects of soil organic carbon mineralization during the formation of a siltation body produced by erosion. *Catena* **2024**, *241*, 108030. [CrossRef]
44. Li, S.; Yao, Y.; Yang, M.; Zhang, Y.; Zhang, S.; Shen, T.; Ding, F.; Li, Z.; Liu, W.; Cui, J.; et al. Effects of different amendments on aggregate stability and microbial communities of coastal saline–alkali soil in the Yellow River Delta. *Land Degrad. Dev.* **2023**, *34*, 1694–1707. [CrossRef]
45. Zhu, R.; Zhang, P.; Li, X.; Duan, Z. How to Remediate Sulfate-Nitrate Salinized Greenhouse Soil? An Optimal Combination of Organic Amendment, Fertilizer and Irrigation. *Sci. Hortic.* **2023**, *321*, 112264. [CrossRef]

46. Abdehvand, Z.Z.; Karimi, D.; Rangzan, K.; Mousavi, S.R. Assessment of soil fertility and nutrient management strategies in calcareous soils of Khuzestan province: A case study using the Nutrient Index Value method. *Env. Monit. Assess.* **2024**, *196*, 503. [CrossRef]
47. Ali, S.; Jan, A.; Manzoor; Sohail, A.; Khan, A.; Khan, M.I.; Inamullah; Zhang, J.; Daur, I. Soil Amendments Strategies to Improve Water-Use Efficiency and Productivity of Maize under Different Irrigation Conditions. *Agric. Water Manag.* **2018**, *210*, 88–95. [CrossRef]

water

MDPI

Article

Effects of Ditch Water and Yellow River Irrigation on Saline–Alkali Characteristics of Soil and Paddy

Liqin Fan [1,2], Jingli Shen [1,2], Xu Wang [1,2,*] and Yonghong Zhang [1,2]

1 Institute of Agricultural Resources and Environment, Ningxia Academy of Agricultural and Forestry Sciences, Yinchuan 750002, China; fanlqnx@126.com (L.F.); lily_s90@163.com (J.S.); zyh8401@163.com (Y.Z.)
2 National Agricultural Environment Observation and Experiment Station in Yinchuan, Yinchuan 750002, China
* Correspondence: wangxu640321@126.com

Abstract: This study examined the agricultural water resource shortage and abundant ditch water resources in the Yinbei region of Ningxia. The effects of ditch water and Yellow River irrigation on the saline–alkali characteristics of soil and paddy were investigated using field monitoring and indoor detection methods in Pingluo County, Ningxia (106°31′ E, 38°51′ N). In addition to monitoring ditch water, four treatment groups were established: direct ditch water irrigation (T1), mixed ditch water and Yellow River water irrigation (T2), alternate ditch water and Yellow River water irrigation (T3), and irrigation solely with Yellow River water (CK). The results show the following: (1) The salinity of ditch water samples collected from the experimental field during the rice growth period was less than 1.60 g/L, and the pH of the samples was lower than 8.62; thus, they were classified as mildly brackish water. The application of ditch water irrigation did not result in soil saline–alkali aggravation and the accumulation of excessive amounts of heavy metals in soils and paddies in Pingluo County, Ningxia. (2) The rice yields for the CK, T1, T2, and T3 treatments were 10,437.5, 8318.4, 9182.1, and 9016.2 kg/hm^2, respectively. Compared with Yellow River irrigation, the rice yields for the T1, T2, and T3 treatments were 20.3, 12.1, and 13.6% lower than that of CK, respectively, with minimal differences observed among them. Hence, under the condition of a water resource shortage in the Yellow River region, ditch water can be appropriately applied for mixed or alternate irrigation to ensure food security. This research has revealed the influences of ditch water irrigation on the saline–alkali properties of soil and the heavy metal contents of paddies.

Keywords: irrigation; ditch water; soil salinity and alkalinity; heavy metals in rice; Ningxia Yinbei region

1. Introduction

Ningxia is located in the northwestern inland region of China, where rainfall is scarce. More than half of the province is located in arid and semi-arid zones, and there is a significant difference between the supply of and demand for water resources [1–3]. Agricultural water intake in Ningxia makes up about 5.86 billion m^3, accounting for 83% of the total water intake in Ningxia [4]. The irrigation area in northern Ningxia is low-lying with strong water retention, making it suitable for rice cultivation [5]. However, it suffers from water shortage problems in respect of the Yellow River region and prominent rice irrigation differences. As an effective measure for alleviating regional water resource problems and ensuring grain production, the rational application of water resources in farmland drainage ditches has attracted increasing attention.

In China, fresh water resources are scarce and unevenly distributed in terms of time and location. The application of poor-quality water such as groundwater and ditch water for irrigation is a hot topic in agricultural research relating to arid regions. The application of shallow underground brackish water for irrigation can not only supply necessary water for crop growth but also decrease groundwater levels [6,7]. Fan (2020) performed ditch soil column-leaching experiments and found that high-salinity water could promote soil flocculation and improve the soil structure. When irrigated water salinity is lower than 3 g/L, the salt in the soil profile is in an equilibrium state, while at salinity levels of above 3 g/L, salt accumulation occurs, mainly in 0–40 cm deep soil layers [8]. While ditch water contains certain amounts of salt, brackish water delivers lower amounts of salt into soil in a short period of time, and an alternative irrigation method involving brackish water and Yellow River water can be used to avoid salinity aggravation in soil [9]. The authors of [9] found that the heavy metal contents of soil after irrigation with reclaimed water did not obviously change compared with those before irrigation; therefore, heavy metal accumulation in soil did not occur in a short time, and the salt content of the surface soil increased while that of the deep soil did not change much. In the coastal saline–alkali areas of Hebei Province, shallow water resources with salt contents lower than 5 g/L could be reasonably employed for the irrigation of winter wheat without causing secondary salinization due to salt accumulation in a short time [10]. However, research has revealed that compared with freshwater irrigation, brackish water irrigation brings in additional salt. Continuous irrigation can cause salt accumulation, damaging the normal growth of crops and affecting their yields over a certain limit [11]. Therefore, the scientific and safe application of brackish water resources is of particular importance. Wang (2019) concluded that irrigation with treated inferior water did not result in heavy metal pollution in rice, and the heavy metal contents of lead, arsenic, and chromium in rice met the requirements of the National Standard for Food Safety (GB 2762-2017) [12,13]. Wang (2022) concluded that different irrigation water sources did not cause soil pollution, and the heavy metal contents of rice grains did not present significant increases. Compliance with the limit requirements of rice pollutants was verified through comparative tests on rice irrigation with clear water from rivers, purified water from ecological ponds, and treated rural domestic sewage [14].

Currently, research in respect of ditch water primarily focuses on monitoring, while studies relating to farmland ditch water emphasize soil infiltration, saline–alkali properties, crop growth, and so on [15]. However, there is a lack of systematic research on water quality monitoring, scientific utilization, and the effects of ditch water on soil and rice in the Pingluo region of the fifth drainage ditch in Ningxia. Hence, in this study, experiments were performed in the rice experimental field of the Pingluo county experimental station (106°31′ E, 38°51′ N) in Ningxia. By monitoring water quality and performing field experiments, the effects of the variations in ditch water quality, direct ditch water irrigation, mixed ditch water and Yellow River water irrigation, and alternate irrigation on soil salinity, heavy metal contents, and rice heavy metal contents were evaluated. These experiments provide a theoretical basis and technical support for the safe utilization of ditch water and rice planting.

2. Materials and Methods

2.1. Overview of Study Area

The experimental field was located in Pingluo county experimental station (106°31′ E, 38°51′ N), Shizuishan City, Ningxia (Figure 1, drawn with ArcGIS 10.2 software). Pingluo County has a long agricultural history, with favorable basic conditions and a sound agricultural industry system, which makes it a representative self-flow irrigation area in respect of the Yellow River. This area has an arid continental climate and is classified as an arid semi-desert saline area in the middle and upper reaches of the Yellow River. The annual average precipitation and evaporation of the area are 185 and 1841 mm, respectively, with rare and unevenly distributed precipitation, mainly from July to September. In this study, the underground water depth was determined to be 1.3~2.0 m. The fifth drainage ditch on the eastern side of the experimental field is considered the main farmland drainage ditch in Pingluo, and its water resources are rich. Before planting, obvious salt spots were observed on the ground surface of each plot of land. According to a surface analysis, the total salt content in the 0–100 cm soil layer was 0.91~4.2 g/kg, and obvious salt spots were observed in local fields with a total salt content of more than 15 g/kg and a soil pH value of 8.17~9.50. In the 0–20 cm soil layer, the soil alkali–hydrolyzed nitrogen level was 49.7 mg/kg, the available phosphorus level was 29.9 mg/kg, the available potassium level was 317.95 mg/kg, the total nitrogen level was 0.79 g/kg, the total phosphorus level was 0.82 g/kg, the total potassium level was 19.5 g/kg, and the organic matter level was 14.83 g/kg. The soil bulk density was 1.4 g/cm^3. The soil texture was clay loam; no CO_3^{2-} was observed in soil salt ions, the cations mainly consisted of Na^+ and Ca^{2+}, the anions mainly included SO_4^{2-} and HCO_3^-, and the ratio of Cl^-/SO_4^{2-} was below 0.5. The soil in the experimental area was sulfate saline soil.

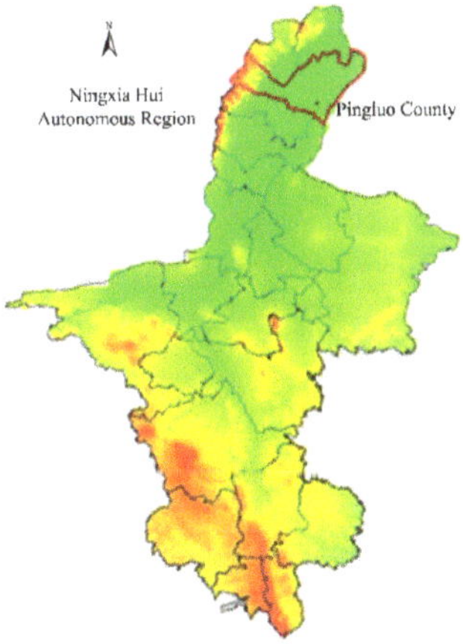

Figure 1. Geographical position of research area (Red represents mountains in Ningxia, yellow represents hills in Ningxia, and green represents plains in Ningxia.).

2.2. Experimental Design and Implementation Process

Ningjing 47 was adopted as the experimental variety. Based on the unified planting of rice, a random block design was prepared to set up four treatment groups: direct ditch water irrigation (T1), mixed ditch water and Yellow River water irrigation (T2), alternate ditch water and Yellow River water irrigation (T3), and irrigation solely with Yellow River water (CK). The experimental plots were 8.0 m long and 2.0 m wide, each with a 16.0 m^2 area. High ridges were built between plots and plots were covered with plastic film. Each treatment was repeated five times. In this research, dry rice planting technology was used. This technology enables rice seeds to be directly sown into the field using special machinery, following the application of a base fertilizer when the field is dry, with subsequent irrigation until the field is sufficiently wet for seed germination. Seeds were sown on 30th April, watered on 4th May, and harvested on 4th October. Border irrigation was adopted as the selected irrigation method. Only ditch water was used in the T1 treatment, 1:1 ditch water and Yellow River water mixed irrigation was used in the T2 treatment, ditch water and Yellow River water alternating irrigation was applied in the T3 treatment, and only Yellow River water was used in the CK treatment during the rice growing period. Fertilization and field management were similar in all regions. During the rice growth period, 225 kg/hm^2 urea, 375 kg/hm^2 compound fertilizer, 225 kg/hm^2 diammonium, and 225 kg/hm^2 ammonium sulfate were used. The total irrigation water quantity for all treatments in the rice growth period except winter irrigation was 1.65×10^4 m^2/hm^2.

2.3. Soil Sample Collection and Determination

In April 2023, soil samples were collected at depths of 0–20, 20–40, 40–60, 60–80, and 80–100 cm from the surface of the experimental area to obtain the soil background value. Soil samples were collected using a ring knife to determine soil bulk density. Ditch water was collected every month from May to August 2023 for inspection, soil and rice samples were collected after rice harvesting in October 2023, and the yield was measured in each district group with uniform growth and an area of 1 m^2. Samples were collected in triplicate.

Debris was removed from the collected soil samples, and the cleaned samples were air dried, milled, and passed through a 2 mm sieve. The soil water content was determined through the drying method, and the soil bulk density was measured using the ring knife method. The supernatant was fully shaken with a soil-to-water ratio of 1:2.5, and the pH was measured using a multi-parameter Mettler Toledo S220 tester (METTLER TOLEDO, Greifensee, Switzerland). The supernatant obtained from the previous stage was fully shaken at a soil-to-water ratio of 1:5 and was analyzed using a DDS-307A soil conductivity meter (INESA, Shanghai, China); the obtained value was converted to the total salt content [16]. The pH and salinity values of water samples were measured using a pH meter and a conductivity meter, respectively. Soil organic matter was determined according to the potassium dichromate method and alkali hydrolysis nitrogen was measured based on the alkali hydrolysis diffusion method. The available phosphorus and available potassium were determined using the sodium bicarbonate extraction–molybdenum antimony anti-colorimetric and ammonium acetate extraction flame luminosity methods, respectively. Finally, the heavy metal mass fraction was determined via inductively coupled plasma optical emission spectrometry.

2.4. Data Processing and Analysis

Microsoft Excel 2010 was applied for data processing and Surfer 10 was employed for 3D Wireframe drawing. SPSS 17.0 was used for significance tests and correlation analyses.

3. Results

3.1. Water Quality Variations of Gully Water and Yellow River Water

The ditch water pH in the rice growth stage from May to August 2023 ranged from 7.60 to 8.62, with the lowest value observed in June ($p < 0.05$), and the salinity value ranged from 0.94 to 1.55 g/L. The ditch water salinity and pH values in May were significantly higher than those in other months (Figure 2). Based on the grading standard of water quality and salinity, the farmland drainage ditch water belonged to the category of weak mineralized water and fresh water and could be directly applied for irrigation. During the experiments, the characteristics of Yellow River water were relatively stable, with salinity ranging from 0.41 to 0.51 g/L and pH ranging from 8.06 to 8.23. The Yellow River water pH and salinity values were not significantly different in the time period from May to August ($p > 0.05$). The irrigation season in this area spans from May to August. During this period, the farmland drained more efficiently, and the quality of ditch water remained relatively good. However, the salinity of ditch water was still higher than that of Yellow River water during the same period (Figure 2). The pH value of Yellow River water was lower than that of ditch water in May, and it was slightly higher than that of ditch water in the other months. The quality of Yellow River water was good, with relatively stable pH and salinity values, thus qualifying it as high-quality irrigation water. The concentrations of suspended matter, zinc (Zn), nickel (Ni), selenium (Se), arsenic (As), mercury (Hg), cadmium (Cd), lead (Pb), and chromium (Cr), as well as the values of the five-day biochemical oxygen demand (BOD), chemical oxygen demand (COD), and anion surface active dose in ditch water were significantly below the thresholds outlined in the Water Quality Standard for Farmland Irrigation (GB 5084-2021) [17] (Table 1). The total phosphorus and nitrogen concentrations in ditch water were 0.042 mg/L and 2.5 mg/L, respectively, which accords with the requirements of the basic project standard of the Surface Water Environmental Quality Standard (GB 3838-2002) [18].

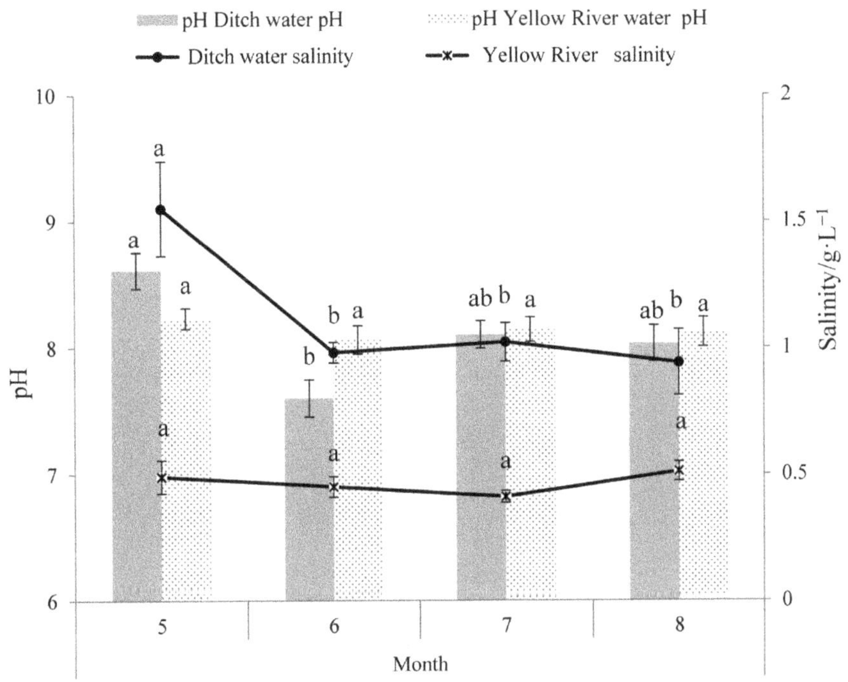

Figure 2. Variations in pH and salinity values of ditch and Yellow River water from May to August. Note: Short lines in the figure represent corresponding average ± standard error. Different letters indicate significant differences among different months in one index at a significance level of 0.05.

Table 1. Water quality physical and chemical properties of ditch water mg/L.

Item	Component Content											Five-Day Biochemical Oxygen Demand (BOD)	Chemical Oxygen Demand (COD)	Anion Surface Active Dose
	Total Nitrogen	Total Phosphorus	Suspended Matter	Zn	Ni	Se	As	Hg	Cd	Pb	Cr			
Ditch water	1.8	0.042	2.5	0.01	0.001	None	0.001	None	None	0.001	0.004	4.5	7.5	None
Threshold value	≤2.0 (Class V)	≤0.2 (Class I)	≤80	≤2	≤0.2	≤0.02	≤0.05	≤0.001	≤0.01	≤0.2	≤0.1	≤60	≤150	≤5

Note: Except for the threshold values of total nitrogen and total phosphorus contents in the Surface Water Environmental Quality Standard (GB 3838-2002), other thresholds are the threshold values of paddy field crops presented in the Water Quality Standard for Farmland Irrigation (GB 5084-2021).

3.2. Effects of Different Treatments on Soil Saline–Alkali Properties

Variations in soil salinity and pH values before planting in April and after harvesting in October are presented in Figures 3 and 4, respectively. Under the effect of the local climate, the salt return was obvious before planting in spring, the surface salt content was the highest (4.6 g/kg), and salt was accumulated on the surface. The soil salt contents of the four treatments were significantly decreased compared with those in spring. The soil salt contents of the 0–20 and 20–40 cm soil layers under CK treatment were 78.7 and 78.9% lower than those before planting, respectively, and that of the 0–100 cm layer was lower than 0.98 g/kg; this was the lowest among all treatments. The soil salinity of the 0–40 cm soil layer treated with T1 was significantly higher than those of other treatments, and that of the soil layer deeper than 40 cm exhibited little difference compared with that before planting ($p > 0.05$). Compared with other treatments, direct ditch water irrigation increased the surface soil salt content to a certain extent. The soil salt content values for the 0–20 cm soil layer treated with T1 were 90.8, 23.1, and 14.7% higher than those of the CK, T2, and

T3 treatments, respectively. The surface soil salt content of the T1 treatment after the first year of rice harvesting was lower than 1.9 g/kg.

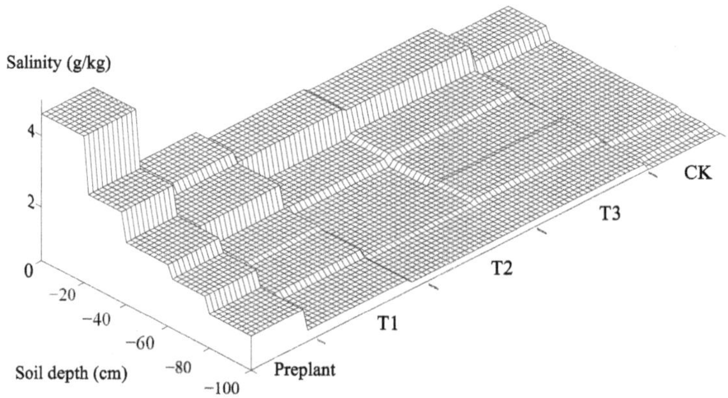

Figure 3. Variations in soil salinity before planting and after harvest.

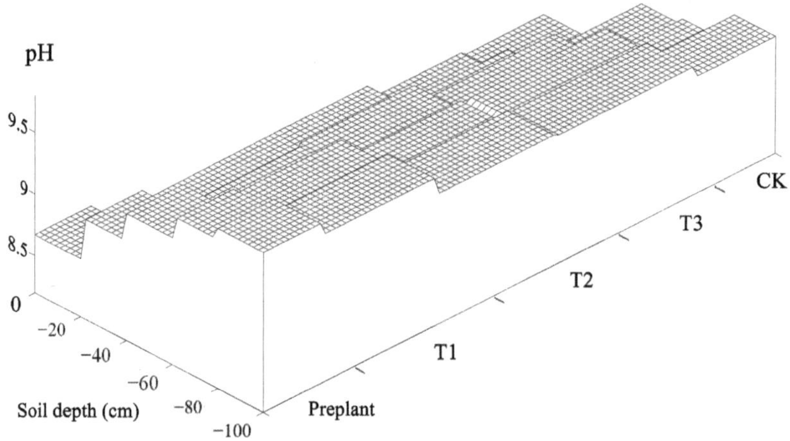

Figure 4. Variations in soil pH before planting and after harvest.

Before planting, the pH value was bounded by the 40 cm soil section, the pH value of the 0–40 cm soil layer was below 9.0, and that of the soil layer deeper than 40 cm was higher than 9.0 (Figure 4). The soil pH value increased with the increase in soil depth. After rice harvesting, the soil pH values of the four treatments in the 0–20 cm soil layer were slightly reduced. The pH values of the soil across the four treatments were between 8.2 and 9.5; the pH values of the surface soil were the lowest, with all values below 8.6. All of the pH values presented increasing trends with increasing soil depth (Figure 4), and there were only small differences in soil pH value values among the four treatments ($p > 0.05$). During the rice growth period, the salt pressure effect was obvious, the average soil salt decrease in the 0–20 cm soil layer for the four treatments was 67.4%, and the pH reduction was less than 6%. The decreased salt and alkali contents in the soil were most obvious in respect of the Yellow River direct irrigation. In addition, direct ditch water irrigation, mixed ditch water and Yellow River water irrigation, and alternate irrigation did not increase the salt and alkali contents in the soil after harvesting. However, direct ditch water irrigation still slightly increased the surface soil salinity.

3.3. Effects of Different Treatments on Soil Heavy Metals

As, Hg, Cd, Cr, and Pb contents in the 0–20 cm soil layer before planting met the thresholds specified in the Secondary Standard for Soil Environmental Quality (GB15618-1995) [19]. The contents of the above heavy metals in the soil treated with T1, T2, and T3 after harvesting were increased. However, they were much lower than the thresholds advised in the Secondary Standard for Soil Environmental Quality (GB15618-1995) (Table 2).

Table 2. Heavy metal contents in soil, mg/kg.

Item	As	Hg	Cd	Cr	Pb
Before planting	15.53 ± 1.43 ab	0.04 ± 0.01 b	0.13 ± 0.01 a	32.01 ± 3.26 b	15.33 ± 0.82 b
T1	16.87 ± 1.65 a	0.13 ± 0.02 a	0.17 ± 0.01 a	39.74 ± 4.06 a	20.16 ± 1.42 a
T2	16.32 ± 0.96 a	0.09 ± 0.01 ab	0.15 ± 0.01 a	36.23 ± 3.12 ab	18.39 ± 1.06 ab
T3	16.57 ± 1.32 a	0.11 ± 0.02 a	0.14 ± 0.01 a	37.76 ± 2.86 ab	19.75 ± 0.97 ab
CK	15.12 ± 1.13 b	0.05 ± 0.01 b	0.12 ± 0.01 a	32.57 ± 1.68 b	16.17 ± 1.32 b
GB15618—1995	25.0	1.0	0.6	250.0	350.0

Note: Different letters indicate significant differences among different scenarios in one index at 0.05 significance level.

No significant differences were observed in Cd contents in any of the treatments compared with levels before planting ($p > 0.05$). The As, Hg, Cd, Cr, and Pb contents in the CK treatment were the lowest, and there were no significant differences compared to those before planting ($p > 0.05$). The Hg, Cr, and Pb contents in the soil treated with T1 were significantly higher than those before planting ($p < 0.05$). Compared with Yellow River irrigation, soil As, Hg, Cr, and Pb contents were increased to a certain extent after ditch water irrigation, but were much lower than the thresholds specified in the Secondary Standard for Soil Environmental Quality. There were no significant differences in the As, Hg, Cd, Cr, and Pb contents between the T2 and T3 treatments ($p > 0.05$). Hence, direct ditch water irrigation, mixed ditch water and Yellow River water irrigation, and alternate irrigation during the rice growth period did not lead to excessive heavy metal contents in the soil.

3.4. Effects of Different Treatments on Heavy Metals and Paddy Yield

After harvesting, the As, Hg, Cd, Cr, and Pb contents in rice corresponding to the four treatments were all lower than the threshold values specified in the National Standard for Food Safety (GB 2762-2017) (Table 3). The Cr content of rice under the T1 treatment was significantly higher than that under the CK treatment ($p < 0.05$), but no significant differences were observed in respect of the As, Hg, Cd, and Pb contents among the four treatments ($p > 0.05$). The rice yields of the T1, T2, and T3 treatments were much lower than that of CK ($p < 0.05$), and the T1 treatment had the lowest rice yield. As mentioned above, the rice yields of the T1, T2, and T3 treatments were 20.3%, 12.1%, and 13.6% lower than that of CK, respectively, while no significant difference was observed between the T2 and T3 treatments ($p > 0.05$). Ditch water irrigation did not result in excessive heavy metal contents in rice, but ditch water irrigation affected rice yield to a certain extent. In addition, direct ditch water irrigation did not decrease the yield. There was little difference between the effects of mixed ditch water and Yellow River water irrigation and alternate irrigation on the rice yield.

Table 3. Heavy metal contents and yields of paddies.

Item	Heavy Metal Content (mg/kg)					Yield (kg/hm^2)
	As	Hg	Cd	Pb	Cr	
T1	0.23 ± 0.09 a	0.006 ± 0.002 a	0.08 ± 0.02 a	0.26 ± 0.07 a	0.37 ± 0.09 a	8318.4 ± 134.7 c
T2	0.15 ± 0.03 a	0.004 ± 0.001 a	0.06 ± 0.01 a	0.21 ± 0.09 a	0.24 ± 0.06 ab	9182.1 ± 128.9 b
T3	0.16 ± 0.04 a	0.005 ± 0.001 a	0.07 ± 0.01 a	0.17 ± 0.04 a	0.21 ± 0.08 ab	9016.2 ± 157.3 b
CK	0.08 ± 0.02 b	0.003 ± 0.001 a	0.04 ± 0.01 a	0.11 ± 0.03 a	0.17 ± 0.07 b	10,437.5 ± 187.5 a
GB2762—2017	0.5	0.02	0.2	0.5	1	

Note: Different letters indicate significant differences among different scenarios in one index at 0.05 significance level.

4. Discussion

Brackish water could be reasonably applied for irrigation in arid and semi-arid areas, and the key to this lies in how to make its use efficient, safe, and reasonable [20,21]. In this research, ditch water salinity in the experimental area varied from 0.66 to 1.19 g/L during the rice growth period. Based on the grading standard of water quality and salinity [22], salinity values lower than 1.0 g/L indicate high-quality irrigation water, while a salinity range of 1.0 to 2.0 g/L indicates weakly mineralized water and could be applied for irrigation. Hence, the ditch water in this study belonged to the category of weakly brackish water and fresh water during the rice growth period. From June to August, ditch water salinity was lower than 1.0 g/L, which indicates high-quality irrigation water.

In the Yinbei irrigation area of Ningxia, brackish water with salinity values from 0.94 to 1.55 g/L has been employed for irrigation, and soil desalination was obvious; however, the pH value, salt content, and alkalinity of the soil were higher than those in respect of Yellow River irrigation [23]. Some research works have revealed that irrigation with brackish water with low salinity effectively reduced soil salinity, presenting no significant difference compared with freshwater irrigation [24]. This result might be related to the number of years of brackish water irrigation. Long-term brackish water irrigation increases the surface soil salt content to a certain extent [8], especially in northwestern inland areas, where there is little rain and severe drought. With the increase in brackish water irrigation years, salt can easily accumulate in the soil surface.

In this study, compared with mixed irrigation and alternate ditch water and Yellow River water irrigation, direct ditch water irrigation increased the soil salinity in the 0–40 cm soil layer, which was consistent with the findings of Fan (2020), suggesting that brackish water irrigation increases the upper soil salinity to a certain extent [8]. Although ditch water contains certain amounts of salt, a limited amount of salt is brought in in a short period of time and salt and alkali aggravation can be avoided by alternating between brackish water and fresh water irrigation [9]. This was also proved by the fact that mixed or alternating irrigation with ditch water and Yellow River water did not significantly affect crop yields, and the effects were relatively minor compared with direct ditch water irrigation. However, the brackish water irrigation threshold in the coastal area of China is typically above 3 g/L [10], which is due to the fact that high rainfall and natural precipitation in coastal areas have a certain leaching impact on soil salinity. Ditch water irrigation had a slight influence on soil pH.

Kazuto (2023) found that the introduction of paddy drainage during the irrigation season significantly alters the DOC components in river waters, and irrigation management is strongly linked to the primary production in agricultural channels [25]. Li (2020) found that fully utilizing ditches and ponds as temporary reservoirs in a typical paddy IDU could reduce nitrogen loads from field edges by 39% and phosphorus loads by 28%. Therefore, natural ditches and ponds are recommended to be included in irrigation [26]. Compared with Yellow River water, ditch water contains certain amounts of total nitrogen and total phosphorus, which can complement soil fertility. However, the long-term effects of direct ditch water irrigation on soil health require further verification through long-term tests. If

Yellow River water is reliably available, it is recommended to implement mixed or alternate irrigation using both Yellow River water and ditch water.

Eldardiry (2013) found that through brackish water irrigation, excessive salt in water affected normal crop growth. Water containing small amounts of salt stimulates crop growth to a certain extent, and crop quality is improved [27]. Ditch water irrigation brings extra salt into the soil, which interacts with the chemical elements present in the soil and changes the physical and chemical properties of the soil. Under alkaline conditions, the irrigation water salinity is low, the soil aggregate structure is destroyed, and clay particle expansion and dispersion reduce the porosity and permeability of soil, while high-salinity irrigation water increases the flocculation of soil and reduces the expansion and dispersion of clay particles [28]. In this experiment, the heavy metal contents of ditch water conformed to farmland irrigation water quality standards; the soil heavy metal contents after direct ditch water irrigation, mixed ditch water and Yellow River water irrigation, and alternate ditch water and Yellow River water irrigation complied with the requirements of the Secondary Standard of Soil Environmental Quality and did not lead to excessive heavy metal contents in soil. Ditch water irrigation did not result in excessive heavy metal contents in rice, but it affected rice yield to a certain extent. The rice yield under direct ditch water irrigation was the lowest, and there was little difference between the mixed ditch water and Yellow River water irrigation and the alternate ditch water and Yellow River water irrigation in terms of rice yield. Therefore, if conditions allowed, mixed ditch water and Yellow River water irrigation and alternate ditch water and Yellow River water irrigation should be given priority. There is a risk that long-term irrigation with ditch water will increase soil salinity and affect crop yields.

5. Conclusions

(1) Ditch water salinity from the experimental field was less than 1.60 g/L and its pH value was lower than 8.62 during the rice growth period from May to August in Pingluo County, Ningxia; the water is therefore classified as mildly brackish water suitable for farmland irrigation.

(2) The application of ditch water irrigation for rice did not lead to soil salinity and alkali aggravation, and the heavy metal concentrations in both soil and rice remained within specified standard values. The heavy metal concentration in soil complied with the requirements of the Secondary Standard of Soil Environmental Quality, while the heavy metal concentration in rice met the requirements of the National Standard of Food Safety.

(3) The rice yields for the CK, T1, T2, and T3 treatments were 10,437.5, 8318.4, 9182.1, and 9016.2 kg/hm^2, respectively. Compared with Yellow River irrigation, the rice yields for T1, T2, and T3 treatments were 20.3%, 12.1%, and 13.6% lower than that for CK, respectively, with minimal differences among them.

Therefore, in the case of a water resource shortage in the Yellow River, farmland drainage ditch water can be effectively utilized for mixed ditch water and Yellow River water irrigation or alternate ditch water and Yellow River water irrigation to ensure food security.

Author Contributions: Investigation, Y.Z.; data curation, J.S. and Y.Z.; writing—original draft, L.F. and J.S.; writing—review and editing, X.W.; visualization, L.F. All authors have read and agreed to the published version of the manuscript.

Funding: This research was funded by the National Key Research and Development Plan (2021YFD1900605-05); Ningxia Hui Autonomous Region key research and development project (2023BCF01050, 2022BBF03032-03); the National Natural Science Foundation (42367043); Agricultural Science and Technology Independent Innovation of Ningxia Hui Autonomous Region (NGSB-2021-11-03); and Observation and Monitoring of Basic Long-Term Scientific and Technological Work in Agriculture (NAES091AE18).

Data Availability Statement: The data are contained within the article.

Conflicts of Interest: The authors declare no conflicts of interest.

References

1. Wang, X.; Xiao, X.; Zou, Z.; Dong, J.; Qin, Y.; Doughty, R.; Menarguez, M.; Chen, B.; Wang, J.; Ye, H.; et al. Gainers and losers of surface and terrestrial water resources in China during 1989–2016. *Nat. Commun.* **2020**, *11*, 3471. [CrossRef] [PubMed]
2. Gu, X.; Zhang, F.; Wang, T.; Xie, X.; Jia, X.; Xu, X. Effects of Nitrogen and Phosphorus Addition on Growth and Leaf Nitrogen Metabolism of Alfalfa in Alkaline Soil in Yinchuan Plain of Hetao Basin. *PeerJ* **2022**, *10*, e13261.
3. Zhu, X.; Liu, J.; Qi, W.; Gao, X.; Zhao, X.; Zhu, B. Future changes and distribution of water resources in China under global warming levels of 1.5–4.5 °C. *J. Hydrol.* **2024**, *53*, 101781. [CrossRef]
4. Lu, C.; Xi, R.; Hei, Z.; Tang, L. Safety Evaluation of Water Environment Carrying Capacity of Five Cities in Ningxia Based on Ecological Footprint of Water Resources. *Asian Agric. Res.* **2022**, *14*, 11–16.
5. Hazem, T.; Guan, H. Hyperspectral remote sensing for extraction of soil salinization in the northern region of Ningxia. *Model. Earth Syst. Environ.* **2020**, *6*, 2487–2493.
6. Kamel, N.; Mohamed, M.; Netig, B. Impacts of Irrigation Regimes with Saline Water on Carrot Productivity and Soil Salinity. *J. Saudi Soc. Agric. Sci.* **2012**, *11*, 19–27.
7. Zhao, J.; Zhang, X.; Yang, Y.; Zang, H.; Yan, P.; Meki, M.; Doro, L.; Sui, P.; Jeong, J.; Zeng, Z. Alternative cropping systems for groundwater irrigation sustainability in the North China Plain. *Agric. Water Manag.* **2021**, *250*, 106867. [CrossRef]
8. Fan, L.; Li, L.; Wu, X.; Wang, X. Effects of Leaching with Different Quality Water and Soil Amendments Application on Soil Water and Salt Transport in Alkalize Solonchaks in Yinbei Irrigation District. *J. Soil Water Conserv.* **2020**, *34*, 369–376. (In Chinese)
9. Zhang, P.; Shen, J. Effect of brackish water irrigation on the movement of water and salt in salinized soil. *Open Geosci.* **2022**, *14*, 404–413. [CrossRef]
10. Lyu, S.; Wu, L.; Wen, X.; Wang, J.; Chen, W. Effects of reclaimed wastewater irrigation on soil-crop systems in China: A review. *Sci. Total Environ.* **2022**, *813*, 152531. [CrossRef]
11. Gao, H.; Fu, T.; Tang, S.; Liu, J. Effects of saline water irrigation on winter wheat and its safe utilization under a subsurface drainage system in coastal saline-alkali land of Hebei Province, China. *Irrig. Sci.* **2023**, *41*, 251–260. [CrossRef]
12. Wei, C.; Li, F.; Yang, P.; Ren, S.; Wang, S.; Wang, Y.; Xu, Z.; Xu, Y.; Wei, R.; Zhang, Y. Effects of Irrigation Water Salinity on Soil Properties, N_2O Emission and Yield of Spring Maize Under Mulched Drip Irrigation. *Water* **2019**, *11*, 1548. [CrossRef]
13. *GB 2762-2017;* National Standard for Food Safety. China Food and Drug Administration. Standards Press of China: Beijing, China, 2017; pp. 2–9. (In Chinese)
14. Wang, Z.; Tian, J. Effects of Fertilizer Application Quotas on Rice Yield, Quality and Heavy Metal Content under Reclaimed Water Irrigation. *Ningxia Eng. Technol.* **2022**, *21*, 326–330. (In Chinese)
15. Wang, L.; Zhou, L.; Hu, J.; Cai, J.; Wang, W.; Xiao, W.; He, M. Effects of reclaimed water irrigation on the distribution of heavy metal in paddy fields. *J. Drain. Irrig. Mach. Eng.* **2022**, *40*, 842–849. (In Chinese)
16. Wang, X.; Shen, J.; Fan, L.; Zhang, Y. Research on Salt Drainage Efficiency and Anti-Siltation Effect of Subsurface Drainage Pipes with Different Filter Materials. *Water* **2024**, *16*, 1432. [CrossRef]
17. *GB 5084-2021;* Water Quality Standard for Farmland Irrigation. Ministry of Ecology and Environment. Standards Press of China: Beijing, China, 2021; pp. 3–4. (In Chinese)
18. *GB 3838-2002;* Surface Water Environmental Quality Standard. State Environmental Protection Administration. Standards Press of China: Beijing, China, 2002; pp. 2–4. (In Chinese)
19. *GB 15618-1995;* Standard for Soil Environmental Quality. Ministry of Ecology and Environment. Standards Press of China: Beijing, China; pp. 2–4. (In Chinese)
20. Liu, C.; Cui, B.; Hu, C.; Wu, H.; Gao, F. Effects of mixed irrigation using brackish water with different salinities and reclaimed water on a soil-crop system. *Water Reuse* **2021**, *11*, 632–648.
21. Wang, Y.; Xiao, Y.; Puig-Bargués, J.; Zhou, B.; Liu, Z.; Muhammad, T.; Liang, H.; Maitusong, M.; Wang, Z.; Li, Y. Assessment of water quality ions in brackish water on drip irrigation system performance applied in saline areas. *Agric. Water Manag.* **2023**, *289*, 108544. [CrossRef]
22. Bao, S. *Agrochemical Analysis of Soil;* China Agriculture Press: Beijing, China, 2018; pp. 178–179. (In Chinese)
23. Zhao, H.; Li, L.; Fan, G.; Xie, S.; Li, F. Effects of aerated brackish water irrigation on growth of *Lycium barbarum* seedlings. *Sci. Hortic.* **2023**, *310*, 111721. [CrossRef]
24. Li, X.; Kang, Y. Water-salt control and response of Chinese rose (*Rosa chinensis*) root on coastal saline soil using drip irrigation with brackish water. *Trans. Chin. Soc. Agric. Eng.* **2019**, *35*, 112–121. (In Chinese)
25. Kazuto, S.; Tomoka, K.; Takafumi, T.; Riku, T.; Fumiaki, N.; Tamihisa, O.; Masaki, N.; Yuji, Y.; Hideki, K. Effects of paddy irrigation-drainage system on water quality and productivity of small rivers in the Himi region of Toyama, Central Japan. *J. Environ. Manag.* **2023**, *342*, 118305.
26. Li, S.; Liu, H.; Zhang, L.; Li, X.; Wang, H.; Zhuang, Y.; Zhang, F.; Zhai, L.; Fan, X.; Hu, W.; et al. Potential nutrient removal function of naturally existed ditches and ponds in paddy regions: Prospect of enhancing water quality by irrigation and drainage management. *Sci. Total Environ.* **2020**, *718*, 137418. [CrossRef] [PubMed]

27. Eldardiry, E.; El-Hady, M.; Zaghloul, A. Relationship Between Soil Physical and Chemical Properties and Hydrophysical Soil Properties Under Reuse of Agricultural Drainage Water. *Am.-Eurasian J. Agric. Environ. Sci.* **2013**, *13*, 1–6.
28. Wang, Y. *Return Flow Water Quality and Reuse in Saline-alkali Paddy Soils*; University of Chinese Academy of Sciences: Changchun, China, 2015; p. 104. (In Chinese)

MDPI AG
Grosspeteranlage 5
4052 Basel
Switzerland
Tel.: +41 61 683 77 34

Water Editorial Office
E-mail: water@mdpi.com
www.mdpi.com/journal/water

www.ingramcontent.com/pod-product-compliance
Lightning Source LLC
Chambersburg PA
CBHW061441210926
51850CB00008B/1569